Advancing Sustainable Engineering Through Next-Generation Thermo-Fluid Systems

Kamalakanta Muduli
https://orcid.org/0000-0002-4245-9149
Papua New Guinea University of Technology, Papua New Guinea

Sachindra Kumar Rout
https://orcid.org/0000-0002-5471-1620
C.V. Raman Global University, India

Bukke Kiran Naik
National Institute of Technology, India

Albert U. Ude
Botswana International University of Science and Technology, Botswana

IGI Global
Scientific Publishing
Publishing Tomorrow's Research Today

Vice President of Editorial Melissa Wagner
Director of Acquisitions Mikaela Felty
Director of Book Development Jocelynn Hessler
Production Manager Mike Brehm
Cover Design Jose Rosado

Published in the United States of America by
IGI Global Scientific Publishing
701 East Chocolate Avenue
Hershey, PA, 17033, USA
Tel: 717-533-8845 | Fax: 717-533-7115
Website: https://www.igi-global.com E-mail: cust@igi-global.com

Library of Congress Cataloging-in-Publication Data

LCCN: 2025026728 (CIP Data Pending)
ISBN13: 979-8-3373-3621-3
Isbn13Softcover: 979-8-3373-3622-0
EISBN13: 979-8-3373-3623-7

British Cataloguing in Publication Data
A Cataloguing in Publication record for this book is available from the British Library.

Table of Contents

Detailed Table of Contents

A two-dimensional computational study is carried out on the solar air heater duct (SAHD) by attaching rivet ribs on the collector plate at constant rib height and relative roughness height of (e =1.4 mm and e/Dh = 0.042). The finding of this problem was done with a computational exercise in ANSYS Fluent R22 by varying dimensionless parameters like Reynolds number (4000-18000), and relative width of rivet shank to height ratio (w/e) of 0.285 to 0.714 by keeping constant pitch distance (P=10 mm). The investigation of the domain was done, with a uniform heat supply of 1000 W/m2 over the collector plate made up of aluminium. From various turbulence models, the RNG k-ε model with enhanced wall treatment (EWT) was chosen for simulating the fluid flow equations (mass, momentum, and energy). The result showed that the Nusselt number (Nur) of roughed plate rises with the increment of Reynolds number (Re), also found average temperature of 361 K with maximum average Nusselt number enhancement factor (Nur / Nus) of 2.53 and thermal enhancement factor (TEF) of 1.78 at Re= 18000 at w/e of 0.36.

The performance of loop heat pipes (LHPs) largely depends on the capillary action within their porous wicks, which drives fluid circulation. In this study, stainless steel wicks with monoporous and biporous structures were tested using methanol, ethanol, and water to evaluate capillary rise behavior. Results show that biporous wicks significantly outperform monoporous ones, enabling faster fluid transport.

range of Rayleigh numbers (10^3 to 10^7), rotational speeds (0 to 4), and aspect ratios (2.5 to 7.5) to quantify the heat transfer enhancement by the rotation of the torus. The temperature contours and the velocity vector plots are presented to elucidate the underlying flow physics and heat transfer mechanisms for both rotating as well as non-rotating cases. Utilizing natural convection to cool rotating components eliminates the need for energy-intensive active cooling systems, thereby enhancing overall energy efficiency and aligning with sustainable engineering principles

P. Srinivas Reddy, CVR College of Engineering, Hyderabad, India
Sk Mohammad Shareef, CVR College of Engineering, Hyderabad, India
Maturi Jeevan Kumar, CVR College of Engineering, Hyderabad, India

Ornithopters, or flying vehicles propelled by flapping wings, have been of great interest in recent years because of their application to micro aerial vehicles (MAVs). There are a host of new problems in vehicle dynamics and control that can be explored because of this renewed interest. An ornithopter is a robotic flight vehicle that generates lift and thrust by flapping its wings. An ornithopter equipped with a wireless camera will be developed to study flapping wing flight better. In this study, it is discussed the essential aspects of developing an ornithopter as well as its operation mechanism and various applications. This project begins with the design of the ornithopter mechanism, design of the mechanical, electrical, and control systems, followed by the manufacturing process. An ornithopter's data is obtained through kinematic and experimental analysis. Based on the ornithopter design, the wingspan is 126 cm, the weight is 370 grams, and the torsional length is 95 cm. This mechanism can achieve a maximum flapping frequency of 6-7 Hz based on the calculation results.

Shantanu Kumar, Indian Institute of Technology, Delhi, India
K. Ravi Kumar, Indian Institute of Technology, Delhi, India
Kavya Dashora, Indian Institute of Technology, Delhi, India

Tea being a non-alcoholic drink places second among all the drinks. Black tea is the highest consumed tea worldwide. Black tea is of two types: Orthodox and Crush-Tea-Curl. Drying is a crucial step in black tea processing. It reduces moisture content. Tea drying especially arrests the oxidation reaction started during oxidation. In the current study, cylindrical tea particle is considered under a natural hot air-drying environment. Orthodox tea particle can be considered cylindrical in shape. A finite difference method-based explicit scheme is used to discretize the governing and boundary conditions. Drying temperatures of 160°C, and 170°C give acceptable moisture content in the drying duration of 1800 s (approx.) and 1500 s – 1800 s, respectively. Drying at 160°C and 170°C gives mean moisture contents of 0.063 g/g (d.b.) and 0.038 g/g (d.b.), respectively in 1800 s. Drying at higher temperatures shows faster moisture ratio reduction as compared to lower drying temperatures. Moisture ratios corresponding to 160°C and 170°C are found to be 0.017 and 0.003, respectively for 1800 s.

Umamaheshz Boddepalli, Indian Institute of Technology, Kharagpur, India
Rohit Roy, Indian Institute of Technology, Kharagpur, India
Sourav Mitra, Indian Institute of Technology, Kharagpur, India
Ramgopal Maddali, Indian Institute of Technology, Kharagpur, India

The present theoretical study reports the characteristics of refrigerant blends of hydrocarbons and CO2.The thermodynamic, environmental and flammability characteristics of these blends are obtained through suitable mixture models. The safety class of blends of propane, propylene, isobutane and dimethyl ether are categorised as per the ASHRAE conditions. This study also presents the thermodynamic performance of these blends in a simple VCRC using standard AHRI cooling conditions with a constant cooling capacity of 3 Ton. The results indicated that the (COP) of the zeotropic blends is low compared to the pure hydrocarbons with a constant temperature heat exchange fluid. This paper also analysed the environmental performance (GWP, TEWI) of the selected pure natural refrigerants and their blends. The effect of emission factor on indirect warming impact is also discussed. The results describe that among the pure refrigerants R-600a performance is best and CO2 performance is least. However, none of the blend performance is better than the R-32. This study also included the analysis of thermodynamic properties of the mixtures.

S. K. Mohammad Shareef, CVR College of Engineering, Hyderabad, India
P. Srinivas Reddy, CVR College of Engineering, Hyderabad, India
B. Abhiram, CVR College of Engineering, Hyderabad, India

Conventional fossil fuels are rapidly depleting due to population growth and rising energy consumption, necessitating the search for renewable energy sources. Among these, biodiesel has gained attention for its potential as a sustainable substitute when blended with diesel in optimal ratios. Recent advancements in nanotechnology have introduced nano-additives as catalysts to enhance engine efficiency and reduce emissions. This study focuses on improving diesel engine performance and reducing emissions by blending alcohol with diesel in varying proportions and incorporating nanoparticles to enhance thermal conductivity. A diesel engine is used to investigate combustion characteristics, engine performance, and emissions. Key challenges, such as fuel burning rate, ignition lag, and emissions like NOx, HC, and CO, are addressed. The study highlights the potential of nanoparticles to improve thermal efficiency and explores the comparative benefits of alcohol-diesel blends, contributing to the development of sustainable energy solutions.

Flow measurement is a critical component across numerous industrial sectors. The orifice meter has become a preferred alternative to traditional wellhead chokes, offering enhanced control and precision in measuring mass flow rate. It is suitable for fluid flow measurements, utilizing in pressure drop and the discharge coefficient. This chapter explores orifice meters in detail, focusing on the impact of various shapes, sizes, classifications, and operational parameters on their performance for high viscous fluid flow applications.This chapter considered different orifices such as sector, crescent, segmental to determine the flow charteristics on the effect of input parameters. It examines how changes in shape factor, width factor, and non-dimensional flow coefficient affect velocity, pressure drop and discharge coefficient of upside and downside of the flow. This study presents vital approaches into the more efficient design for high viscous flow application of orifice meters.

There are several things that can power with lithium-ion batteries thanks to their high energy density and long cycle life. However, they face several challenges, including thermal management issues, which can lead to performance degradation and safety concerns. This can result in capacity loss, reduced lifespan, and potential safety hazards. The right thermal management prevents overheating and maintains optimal operating temperatures. Active and passive cooling are used to enhance thermal management of batteries. Passive cooling methods, such as phase change materials (PCMs), are commonly hampered by low thermal conductivity, making them less efficient at dissipating heat quickly. To optimize the thermal management of Li-ion batteries, a numerical simulation was performed using ANSYS-FLUENT with different materials for fins and various fin numbers at a discharge rate of 8C. The results of this study show that the temperature reduction achieved by installing copper fins with fewer numbers is comparable to the temperature reduction achieved by using a higher number of aluminum fins.

In light of increasing environmental and regulatory concerns, there is a growing emphasis on the use of natural refrigerants. In the present study theoretical performance of a cold storage that uses a NH3 based VCRS with a CO2 based natural circulation loop (NCL) is analyzed. The secondary loop is a rectangular, two-phase, buoyancy driven, NCL with adiabatic riser and downcomer. As addition of secondary loop introduces, energy and exergy analyses of system are carried out to estimate the performance. The key operating parameters such as evaporating and condensing temperatures, temperature differences in evaporator, condenser and intermediate heat exchanger are found to have a significant effect on COP, second law efficiency and exergy destruction per unit refrigeration capacity of the system. Based on the data generated, suitable correlations are developed between the operating and performance parameters. Results show that despite the minor penalty over single-stage ammonia refrigeration system, indirect system provides a safer working space by reducing charge of ammonia.

The rising interest of renewable resources due to the increasing demand of sustainable energy solutions world-wide has heightened the development of thermo-fluid systems. These systems are very significant in facilitating fluid conversion, storage and distribution of energy in adherence to the principles of thermodynamics and fluid mechanics. Due to the increasing rate of climate change and exhaustion of materials, renewable sources have become viable to achieve such as solar, wind, biomass and geothermal sources which have been encouraged by advances in computational fluid dynamics (CFD), and material science, as well as hybrid systems countering intermittency and increasing reliability. In recent research, the combination of multiple sources of renewable energy is emphasized, with solutions providing a move toward operational resilience and system efficiency. This chapter offers a wholesome investigation of the applications of renewable energy within thermo-fluid systems with an emphasis on principles, development and hybrid methods.

The present article focuses on the simulation of a hybrid desiccant wheel based air-conditioning unit coupled with a vortex tube cooler. The desiccant wheel and the vortex tubes are modelled from empirical correlations developed through regression analysis from experimental data. The sensible cooling of the supply air is carried out by the vortex tube cooler after it is dehumidified in the desiccant wheel and releasing the adsorption heat in the energy recovery wheel. The variation of the output parameters with respect to some of the factors are discussed. The variation of temperature and specific humidity of the supply air are illustrated and discussed with respect to the weather conditions of Bhubaneswar city (Odisha, India) during the whole year. The output supply air temperature and humidity levels suggest the feasibility of implementing the vortex tube cooler in conjunction with the desiccant wheel.

Latent Thermal Energy Storage (LTES) in a compact, solar-powered vapour compression air conditioner system with ice as the phase change material is studied numerically. In addition to using propane (R290) as a refrigerant, the system has two VCR systems: a high-capacity unit for charging the LTES and a low-capacity one for chilling during the day. With interior fins to improve heat transfer, a CFD model replicates the storage tank. Findings indicate that the solidification rate does not significantly increase beyond 32 fins. The cooling load of a typical student room is determined, and the necessary power and associated PV panel area are computed using the thermodynamic cycle. The results show that full solar operation is not practical due to the small roof space and that some grid electricity is required. The study backs the development of environmentally friendly cooling solutions for tropical regions.

The steady flow characteristics of circular and elliptical cylinders have been extensively examined at Reynolds numbers (Re) well beyond the critical value, where flow transitions from steady to unsteady. However, studies concerning the steady flow over square cylinders have been mainly confined to Reynolds numbers up to 40. While the flow becomes unsteady beyond the critical Re, it remains valuable to numerically investigate the characteristics of steady flow, as this can provide insights into flow stability. In the present work, steady two-dimensional flow past a stationary square cylinder has been computed for Reynolds numbers up to 200. The main characteristics of steady flow, such as bubble length, bubble width, drag coefficient on the cylinder surface, and maximum vorticity on the surface of the cylinder, are investigated. Results indicate that bubble length and bubble width are linear functions of Re up to Re = 80, after which it exhibits nonlinear behaviour well fitted by a second-order polynomial. The drag coefficient shows linear variation with Re^(−0.66).

Theoretical investigations have been carried out to evaluate the thermodynamic and environmental properties of different CO_2 (R744) blends with low Global Warming Potential (GWP) Hydrofluoroolefins (HFOs) to assess the suitability of these blends for refrigeration and air conditioning applications. The results show that compared to other available refrigerants, the selected blends offer significantly low GWP values but high indirect emissions due to their poor coefficient of performance (COP). The binary blend yields good thermodynamic characteristics, including large volumetric capacity and low operating pressures and a subcritical cycle of operation. The temperature glide characteristics show the zeotropic behaviour of the blends.

Preface

The field of thermo-fluid engineering stands at a defining crossroads where innovation, sustainability, and efficiency converge to address the urgent global challenges of climate change, resource depletion, and energy transition. *Advancing Sustainable Engineering Through Next-Generation Thermo-Fluid Systems* emerges from this intersection, offering a timely and comprehensive exploration of how advanced thermal and fluid sciences can drive a more sustainable and energy-efficient future.

Thermo-fluid systems are central to countless modern technologies, from power generation and renewable energy systems to industrial processing, HVAC applications, and transportation. As the world shifts toward decarbonization and environmental responsibility, the need to redesign these systems through the lens of sustainability has never been greater. This book seeks to bridge the gap between conventional approaches and next-generation innovations by highlighting research that combines scientific rigor with practical impact. It brings together contributions that explore emerging frontiers such as AI-driven modeling, hybrid energy systems, bio-inspired designs, and materials engineering, all directed toward achieving efficiency and environmental harmony.

Our editorial vision emphasizes a future-oriented perspective, focusing on innovations that will shape the next era of thermo-fluid technology. Beyond individual components and processes, we adopt a holistic approach that integrates thermo-fluid systems with broader sustainability frameworks, including renewable energy integration, lifecycle optimization, and environmental resilience. The book also maintains a strong practical orientation, linking advanced theory with real-world case studies that demonstrate tangible solutions for industry and research applications.

The collective expertise of contributors from diverse academic and industrial backgrounds ensures that this volume serves as both a scholarly reference and a practical guide. It is designed to meet the needs of academicians, researchers, and students engaged in cutting-edge studies, as well as engineers, consultants, and policymakers seeking actionable insights into sustainable energy design and implementation. By presenting complex concepts in an accessible format, the book aims to foster knowledge transfer across disciplines and experience levels.

The chapters encompass a broad spectrum of topics, from advanced heat transfer and computational fluid dynamics (CFD) to smart materials, hydrogen energy systems, and waste heat recovery technologies. Each contribution provides not only technical depth but also a vision for how thermo-fluid science can be harnessed to achieve cleaner, smarter, and more sustainable engineering outcomes.

CHAPTER OVERVIEWS

Chapter 1 introduces a computational investigation into the performance of solar air heater ducts enhanced with rivet ribs. The study provides valuable insights into how surface roughness and rib geometry influence fluid flow and thermal efficiency. By employing advanced simulation models, the research demonstrates significant improvements in Nusselt number and thermal enhancement factors, emphasizing the potential for optimized solar air heaters to contribute to sustainable energy harvesting and efficient heat utilization.

Chapter 2 explores the capillary rise behavior of biporous and monoporous stainless steel wicks for loop heat pipe applications, comparing their performance with different working fluids. Through infrared imaging and microscopic analysis, the study reveals that biporous structures markedly improve fluid transport and thermal performance. The findings contribute to the development of high-performance wicking materials essential for compact and efficient thermal management systems in modern electronics and spacecraft.

Chapter 3 focuses on the thermal behavior of lithium-ion batteries for electric vehicles using mini-channel cooling. By examining the influence of geometric and thermal parameters, the study identifies an optimal aspect ratio that effectively reduces battery temperature while maintaining efficiency. This chapter highlights how precision cooling design supports safe, reliable, and high-performance energy storage systems critical to sustainable mobility and electrification.

Chapter 4 presents a computational study on natural and mixed convection heat transfer from a rotating solid torus suspended in air. The research elucidates how buoyancy and rotation interact to enhance heat transfer, proposing a passive cooling mechanism that minimizes energy consumption. This sustainable approach to thermal management underscores how natural convection principles can replace conventional cooling methods in various industrial applications.

Chapter 5 investigates the design and fabrication of an ornithopter, a bio-inspired aerial vehicle propelled by flapping wings. The work integrates mechanical, electrical, and control systems engineering to replicate avian flight dynamics. Through kinematic and experimental analysis, the project showcases the engineering challenges and innovations involved in developing lightweight, agile, and energy-efficient flight systems for surveillance and research applications.

Chapter 6 examines the drying characteristics of orthodox fermented tea particles under convective drying conditions. Employing finite difference modeling, the study evaluates the impact of temperature and time on moisture reduction, identifying optimal drying parameters for quality preservation. This research bridges thermal science and agricultural processing, offering insights into energy-efficient drying techniques that enhance product quality and sustainability in the food industry.

Chapter 7 evaluates low global warming potential hydrocarbon–CO2 blends for refrigeration systems. The theoretical analysis investigates thermodynamic, environmental, and safety characteristics of various mixtures, revealing trade-offs between performance and environmental impact. The findings support ongoing efforts to identify environmentally responsible refrigerants that balance energy efficiency with minimal ecological footprint.

Chapter 8 experimentally analyzes the performance and emissions of phytol–diesel blends in a compression ignition engine. By incorporating nanoparticles and alcohol additives, the study achieves improvements in thermal efficiency and significant emission reductions. The work provides a promising direction for the adoption of biofuel blends as a sustainable alternative to conventional diesel, supporting cleaner combustion technologies for the energy sector.

Chapter 9 explores geometry-driven optimization of orifice plates used for measuring viscous fluid flow. The chapter details how variations in shape, width, and flow coefficients affect velocity and pressure characteristics, presenting computational strategies for improving design efficiency. The findings have broad applicability in industrial metering and control systems, where precision fluid measurement underpins process optimization and energy conservation.

Chapter 10 presents a comparative analysis of fin materials and convection effects for thermal management of electric vehicle batteries. Using numerical simulation, the study demonstrates how optimized fin configurations enhance heat dissipation while maintaining compactness. The research contributes to safer, more efficient battery systems and reinforces the importance of material selection in the design of next-generation energy storage technologies.

Chapter 11 analyzes the performance of a low-charge ammonia system integrated with a CO2-based natural circulation loop for cold storage applications. The study demonstrates that such systems can achieve effective cooling with reduced refrigerant charge, enhancing safety while maintaining performance. This hybrid approach supports the transition to natural refrigerants and reflects a commitment to environmentally responsible refrigeration solutions.

Chapter 12 provides a comprehensive discussion on renewable energy applications within thermo-fluid systems. The chapter synthesizes advances in solar, wind, biomass, and geothermal technologies, illustrating how hybrid configurations enhance resilience and efficiency. Through the integration of computational tools and material innovations, the study underscores how renewable energy can be effectively harnessed to meet the global demand for sustainable power.

Chapter 13 simulates a desiccant-based air-conditioning system incorporating a vortex tube as a cooling source. The analysis demonstrates how coupling desiccant dehumidification with vortex tube cooling can achieve efficient thermal comfort in humid climates without reliance on conventional refrigerants. The findings highlight the potential of hybrid cooling systems to reduce energy consumption and environmental impact in building applications.

Chapter 14 investigates solar-based air conditioning systems enhanced with latent thermal energy storage using ice as the phase change material. The numerical study demonstrates the trade-offs between compact system design, energy storage efficiency, and solar dependence. By examining realistic operational constraints, the research contributes to the development of viable solar-assisted cooling solutions for energy-constrained regions.

Chapter 15 examines the steady flow behavior around a square cylinder, providing numerical insights into drag, vorticity, and flow stability at moderate Reynolds numbers. The study extends existing fluid dynamics understanding beyond traditional circular geometries and aids in predicting flow-induced forces in structural and aerodynamic applications. These findings have implications for optimizing design elements in both civil and mechanical engineering systems.

Chapter 16 concludes the volume with a theoretical study of zeotropic mixtures of carbon dioxide and hydrofluoroolefins for refrigeration and air-conditioning use. The research evaluates thermodynamic and environmental properties, identifying blends that balance low global warming potential with favorable performance characteristics. This work reinforces the broader theme of sustainability by advancing refrigerant technologies that align with global environmental objectives.

As editors, we envision this book as both a reflection of current progress and a catalyst for future exploration. It underscores the vital role of thermo-fluid systems in global sustainability and encourages continuous innovation at the intersection of science, engineering, and environmental stewardship.

We extend our sincere gratitude to all contributing authors, reviewers, and collaborators for their dedication and insight. Their efforts have made this volume a meaningful contribution to the ongoing dialogue on sustainable engineering.

Kamalakanta Muduli
Papua New Guinea University of Technology, India

Sachindra Rout
C. V. Raman Global University, India

Bukke Naik
National Institute of Technology, India

Albert Ude
Botswana International University of Science and Technology, Botswana

Chapter 1
A Computational Study of Rivet Ribs in Solar Air Heater Ducts

Sajal Chanda
National Institute of Technology, Rourkela, India

Kashinath Dhamudia
National Institute of Technology, Rourkela, India

Jnana Ranjan Senapati
https://orcid.org/0000-0002-2672-3763
National Institute of Technology, Rourkela, India

ABSTRACT

A two-dimensional computational study is carried out on the solar air heater duct (SAHD) by attaching rivet ribs on the collector plate at constant rib height and relative roughness height of (e =1.4 mm and e/Dh = 0.042). The finding of this problem was done with a computational exercise in ANSYS Fluent R22 by varying dimensionless parameters like Reynolds number (4000-18000), and relative width of rivet shank to height ratio (w/e) of 0.285 to 0.714 by keeping constant pitch distance (P=10 mm). The investigation of the domain was done, with a uniform heat supply of 1000 W/m2 over the collector plate made up of aluminium. From various turbulence models, the RNG k-ε model with enhanced wall treatment (EWT) was chosen for simulating the fluid flow equations (mass, momentum, and energy). The result showed that the Nusselt number (Nur) of roughed plate rises with the increment of Reynolds number (Re), also found average temperature of 361 K with maximum average Nusselt number enhancement factor (Nur / Nus) of 2.53 and thermal enhancement factor (TEF) of 1.78 at Re= 18000 at w/e of 0.36.

DOI: 10.4018/979-8-3373-3621-3.ch001

1. INTRODUCTION

The importance of various forms of energy has grown significantly in recent decades due to essential contribution in driving industrialization and fostering global economic growth. Solar energy, in particular, is seen as a vital solution for addressing climate change and satisfying the increasing energy demands within the context of sustainable development. Solar radiation is a clean, limitless energy source that is available for free. During the 1970s, intensification in oil prices and the pollution caused by non-renewable energy challenged the research community to devote sustainable research work. Despite being challenging and expensive, switching to non-traditional energy sources such as solar energy and maximizing its benefits is essential for society. Indeed, solar power offers various applications, with one typical example being heating air by solar air heater duct (SAHD). It is used extensively in process industries, drying applications, and room heating. One major shortcoming of SAHD is the relatively low heat removal from collector plate to the working fluid. Air has a low density with minimal heat transfer coefficient. Indeed, there are two primary means to maximizing heat removal in SAHDs:

i) With an increment of the active surface area of the collector plate by adding ribs or grooves, there will be enhancement heat transfer.
ii) Creating swirl or turbulence within the fluid flow domain promotes better blending, that enriches heat transferal efficiency.

The enhancement in heat flow is done by maximizing the heat transmission area of SAHD with the association of ribs, which creates turbulence in fluids. However, this approach may also lead to undesired pressure losses in the fluid domain, which can be challenging to manage. It's crucial to strike a delicate balance between improving heat transfer efficiency and minimizing pressure losses when designing solar air heaters.

2. LITERATURE REVIEW AND OBJECTIVE

The researchers used experimental methods to study solar air heaters by attaching roughed struts in absorber plates. (Prasad, B.N and Saini, 1988) conducted research to explore the impact of ribbed surface over SADH. Their findings indicated that both Nusselt number and friction factor decrease when (*P/e*) drops below 8 to 10. This effect is attributed to a deficiency of a reattachment-free shear layer, resulting in flow separation. (Ahn, 2001) studied heat removal and friction factor in rectangular ducts with ribbed surfaces. His findings indicated that triangular ribs raise heat removal more effectively than square ribs. Additionally, he concluded that ribbed channels offer superior performance compared to smooth channels. (S. K. Sharma & Kalamkar, 2017) did a study on the SAH with thin ribs, where the blockage effect was maintained at e/H=0.1. It was observed that the optimal TEF was achieved when $8 \leq P/e \leq 10$. Performance outside this range was suboptimal due to reduced heat transfer, which was attributed to low reattachment of the free shear. (Gawande et al., 2016) executed experiment and CFD investigation with reverse L-shaped ribs at a constant e/D_h of 0.042, and found at Re = 15000, Nusselt number has maximum augmentation of 2.87 times more w.r.t smooth duct. (Hans et al., 2010) inspected V-ribs over the collector plate and discovered that this artificial roughness resulted in an enhancement of the Nusselt number and friction factor by approximately 6 and 5 times, respectively, compared to a smooth duct. (Karwa et al., 2001) explored transverse, rectangular, chamfered ribs. They found an enhancement of thermal efficiency from 10 to 40% due to a roughened surface on the absorber plate, whereas the Nusselt numbers increased by 50–120% at an e/D_h of 0.0441. (Alam et al., 2014) used V-shaped perforated blocks in a rectangular duct to experiment by varying (e/H) from 0.4 to 1. They conclude that 0.8 has the maximum TEF. (Skullong et al., 2015) done a proportional study on rib size arrangement in an SAH for forced convective heat transfer and concluded thin ribs have a higher heat removal rate. By using six different baffles, (S. Sharma et al., 2021) carried out a computational with an experimental assessment of SAH. The maximum TEF is 2.05 at Re of 15000. (Mahanand & Senapati, 2020) has done a comparative study to find the TEF of a SAH inverted-T ribs over heated surface at constant e/D_h = 0.042. They found at 15000 Reynolds number, P/e = 7.14 has a maximum value of TEF of 1.86. (Saini & Verma, 2008) researched dimple-shaped artificial ribs and resolute maximum Nusselt number and friction factor values were achieved at e/D_h of 0.0379 and 0.0289, respectively, while maintaining a constant P/e of 10. (Singh et al., 2019) did a comparative study with multiple broken transverse and square-waved ribs. Their findings reveal a maximum TEF of 2.10 and 1.62 at Re of 15000 at e/D_h = 0.043. (Singh Yadav & Bhagoria, 2015) investigated the SAHD featuring square section transverse ribs, examining a range of Re from 3800 to 18000 and P/e ratios in the range of 7.14 to 17.86. Their study revealed that the highest TEF = 1.82 was reached at a Re of 18000 and a P/e ratio of 7.14. (Kumar et al., 2019) investigated carried on triangular duct SAH with square ribs, with variations in the (e/D_h) and (P/e). Their research revealed that the maximum TEF of 1.97 is achieved at a (P/e) of 10.In review, (Yadav & Bhagoria, 2013) examined the CFD approach, specifically focusing on the RNG k-ε turbulence model. They concluded that when considering 2D analysis models, the results tend to closely align with experimental findings compared to 3D simulations. Besides these literature, several studies on different rib turbulators in SAHD can be found by reviewing various research works by (Bhushan & Singh, 2010) and (Mund et al., 2021)) have been performed to assess the performance of SAHD featuring ribs on the collector plate. The survey shows that while many researchers have used various artificial tabulator shapes to enhance TEF, none have explored a rivet with an elliptical head and smooth, rounded body with varying shank width. This design induces secondary flow structures, improving fluid mixing, temperature uniformity, and thermal performance, and, to our knowledge, no published CFD study has examined this rib geometry.

The analysis fluid domain has inlet and exit section with an absorber plate (as test section) where, artificial rivet ribs are attached as shown in Figure 1 The dimensions of SAHD are chosen as per ASHRAE 93-2010 standards (Anon, 1977). The fluid is in the fully developed condition at the collector plate. For passing air, a height H =20 mm is taken for this model.

In the present study, we comprehend results by varying relative width of rivet shank to height ratio and Reynolds number to get optimal rib configuration.

Figure 1. Schematic of two-dimensional computational analysis with boundary conditions flow domain.

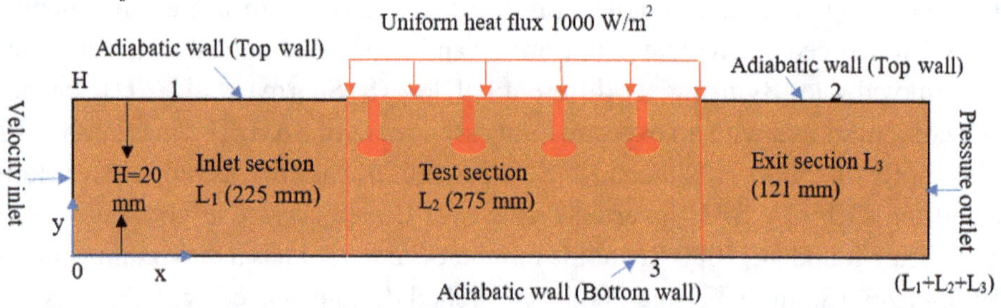

2.1 Boundary Condition and Solution Method

For this investigation, air at 300 K is used as the analysis fluid. Here, Reynolds numbers varied from 4000 to 18000, considering adiabatic boundary conditions for the wall (1,2 and 3), inlet (velocity), outlet (pressure), and uniform heat flux at absorber plate as shown in Figure 1. The steady-state conformation was achieved by solving basic fluid flow equations, i.e., energy, momentum, and continuity. The (SIMPLE) algorithm was adopted with a second-order upwind scheme. The converging criteria of 10^{-6} were set for continuity, momentum, and turbulence equations, and 10^{-9} chosen energy equation.

2.2 Meshing and Grid Independence Test

A non-uniform meshing is adopted for smooth and rough duct, shown in Figure: 2. The grid independence was achieved by varying elements from 359117 to 231784, and node size 304742 has the best result with a % difference of 0.917.

Figure 2. Meshing rough duct SAHD.

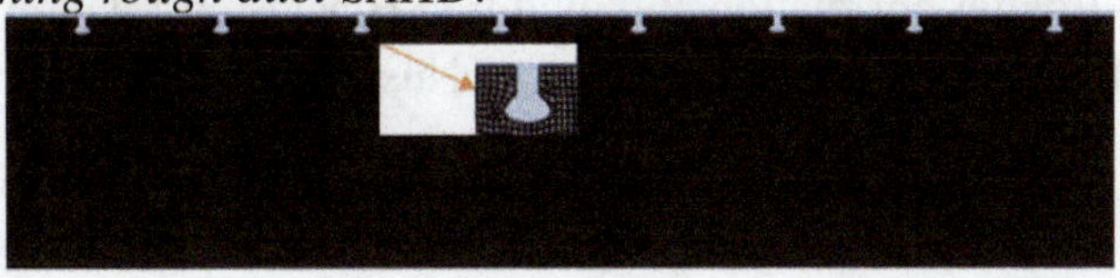

2.3 Validation and Model Selection

The computational results for the rough duct were obtained using the mathematical relations provided in reference (Mahanand & Senapati, 2020) and the abbreviations are listed in nomenclature .

$$h = \frac{Q}{A_P} / \left(T_a - T_m \right) \quad (1)$$

Where, $\frac{Q}{A_P} = 1000\ W/m^2$,

h= convective heat transfer

T_a= avgerare absorber plate temperature and

$T_m = \frac{T_i + T_o}{2}$ mean fluid temperature

The Nusselt number was calculated by

$$Nu_r = \frac{hD_h}{k} \quad (2)$$

The Nusselt number is ratio of convective to conductive heat transfer across fluid boundary, and it helps to calculate how well heat is moving from hot medium to cooler medium, also give better understanding the natural phenomenon in heat exchangers. Present analysis Nusselt number evaluated by considering the duct's width (W) and depth (H) 100 mm and 20 mm respectively, to calculate D_h as follows.

$$D_h = \frac{4W \times H}{2(W+H)}, \tag{3}$$

The 'f_r' is determined

$$f_r = \frac{D_h \Delta p}{2\rho L_2 V^2} \tag{4}$$

The Dittus–Boelter correlation and modified Blasius correlation were used by (Dhamudia & Senapati, 2025) to estimate the Nusselt number and friction factor.

$$N_u = 0.023 R_e^{0.8} P_r^{0.4} \tag{5}$$

$$f = 0.85 R_e^{-0.25} \tag{6}$$

To reach maximum accuracy, five turbulence models were analyzed by numerical simulation in a duct with no rough surface. A rigorous analysis was done on Nusselt number by varying Re from (4000 to 18000) with turbulence models and Dittus-Boelter correlation. It is observed the average Nusselt number variation is 4.6 to 10.8%, where RNG k–ε with EWT has the lowest deviation. the validation of rough and smooth RNG k–ε with EWT duct turbulence model with the experimental result of square rib from the literature (Skullong et al., 2015). Figure.3 illustrated maximum deviation of ±5% and both rough ducts has higher Nusselt number compare to smooth duct.

Figure 3. Validation of numerical model with experimental results of square ribs of P/e = 6.67.

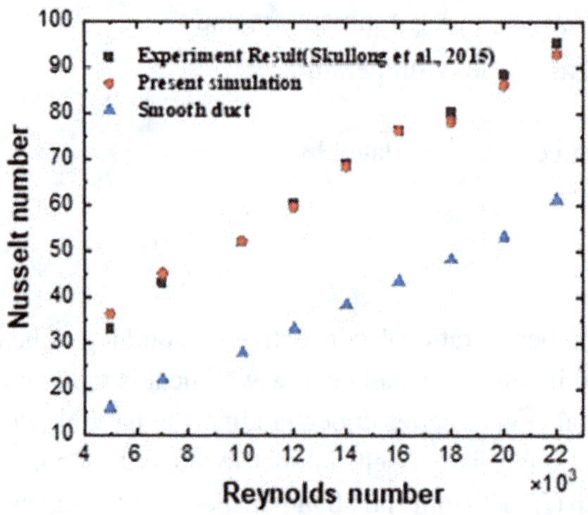

3. RESULTS AND DISCUSSION

The *w/e* is a lateral to vertical dimension of surface protrusion. A lower *w/e* reduces flow resistance and raises flow intensity, which generally promotes '*h*'. At the same time, when the rib width becomes too small, the separated flow on the downstream side reattaches less effectively, which can lower the local heat-transfer coefficient. Therefore, selecting an appropriate rib width is important. Figure 4 compares the variation of the (*h*) with velocity for different w/e values. It shows that '*h*' increases steadily as velocity rises and reaches its highest value when *w/e* = 0.36, indicating this ratio provides the best balance between pressure drop and fluid reattachment.

Figure 4. Heat transfer coefficient vs Velocity.

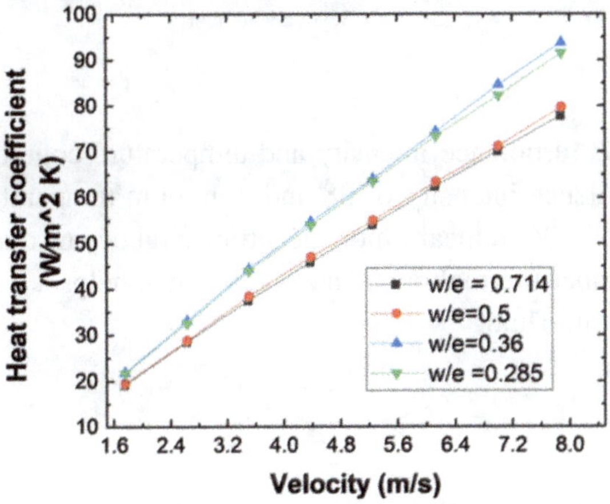

A comparison was made by varying 'Nu_r' vs Re for different cases shown in Figure: 5. It shows enhancement of Nusselt happens with rise in Reynolds numbers for every case. This upward trend arises from the accompanying rise in the (*h*) as the flow becomes more vigorous.

The comparison of average Nusselt number enhancement factor (Nu_r/Nu_s) and average friction factor enhancement ratio (f_r/f_s) was carried out with Reynolds number which is shown in Table 1, The results show that (Nu_r/Nu_s) gradually increases with the increase of Reynolds number from 4000 to 18000. The maximum value of 2.53 was achieved at Re = 18000. It shows higher heat transfer enhancement due to enhancement in turbulence intensity. However, (f_r/f_s) is increase at width of rivet shank of 0.7 mm and 1 mm this cause due to generation of more pressure drop near the ribs.

Figure 5. Nu_r vs Reynolds number.

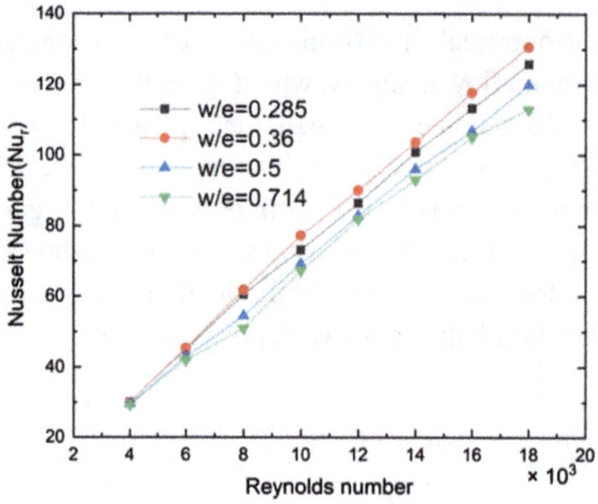

Figure 6 shows turbulence intensity and temperature contours at Re= 18000. It illustrates turbulence intensity of 38 and minimum temperature of 361 K was achieved for w/e of 0.36. It means more heat transferal occurs due to proper mixing of fluid, leads thinner thermal boundary layer, which helps a better heat augmentation of surface with fluid.

Figure 6. Turbulence intensity and temperature contours at Re =18000 and w/e of 0.285,0.36,0.5 and 0.714

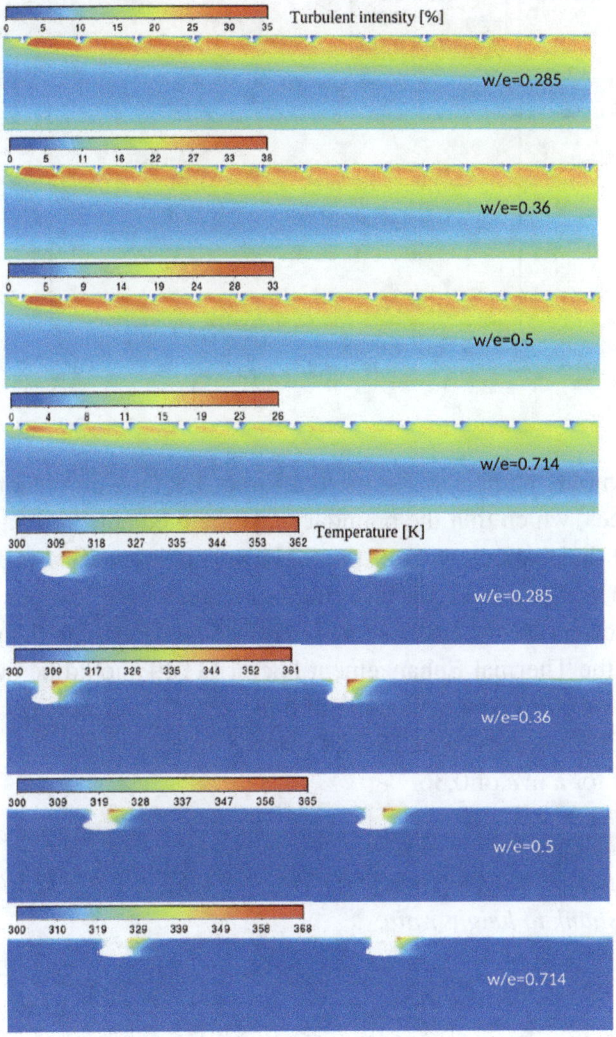

When artificial roughness is attached in SAHD, the performance parameter is defined as

$$\text{Thermal enhancement factor}\left(\text{TEF}\right) = \frac{\left(\frac{Nu_r}{Nu_s}\right)}{\left(\frac{f_r}{f_s}\right)^{1/3}}. \quad (7)$$

Figure 7. TEF vs Reynolds number.

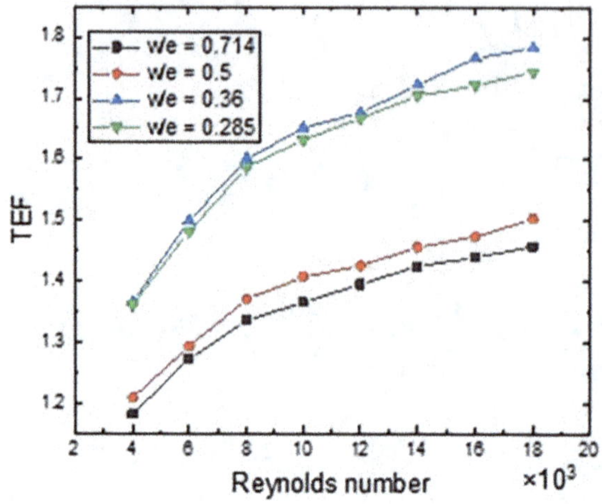

As the Reynolds number rises, inertial forces become dominant and generate stronger vortices, which thin the boundary layer and enhance mixing between the core fluid and the channel walls of the SAHD. This intensified mixing improves the convective heat-transfer rate. In addition, increasing Re leads to greater turbulence, which lowers the friction factor. Therefore, to determine the most effective rib geometry, the Thermal Enhancement Factor (TEF) should be evaluated using equation 7 where (f_r/f_s) play major role in minor modification of geometry. Figure 7 shows that TEF increases with Re and reaches an optimum value of about 1.783 at Re = 18000 for a *w/e* of 0.36.

Table 1. Variation of (Nu_r/Nu_s) and (f_r/f_s) with Reynolds number at different relative width of rivet shank to height ratio.

	(Nu_r/Nu_s)				(f_r/f_s)			
Re	**w/e= 0.714**	**w/e= 0.5**	**w /e= 0.36**	**w /e= 0.285**	**w/e= 0.714**	**w/e= 0.5**	**w /e= 0.36**	**w /e= 0.285**
4000	1.72	1.75	1.96	1.95	**3.089**	3.039	2.962	2.948
6000	1.85	1.87	2.15	2.12	3.076	3.026	2.951	2.933
8000	1.94	1.98	2.29	2.27	3.068	3.019	2.931	2.928
10000	1.98	2.03	2.36	2.33	3.051	3.007	2.920	2.912
12000	2.02	2.05	2.39	2.37	3.036	2.976	2.894	2.874
14000	2.06	2.09	2.45	2.42	3.027	2.957	2.877	2.858
16000	2.08	2.11	2.51	2.44	3.015	2.935	2.869	2.847
18000	2.10	2.15	**2.53**	2.47	2.985	2.925	2.854	2.839

4. CONCLUSION

The article presents the variation of relative width of rivet shank to height ratio at constant pitch length (p = 10 mm) in SAHD.

I. The results illustrated the '(Nu_r/Nu_s)' increases with Re. For w/e = 0.36 has highest value of 2.53 achieved at Re of 18000.
II. The highest average enhancement friction factor of 3.089 occurs at Reynolds number of 4000 at w/e = 0.714.
III. The highest heat augmentation and the highest TEF of 1.783 is attained at *w/e* = 0.36 and *Re* of 18000.
IV. The novelty of the present study lies in using a rivet with an elliptical head and a smooth, rounded body at an optimum w/e, which encourages the formation of additional secondary flow structures. This enhances fluid mixing across the cross-section, resulting in a more uniform temperature distribution and improved thermal performance.

Box 1. Nomenclature

A_p	Area of absorber plate,	(m^2)	**Greek letters**		
T_a	Average absorber plate temperature,	K)	ρ	Air density	(Kg / m^3)
H	Convective heat transfer coefficient	(W/ m^2K)	k	Thermal conductivity,	(W/m K)
H	Depth of duct,	(mm)	P_r	Prandtl number	
f_r	Friction factor for rough duct.		R_e	Reynolds number	
f	friction factor for smooth duct				
D_h	Hydraulic diameter,	(mm)	**Dimensionless parameters**		
T_i	Inlet air temperature,	(K)	e/D_h	Relative roughness height	
T_m	Mean fluid temperature,	(K)	f/f_s	average friction factor enhancement ratio	
T_o	Out let air temperature,	(K)	Nu_r/Nu_s	average Nusselt enhancement factor	
P	Pitch distance	(mm)	P/e	Relative roughness pitch.	
ΔP	Pressure drop across test section	(N/ mm^2)	Re	Reynolds number	
p/e	Relative roughness pitch.				
w/e	Relative width of rivet shank to height ratio				
E	Rib height,	(mm)			
Nu_r	Rough duct Nusselt number				
Nu_s	Smooth duct Nusselt number				
TEF	Thermal enhancement factor				

continued on following page

Box 1. Continued

W	Width of duct,	(mm)			
w	Width of rivet shank	(mm)			

ACKNOWLEDGEMENT

I would like to express my sincere gratitude to NIT, Rourkela to provide all research facilities and expertise to insights my work for successful completion.

REFERENCES

Ahn, S. W. (2001). The effects of roughness types on friction factors and heat transfer in roughened rectangular duct. *International Communications in Heat and Mass Transfer*, *28*(7), 933–942. https://doi.org/https://doi.org/10.1016/S0735-1933(01)00297-4

Alam, T., Saini, R. P., & Saini, J. S. (2014). Experimental investigation on heat transfer enhancement due to V-shaped perforated blocks in a rectangular duct of solar air heater. *Energy Conversion and Management*, *81*, 374–383. DOI: 10.1016/j.enconman.2014.02.044

Anon. (1977). Methods of Testing To Determine the Thermal Performance of Solar Collectors. *ANSI Stand B198. 1, 2010*(Ra).

Bhushan, B., & Singh, R. (2010). A review on methodology of artificial roughness used in duct of solar air heaters. *Energy*, *35*(1), 202–212. DOI: 10.1016/j.energy.2009.09.010

Dhamudia, K., & Senapati, J. R. (2025). A computational study for enhancement in thermal performance of plus(+)-shaped ribbed solar air heater duct. *International Communications in Heat and Mass Transfer*, *161*. Advance online publication. DOI: 10.1016/j.icheatmasstransfer.2024.108411

Gawande, V. B., Dhoble, A. S., Zodpe, D. B., & Chamoli, S. (2016). Experimental and CFD investigation of convection heat transfer in solar air heater with reverse L-shaped ribs. *Solar Energy*, *131*, 275–295. DOI: 10.1016/j.solener.2016.02.040

Hans, V. S., Saini, R. P., & Saini, J. S. (2010). Heat transfer and friction factor correlations for a solar air heater duct roughened artificially with multiple v-ribs. *Solar Energy*, *84*(6), 898–911. DOI: 10.1016/j.solener.2010.02.004

Karwa, R., Solanki, S. C., & Saini, J. S. (2001). Thermo-hydraulic performance of solar air heaters having integral chamfered rib roughness on absorber plates. *Energy*, *26*(2), 161–176. DOI: 10.1016/S0360-5442(00)00062-1

Kumar, R., Kumar, A., & Goel, V. (2019). Performance improvement and development of correlation for friction factor and heat transfer using computational fluid dynamics for ribbed triangular duct solar air heater. *Renewable Energy*, *131*(February), 788–799. DOI: 10.1016/j.renene.2018.07.078

Mahanand, Y., & Senapati, J. R. (2020). Thermal enhancement study of a transverse inverted-T shaped ribbed solar air heater. *International Communications in Heat and Mass Transfer*, *119*(October), 104922. DOI: 10.1016/j.icheatmasstransfer.2020.104922

Mund, C., Rathore, S. K., & Sahoo, R. K. (2021). A review of solar air collectors about various modifications for performance enhancement. *Solar Energy*, *228*(August), 140–167. DOI: 10.1016/j.solener.2021.08.040

Prasad, B.N and Saini, J. S. (1988). *Effect of artificial roughness on heat transfer and friction factor in a solar air heater. 41*(6), 555–560.

Saini, R. P., & Verma, J. (2008). Heat transfer and friction factor correlations for a duct having dimple-shape artificial roughness for solar air heaters. *Energy*, *33*(8), 1277–1287. DOI: 10.1016/j.energy.2008.02.017

Sharma, S., Das, R. K., & Kulkarni, K. (2021). Computational and experimental assessment of solar air heater roughened with six different baffles. *Case Studies in Thermal Engineering*, *27*(July). Advance online publication. DOI: 10.1016/j.csite.2021.101350

Sharma, S. K., & Kalamkar, V. R. (2017). Experimental and numerical investigation of forced convective heat transfer in solar air heater with thin ribs. *Solar Energy*, *147*, 277–291. DOI: 10.1016/j.solener.2017.03.042

Singh, I., Vardhan, S., Singh, S., & Singh, A. (2019). Experimental and CFD analysis of solar air heater duct roughened with multiple broken transverse ribs: A comparative study. *Solar Energy*, *188*(February), 519–532. DOI: 10.1016/j.solener.2019.06.022

Singh Yadav, A., & Bhagoria, J. L. (2015). Numerical investigation of flow through an artificially roughened solar air heater. *International Journal of Ambient Energy*, *36*(2), 87–100. DOI: 10.1080/01430750.2013.823107

Skullong, S., Thianpong, C., & Promvonge, P. (2015). Effects of rib size and arrangement on forced convective heat transfer in a solar air heater channel. *Heat and Mass Transfer/Waerme- Und Stoffuebertragung, 51*(10), 1475–1485. https://doi.org/ DOI: 10.1007/s00231-015-1515-5

Yadav, A. S., & Bhagoria, J. L. (2013). Heat transfer and fluid flow analysis of solar air heater: A review of CFD approach. *Renewable & Sustainable Energy Reviews*, *23*, 60–79. DOI: 10.1016/j.rser.2013.02.035

Chapter 2
Capillary Rise Behaviour in Biporous and Monoporous Stainless Steel Wicks for Loop Heat Pipe Applications:
A Comparative Study on Fluid Wicking Dynamics Using Water, Ethanol, and Methanol

Seshraj Pratihari
https://orcid.org/0009-0009-3606-7569
Indian Institute of Technology, Guwahati, India

Toni Kumari
Indian Institute of Technology, Guwahati, India

Manmohan Pandey
Indian Institute of Technology, Guwahati, India

Marco Marengo
University of Pavia, Italy

ABSTRACT

The performance of loop heat pipes (LHPs) largely depends on the capillary action within their porous wicks, which drives fluid circulation. In this study, stainless steel wicks with monoporous and biporous structures were tested using methanol,

DOI: 10.4018/979-8-3373-3621-3.ch002

ethanol, and water to evaluate capillary rise behavior. Results show that biporous wicks significantly outperform monoporous ones, enabling faster fluid transport. This performance improves further with higher biporosity, achieved by adjusting pore former content during fabrication. Infrared imaging confirmed the faster fluid movement in biporous wicks, and SEM analysis revealed distinct differences in pore structure and distribution. These findings suggest that biporous wicks can substantially enhance the fluid pumping and thermal performance of LHPs, offering valuable direction for future wick design.

1. INTRODUCTION

Loop heat pipes (LHPs) have played an essential role in thermal management for aerospace systems, high-power electronics, and energy-efficient applications owing to their passive operation and high heat transfer capability (Ku, 1999; Ambirajan et al., 2012. The basis of LHP performance is the porous wick structure of the evaporator: it relies on capillary action to drive fluid circulation, compensating for pressure differences across the working fluid's liquid-vapor interface (Ku, 2016; Altschuh et al., 2022). Although traditional monoporous wicks offer baseline performance, their mono-distributed pore structure typically imposes a compromise between the capillary pressure and permeability, which directly compromises effective heat transport and start-up assurance (Shen et al., 2024; Yang et al., 2024).

Biporous designs that take advantage of a bimodal pore distribution have been used in recent wick designs to decouple these opposing parameters. Both silicon-based and aluminium-based wicks demonstrate how biporous structures increase capillary pressure by promoting permeability through interconnected macropore networks and using fine pores for meniscus initiation (Shen et al., 2024; Yang et al., 2024). The different pore structures will significantly affect the capillary rise behaviour and fluid transport capabilities of the respective wick. What contributes to overall better performance in loop heat pipe applications, is that the larger pore sizes tend to assist in facilitating better permeability and better fluid distribution while the smaller pores help to hold the capillary pressure and this entire combination would tend to provide an overall improved performance.

However, there have been surprisingly few systematic comparisons of stainless steel wicks, which are a tempting alternative due to their mechanical stability, resistance to corrosion, and compatibility with aggressive working fluids.

In comparison, stainless steel wicks have many more benefits for LHP systems, such as high strength to weight ratio, high thermal stability, and low degradation in extreme environments. These attributes position stainless steel as an ideal material for use in long-term, reliable operation across a range of thermal management applications. In addition, the precise control of pore structure during the sintering of stainless-steel powders provide unprecedented opportunities for tuning wick performance. The capillary rise function of several stainless-steel wicks featuring controlled monoporous and biporous structures were studied using methanol, ethanol, and water as working fluids. We align pore former concentration during fabrication to capillary kinetics, infrared ((IR)) fluid-front tracking, and Scanning Electron Microscope (SEM) based morphological analysis to close three major gaps:

Performance Benchmarking: Quantitative comparison of monoporous vs biporous stainless steel wicks with the same fluids and same geometry.
Fabrication-Property Relationships: Dosage effect of pore former on biporous nature improvement of stainless steel wicks and its implications for capillary pumping.
Fluid Transport Visualization: Direct observation of fluid redistribution dynamics through IR thermography in stainless steel wick structures (Akafuah et al., 2010).

These results yield specific recommendations for enhanced stainless steel wick designs for future LHPs targeting rapid thermal response and large heat flux dissipation. Reconciling the complex relationships across pore hierarchy, surface morphology, and fluid transport kinetics in stainless steel wicks, this work significantly promotes the fundamental understanding of capillary-driven phase change systems and provides practical directions toward the manufacture of thermal management solutions achieving high performance (Zhao et al., 2024; Nashine et al., 2025).

While biporous wicks have been extensively explored in aluminium and silicon wicks, systematic studies on stainless steel wicks remain scarce despite their high thermal stability, mechanical robustness, and fluid compatibility. This study fills that gap by providing the first direct quantitative comparison of monoporous and biporous stainless steel wicks across multiple working fluids, linking fabrication parameters to fluid transport performance.

2. MATERIALS AND WICK FABRICATION

The fabrication of monoporous and biporous stainless steel wicks is carried out through a controlled process. First, stainless steel powder is measured to the required volume for monoporous wicks. The powder is then mixed with a binder, which in this case is silicone oil, at about fifteen percent of the wick material volume. This mixture serves as the base for all the wick structures studied (Kumari & Pandey, 2022).

Biporous wicks are created with one additional step that uses sodium chloride (NaCl) as pore former. NaCl is blended with the stainless steel powder and silicon oil mixture, and the amount of NaCl is varied between 20% and 40% by volume to generate variable levels of biporous nature from 20% to 40%. This variation in NaCl content allows for the creation of wicks with distinct biporous characteristics.

This study focused on the comparative study between monoporous wicks and biporous with various amount of biporous nature. In particular, the study looked at:

- Monoporous wicks (without addition of NaCl)
- 20% biporous wicks
- 30% biporous wicks
- 40% biporous wicks

The percentages stated for the biporous wicks demonstrate the proportion of added NaCl to the mixture of stainless steel and silicon oil. By systematically varying the amount of pore former in the wicks, we are able to explore the impact of increased biporous nature on the capillary rise behaviour and performance of the stainless steel wicks.

Figure 1. Wick Fabrication Procedure

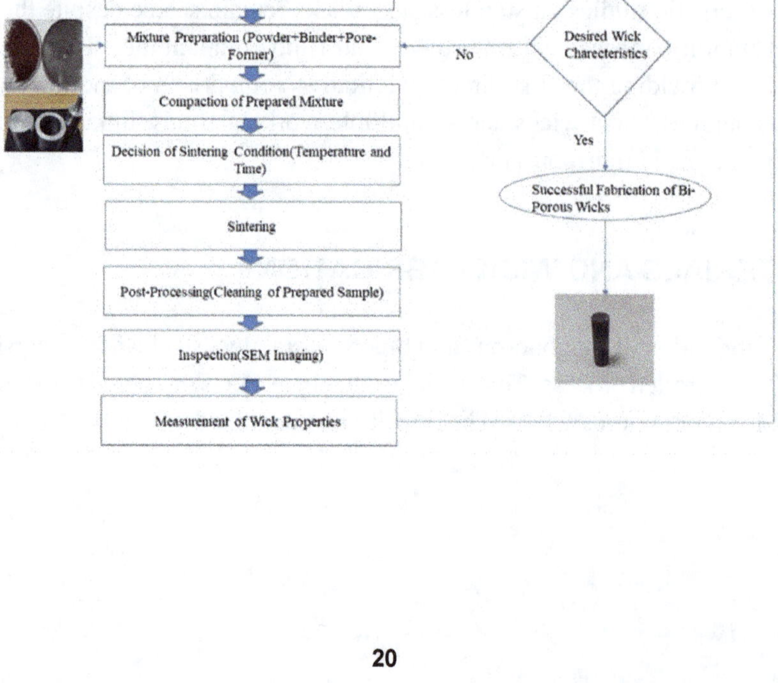

Figure 2a. Monoporous Wick

Figure 2b. 20% Biporous Wick

Figure 2c. 30% Biporous Wick

Figure 2d. 40% Biporous Wick

Figures 2(a) to 2(d) provided above show monoporous and biporous wicks with varying degrees of biporous nature.

2.1 Wick Comparison

The series of figures from figure 2(a) to 2(d) showcases four distinct wick structures:

1. **Monoporous Wick**: The sintering of stainless steel powder combined solely with the silicon oil binder gives this wick a consistent pore structure throughout. The pores are uniformly spaced and rather tiny.
2. **20% Biporous Wick**: With the addition of 20% NaCl as a pore former, this wick begins to show a bimodal pore distribution. Some larger pores are visible, created by the spaces left after NaCl removal, interspersed with the smaller pores of the base structure.
3. **30% Biporous Wick**: As the NaCl content increases to 30%, a more pronounced biporous structure is evident. The larger pores become more numerous and potentially more interconnected, while still maintaining a significant presence of smaller pores.
4. **40% Biporous Wick**: This wick, with the highest NaCl content, displays the most pronounced biporous character. The larger pores are more dominant and likely form a more extensive network throughout the wick structure.

These images are visually representative of the progressive increase of biporous nature from the monoporous wick at left to the 40% biporous wick at right.

3. EXPERIMENTAL METHODOLOGY

3.1 Experimental Setup

Figure 6 shows the schematic of capillary rise experiment whereas figure 7 shows capillary rise setup. The experimental setup for investigating capillary rise behavior in stainless steel wicks consists of several key components, each serving a specific purpose in the measurement and analysis process:

Figure 3. Schematic of Capillary Rise Experiment

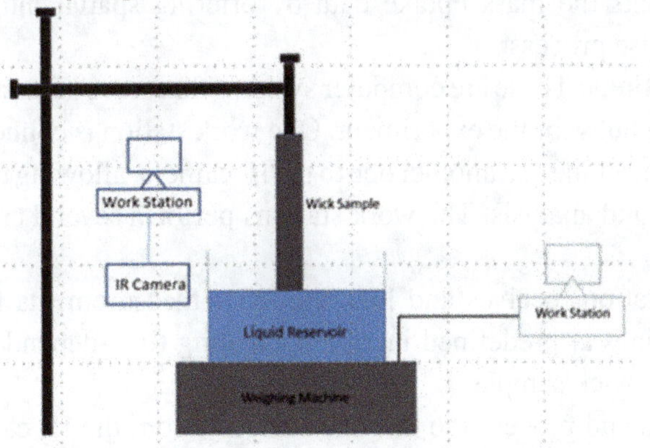

Figure 4. Capillary Rise Setup

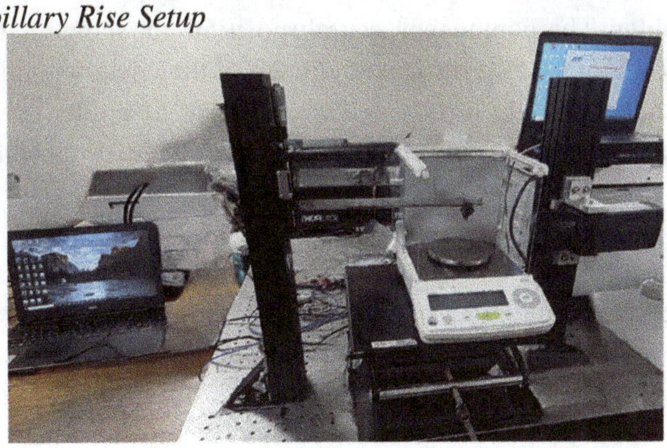

Digital Weighing Machine: This high-precision instrument is utilized for weighing the mass of fluid absorbed by the wick in a time-variant condition. Using this principle, the liquid flowing through the porous structure can be measured qualitatively by observing the increase in mass as the liquid rises through its pores. Data from the weighing machine are generally saved at certain time intervals to formulate a mass uptake profile that is time dependent.
IR Camera: An IR camera is employed to visualize the fluid front progression within the wick. This non-invasive technique captures the temperature differences between the dry and wetted portions of the wick, providing real-time, two-dimensional mapping of the liquid distribution. The IR imagery complements the mass uptake data by offering spatial information on the capillary rise process.
Work Stations: These are computer systems that serve as data acquisition and processing hubs for the experiment. One work station is connected to a digital weighing machine and another one to the IR camera, allowing for real-time data collection and analysis. The work stations perform several crucial functions:

One work station receives and logs the mass measurements from the digital weighing machines at predefined intervals, creating time-dependent mass uptake profiles for each wick sample.

They capture and process the thermal imagery from the IR cameras, enabling real-time visualization of the fluid front progression within the wicks.

Liquid Reservoir: A container holding the working fluid (methanol, ethanol, or water) is positioned beneath the wick.
Wick Holder: This component securely positions the stainless steel wick sample. It allows for precise vertical alignment of the wick, with its bottom edge just touching the liquid surface in the reservoir. The holder is typically made of a material that does not interfere with the IR measurements and ensures that only the wick material is in contact with the working fluid.

This integrated setup enables simultaneous measurement of mass uptake and visual tracking of the capillary rise process, providing comprehensive data on the fluid transport characteristics of both monoporous and biporous stainless steel wicks.

In this study, a high-precision digital weighing machine (SHIMADZU-TW423L) plays a crucial role in measuring the mass of fluid absorbed by the wick over time. As the liquid rises through the porous structure, its increasing mass is recorded, allowing for a qualitative assessment of the fluid's movement. The weighing machine used in this experiment has a maximum capacity of 420 gram and a minimum display of 0.001 gram.

The weighing machine captures data at specific time intervals, forming a time-dependent mass uptake profile that helps characterize the capillary rise behavior of the wicks. To complement these measurements, an IR camera is employed to visualize the progression of the fluid front within the wick. The IR camera used in this experiment is FLIR A655SC having resolution 640 x 480, pixel spectral range of 7.5-14um, thermal sensitivity of 0.03°C at 30°C, and an accuracy of ±2°C or ±2% of reading.

Two dedicated computer serve as data acquisition and processing hubs, ensuring seamless real-time monitoring of the experiment. One PC is linked to the digital weighing machine, continuously logging mass measurements and generating uptake profiles. Meanwhile, the second PC processes thermal imagery from the IR camera, allowing for immediate visualization of the fluid's movement within the wick.

Together, these systems enable a synchronized approach to data collection, merging quantitative mass readings with qualitative spatial analysis.

At the heart of the experimental setup, a liquid reservoir holds the working fluid (methanol, ethanol, or water) positioned directly beneath the wick. To ensure consistency in measurements, a wick holder securely maintains the stainless steel wick in a precise vertical alignment, with its bottom edge just touching the liquid surface. Made from materials that do not interfere with IR imaging, the holder ensures that only the wick material interacts with the fluid, preventing any unintended influences on the results. This carefully integrated system allows for the simultaneous measurement of mass uptake and fluid front progression, providing a comprehensive understanding of fluid transport in both monoporous and biporous stainless steel wicks.

3.2 Capillary Rise Experiments

The capillary rise experiments were designed to quantitatively assess the fluid transport capabilities of the fabricated monoporous and biporous stainless steel wicks. These experiments provide crucial insights into how the varying pore structures affect the wicks' ability to draw and distribute working fluids, a key factor in loop heat pipe performance.

Procedure

- The bottom edge of each wick sample (monoporous, 20%, 30% and 40% biporous) was clipped into the wick holder, so that it barely touched the liquid surface in the reservoir.
- Three working fluids were investigated in this work: methanol, ethanol, and water. These fluids were selected due to their different respective physical characteristics (surface tension, viscosity) in order to provide

a holistic view of the performance of the wicks with respect to different aspects of the physical characteristics of fluids.

- As the experiment began, the digital weighing machine recorded the mass increase at regular intervals as the fluid rose through the wick structure.
- At the same time, the IR camera operated on the wick, generating thermal images to visualize the evolution of the fluid front.

To examine the wicks' internal structure and surface morphology, high-resolution SEM imaging (Zeiss FE-SEM machine) was done. This method made it possible to visualise the pore networks in great detail, including the distribution of pore sizes, interconnectivity, and the overall effect of changing the NaCl content on the biporous structure. SEM pictures of the wick microstructures were taken at various magnifications to give both a broad and a more in-depth view. During capillary rise experiments, the dynamic fluid front progression was visualised using an IR camera. Real-time thermal images were obtained using this non-invasive method, which made it possible to identify preferred flow paths within the wick structures and observe patterns of fluid distribution.

4. RESULTS AND DISCUSSION

It is clear from the capillary rise experiment that the wicking ability improves under all fluid conditions as the wick's biporous nature increases. Figures 5(a) to 5(c) unequivocally demonstrate that water performs worse than methanol and ethanol.

Figure 5a. Capillary Rise Rate of Methanol

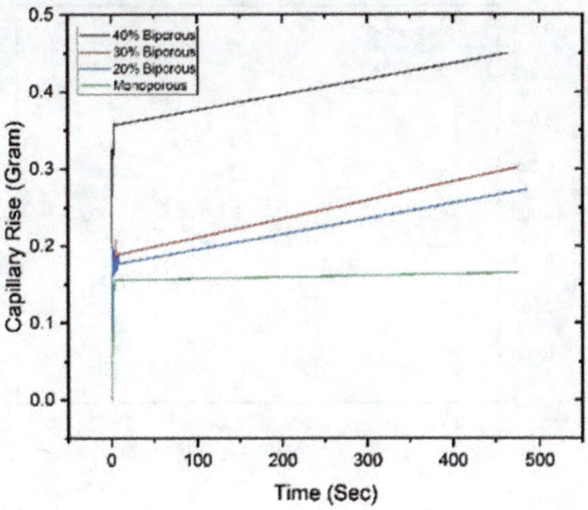

Figure 5b. Capillary Rise Rate of Ethanol

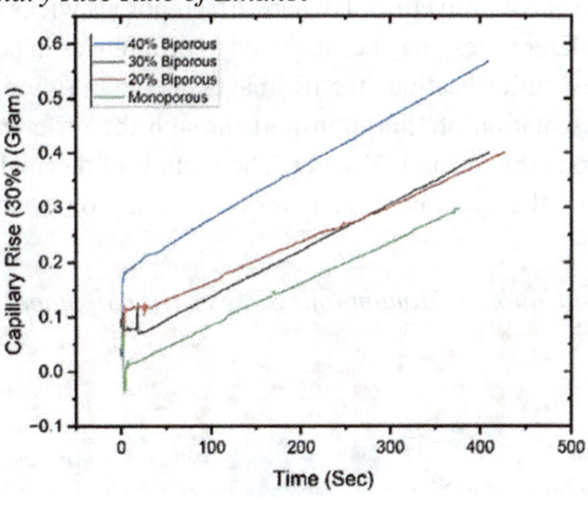

Figure 5c. Capillary Rise Rate of DI Water

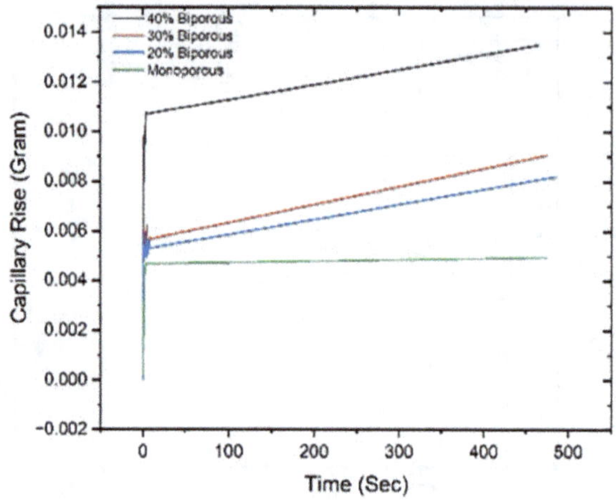

The capillary rise for 40% biporous and monoporous wicks in the case of ethanol exhibits a usual pattern. For 20% and 30% biporous wicks, however, the 30% biporous wick performs better at first, but eventually the 20% biporous wick overtakes it. The crossover behaviour in ethanol, where 20% biporous wicks outperform 30% biporous wicks at later times, may be attributed to differences in pore connectivity.

In the IR Visualization section, the IR images shown in figure 6-17 are a powerful visual representation of fluid transport through the different stainless steel wicks for methanol, ethanol, and DI water. These kinds of thermal images provide valuable insights on the dynamic behaviour of the fluid progression in the wicks.

Figure 6. IR Visualization of Monoporous SS Wick with Methanol

Figure 7. IR Visualization of 20% Biporous SS Wick with Methanol

Figure 8. IR Visualization of 30% Biporous SS Wick with Methanol

Figure 9. IR Visualization of 40% Biporous SS Wick with Methanol

Figure 10. IR Visualization of Monoporous SS Wick with Ethanol

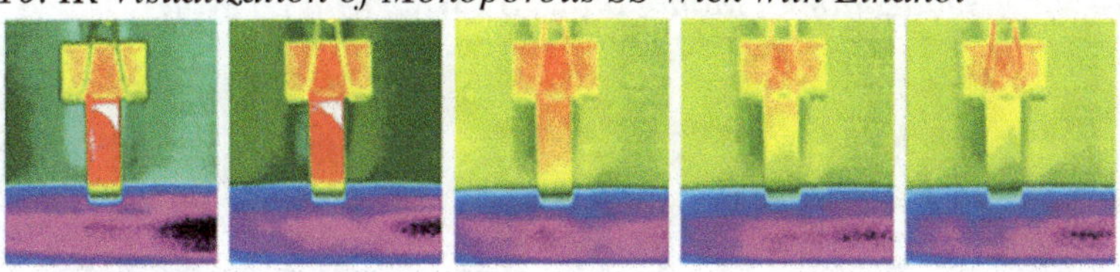

Figure 11. IR Visualization of 20% Biporous SS Wick with Ethanol

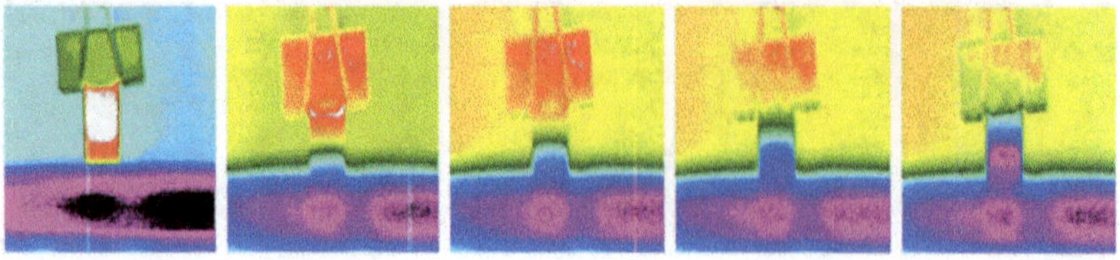

Figure 12. IR Visualization of 30% Biporous SS Wick with Ethanol

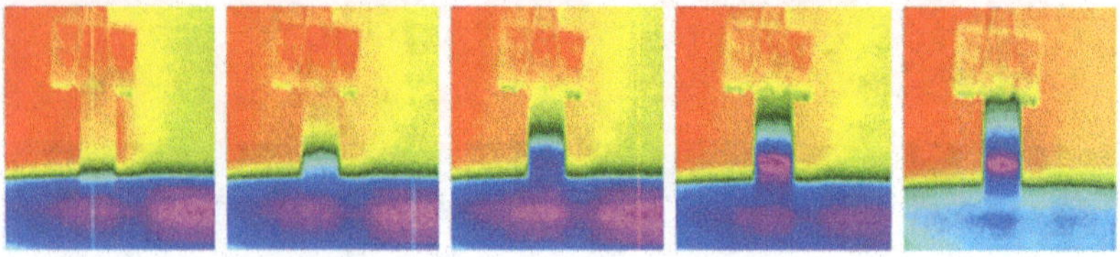

Figure 13. IR Visualization of 40% Biporous SS Wick with Ethanol

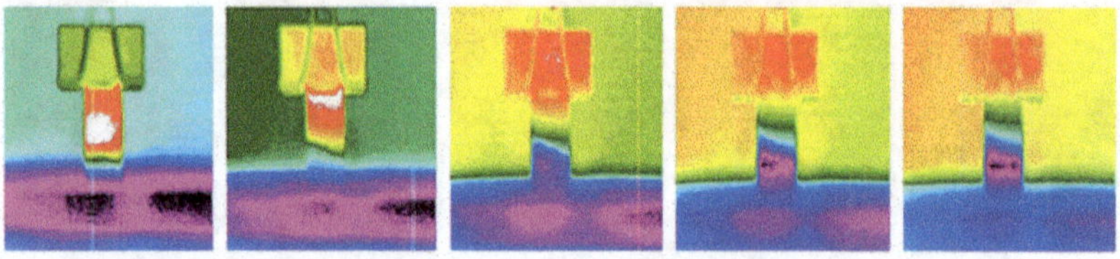

Figure 14. IR Visualization of Monoporous SS Wick with DI Water

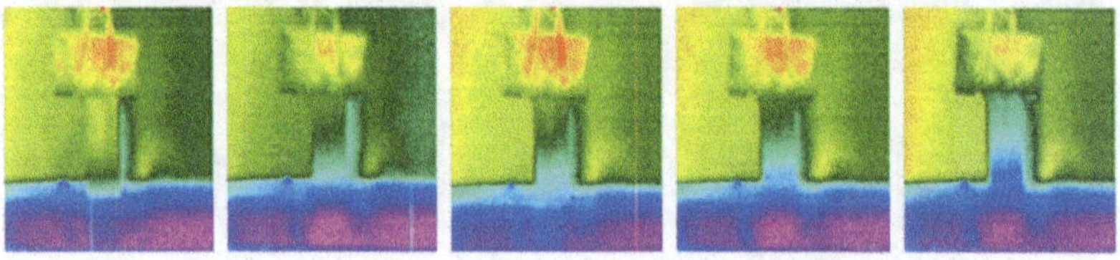

Figure 15. IR Visualization of 20% Biporous SS Wick with DI Water

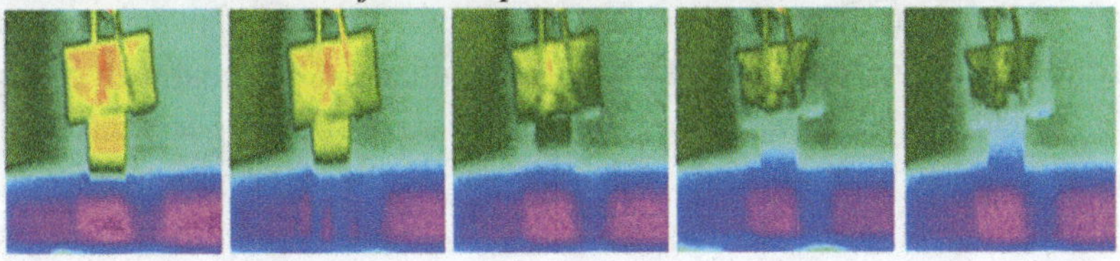

Figure 16. IR Visualization of 30% Biporous SS Wick with DI Water

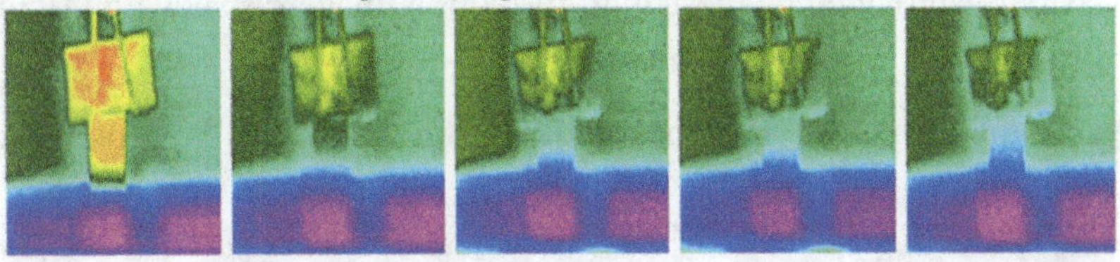

Figure 17. IR Visualization of 40% Biporous SS Wick with DI Water

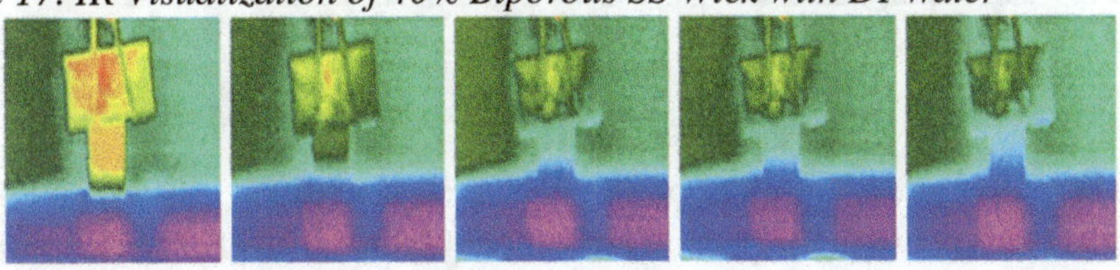

In all three fluids, the advance of the fluid front is rather uniform but slower in the case of monoporous wicks. The fluid distribution pattern in the biporous wicks, on the other hand, is more intricate and quicker. We find that the speed and homogeneity of fluid transport improve as the biporous nature level rises from 20% to 40%. Among all tested fluids, the 40% biporous wicks continuously show the fastest and most uniformly distributed fluid front. Interestingly, the pictures show unique fluid transport properties for every working fluid. Although its fluid fronts are less distinct, methanol exhibits the fastest overall progression due to its lower surface tension. Ethanol exhibits a harmony between front definition and speed. Despite having a higher surface tension, DI water shows distinct fluid fronts and best demonstrates the improved capillary performance of the biporous structures.

In addition to offering a clear, qualitative understanding of how biporous nature improves fluid transport in stainless steel wicks for loop heat pipe applications, these visualisations support the quantitative data from the capillary rise experiments. Faster and more uniform capillary rise directly translates to improved evaporator priming and startup reliability in loop heat pipes. The enhanced permeability of biporous wicks also reduces liquid–vapor pressure drop, supporting higher maximum heat load capacity. These findings suggest that increasing the biporous nature of stainless steel wicks could enable faster startup and sustain higher heat fluxes in practical LHP systems compared to conventional monoporous structures.

The microstructure and surface morphology of the monoporous and biporous stainless steel wicks are thoroughly examined visually in the SEM images. Important information about the pore distribution and structure underlying the observed capillary rise behaviour is provided by these high-resolution images as shown in figure 18.

Figure 18. SEM Images of Monoporous, 20% Biporous, 30% Biporous, and 40% Biporous SS Wicks

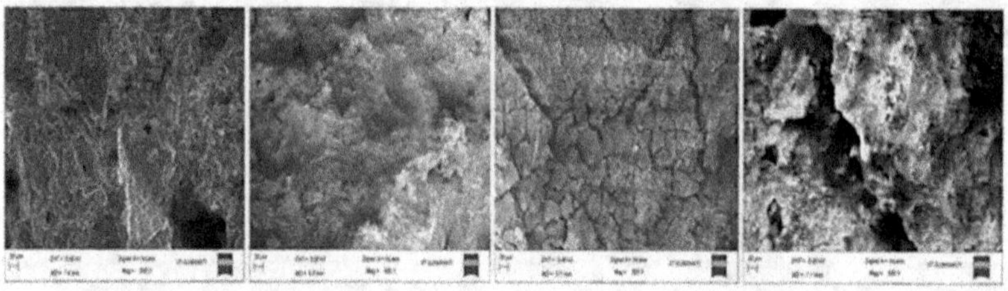

The pore structure of the monoporous wick is comparatively uniform, with interconnected voids between the particles of sintered stainless steel. The pore structure clearly changes as we move on to the biporous wicks. The removal of NaCl particles from the 20% biporous wick results in the formation of larger pores scattered throughout the finer pore network. These larger pores become more numerous and better connected in the 30% biporous sample. The bimodal pore distribution is most noticeable in the 40% biporous wick, which has a network of large, connected pores in addition to the base structure's smaller pores.

5. CONCLUSION

In conclusion, this study offers strong proof that, in loop heat pipe, biporous stainless steel wicks perform better in terms of capillary rise than their monoporous counterparts. By means of a thorough investigation employing capillary rise experiments, IR thermography, and scanning electron microscopy, we have shown that raising the biporous nature level from 20% to 40% consistently improves fluid transport properties for a range of working fluids. Further increasing the biporous nature of wick would deteriorate its strength.

According to SEM analysis, the hierarchical pore structure in biporous wicks promotes strong capillary pressure and quick fluid distribution, improving overall performance.

The IR visualisation results show that fluid front progression in biporous wicks is more uniform and efficient, especially in the 40% biporous samples. These findings demonstrate the adaptability of biporous wicks across a range of fluid properties and are valid for methanol, ethanol, and water. Strong capillary transport is provided by wicks with a 40% biporous structure; however, strength is diminished with a larger pore concentration. Permeability and mechanical stability need to be optimised for reliable loop heat pipe operation.

These results have important implications for the design and optimization of loop heat pipes. The overall thermal performance and efficiency of these systems can be raised by taking advantage of the improved fluid transport capabilities of biporous wicks. The integration of these optimised wick structures into complete LHP systems and an assessment of their long-term performance under varied operating conditions should be the main goals of future research. Furthermore, more research into the connection between capillary performance and pore size distribution may result in even better wick designs.

Future research should focus on integrating these optimized wicks into complete loop heat pipe prototypes to evaluate their performance during startup, steady state, and maximum heat load in practical operating conditions. It is also important to study their long-term reliability when subjected to repeated thermal stresses and in environments such as microgravity.

ACKNOWLEDGEMENT

This work was partially funded through Grant No. CRG/2020/006333 of SERB, DST, Govt. of India. The SEM imaging was conducted at the Central Instruments Facility of IIT Guwahati.

REFERENCES

Akafuah, N. K., Salazar, A. J., & Saito, K. (2010). Infrared thermography-based visualization of droplet transport in liquid sprays. *Infrared Physics & Technology*, *53*(3), 218–226.

Altschuh, P., Kunz, W., Bremerich, M., Reiter, A., Selzer, M., & Nestler, B. (2022). Wicking In porous polymeric membranes: Determination of an effective capillary radius to predict the flow behavior in lateral flow assays. *Membranes*, *12*(7), 638. PMID: 35877842

Ambirajan, A., Adoni, A. A., Vaidya, J. S., Rajendran, A. A., Kumar, D., & Dutta, P. (2012). Loop heat pipes: A review of fundamentals, operation, and design. *Heat Transfer Engineering*, *33*(4-5), 387–405.

Ku, J. (1999). Operating characteristics of loop heat pipes. *SAE Transactions*, •••, 503–519.

Ku, J. (2016). "Loop heat pipe startup behaviors." In *International Conference on Environmental Systems*, no. ICES-2016-24.

Kumari, T., & Pandey, M. (2022). "Study of Physical Characteristics of a Bi-porous Composite Capillary Wick for a Flat Miniature Loop Heat Pipe." In *Conference on Fluid Mechanics and Fluid Power*, pp. 211-221. Singapore: Springer Nature Singapore.

Nashine, C., Pandey, M., & Baraya, K. K. (2025). Experimental studies on the transient characteristics and start-up behaviour of a miniature loop heat pipe. *Applied Thermal Engineering*, *259*, 124814.

Shen, H., Suzuki, A., Takata, N., & Kobashi, M. (2024). Enhancing capillary pressure of porous aluminum wicks by controlling bi-porous structure using different-sized NaCl space holders. *Materials (Basel)*, *17*(19), 4729. PMID: 39410300

Yang, X., Cui, Q., You, Z., Liu, J., Zhang, Y., & Wei, J. (2024). Experimental and theoretical study on performance of bi-porous wick for passive phase-change devices. *Physics of Fluids*, *36*(12).

Zhao, Y., Wei, M., & Dan, D. (2024). Modeling, design, and optimization of loop heat pipes. *Energies*, *17*(16), 3971.

Chapter 3

Computational Analysis of the Thermal Behavior of Lithium-Ion Batteries for Electric Vehicle (EV) Technology Using Mini-Channels

Prateek Debadarsi Roul
National Institute of Technology, Rourkela, India

Affilkumar Bariya
National Institute of Technology, Rourkela, India

Suman Ghosh
National Institute of Technology, Rourkela, India

Jnana Ranjan Senapati
https://orcid.org/0000-0002-2672-3763
National Institute of Technology, Rourkela, India

ABSTRACT

Electric vehicle (EV) batteries generate heat during charging and discharging, which can gradually impact their efficiency and reliability. This study examines the role of thermal and mathematical factors in managing battery temperature through a cooling liquid system flowing within small channels integrated into the battery's surface. Key design elements such as mini-channel aspect ratio, placement, coolant

DOI: 10.4018/979-8-3373-3621-3.ch003

flow rate, and initial coolant temperature significantly influence thermal regulation. In this research, the aspect ratio of the mini-channel was systematically varied while maintaining other parameters constant. The tested aspect ratios were 5.3, 5.8, 6.3, 6.8, and 7.3. Findings indicate that increasing the aspect ratio reduces the battery's peak temperature until an optimal value is reached. The coolant inlet conditions used in the study included a temperature of 299.98 K, a velocity of 0.33 m/s, and an inclination angle of 11.3°.

1. INTRODUCTION

Electric vehicles (EVs) have gained significant attention in recent years as a sustainable alternative to traditional internal combustion engine vehicles, mainly because of their potential to reduce greenhouse gas emissions and urban noise pollution. The core component of an EV is its lithium-ion battery module, known for high energy density, long cycle life, and relatively low self-discharge rates. However, lithium-ion batteries generate heat during charging and discharging, which can adversely impact their efficiency, reliability, and safety if not managed effectively.

Maintaining the battery temperature within an optimal operational range (typically between 287 K and 307 K) is critical, as excessive heat can accelerate capacity degradation, while low temperatures increase internal resistance. Temperature uniformity across cells is also vital because large gradients (above 5 K) may lead to uneven aging and reduced pack life. Thus, an effective Battery Thermal Management System (BTMS) is essential to control temperature and ensure battery longevity and performance.

Various cooling strategies have been explored including air cooling, liquid cooling, heat pipes, and phase change materials (PCM). Air cooling is simple but limited by low thermal conductivity and requires substantial space. PCMs can absorb latent heat effectively but may fail under high thermal loads when latent capacity depletes. Liquid cooling, particularly using mini- or micro-channel structures integrated into battery modules, offers enhanced heat transfer and temperature uniformity by maximizing contact surface area and promoting localized cooling. Studies have demonstrated that mini-channel design parameters such as shape, size, flow configuration, and aspect ratio significantly influence thermal performance and pressure drop.

Despite advances, optimizing mini-channel geometry for practical automotive BTMS applications remains a challenge. Existing research often lacks systematic analysis of aspect ratio effects or does not balance thermal benefits with hydraulic losses. This study aims to fill this gap by computationally analysing the thermal behaviour of lithium-ion batteries integrated with liquid-cooled mini-channels, focusing on the influence of mini-channel aspect ratio on temperature regulation and pumping power requirements. The outcomes aim to guide the design of more efficient, scalable BTMS solutions tailored for EV battery packs.

2. LITERATURE REVIEW AND OBJECTIVE

Environmental conditions significantly influence the thermal management of lithium-ion batteries. Energy degradation commonly occurs at elevated ambient temperatures. These batteries are designed to function within a temperature range of 252 K to 334 K, with an optimal range between 287 K and 307 K (14°C to 34°C) (Wright et al., 2018). Temperatures exceeding 313 K or dropping below 293 K can negatively affect performance—higher temperatures accelerate capacity degradation, while lower temperatures increase internal resistance. Additionally, for optimal thermal performance, it is crucial to maintain a temperature difference of no more than 5 K between cells (Kim et al., 2019). Ensuring temperature uniformity is essential for the battery's longevity and efficiency. Therefore, managing both heat and performance throughout the entire operating cycle is critical for lithium-ion batteries. An effective Battery Thermal Management System (BTMS) plays a key role in enhancing battery performance (Safdari et al., 2022). BTMS is primarily classified based on the cooling medium used, such as air cooling, liquid cooling, heat pipe cooling, and phase change material (PCM) cooling. Phase Change Materials (PCM) significantly impact the thermal efficiency of batteries, and a well-optimized design helps maintain battery temperature within an acceptable range. However, under extreme operating conditions, if the PCM exhausts its latent heat capacity, the thermal management system may fail (Khateeb et al., 2004),(Al-Hallaj & Selman, 2002) .Air cooling is a commonly used method for regulating temperature in battery modules. However, its thermal efficiency declines at higher ambient temperatures due to its lower thermal conductivity and specific heat capacity(Pesaran et al., 1999). Additionally, higher flow rates present challenges, and air-cooled systems require more space than liquid-cooled alternatives. The design of mini-channels plays a crucial role in improving battery cooling. For example, sine and sawtooth wave-shaped cross-sections have shown better thermal performance compared to traditional rectangular mini-channels. Another study found that expanding-shaped channels with two inlets and one outlet resulted in a 7.2% reduction in pressure drop and a decrease in peak temperature by 0.8 K (Kong et al., 2021). Furthermore, implementing an opposite-flow configuration for the cooling medium within the channels helps reduce temperature variations across the cell. In (Jang et al., 2022), key control parameters for liquid cooling systems were identified, including mass flow rate, coolant inlet temperature, and ambient temperature. This study introduced three designs utilizing heat pipes to enhance the heat transfer surface area of the cooling system, with the most effective design achieving a temperature reduction of 9.4 K. In (Guo & Li, 2022), the performance of parallel-spiral serpentine liquid cooling channels was analyzed, revealing that while a higher flow rate helps reduce the maximum temperature, it comes at the expense of increased pressure drop and temperature variation. A comparison in (Xu et al., 2021) between S-shaped and U-shaped channels indicated that S-shaped channels provide better heat dissipation. Similarly, (Ding et al., 2022) demonstrated that square cooling channels outperform circular ones in thermal performance, though they may slightly increase temperature non-uniformity. Additionally, (Ding et al., 2021) highlighted that the aspect ratio of cooling channels influences the thermal efficiency of liquid cooling systems in batteries. The objective of this research was to develop an efficient and effective cooling system for electric vehicle (EV) batteries, ensuring a balance between battery cooling performance and the energy consumption of the cooling system.

Mini- and micro-channel liquid cooling has emerged as an effective approach to improving thermal management in lithium-ion batteries due to enhanced heat transfer and improved temperature uniformity over traditional methods. Diverse channel geometries—such as rectangular, wavy, and divergent shapes—have been explored to optimize the balance between heat removal efficiency and pressure drop. Key flow parameters and configurations have also been studied to reduce thermal gradients and hotspots within battery packs.

Nonetheless, a comprehensive study focusing specifically on the influence of mini-channel aspect ratio remains scarce. Previous works tend to investigate geometric design or hydraulic performance individually without fully examining how aspect ratio impacts both cooling capability and pumping power simultaneously. This gap hinders the ability to achieve an optimized design that meets thermal management needs without excessive energy consumption.

The present research addresses this deficiency by performing in-depth numerical simulations to evaluate varying aspect ratios, aiming to deliver design insights that balance effective cooling and energy efficiency for electric vehicle battery thermal systems.

3. MATERIALS AND METHODS

3.1 Material Specification

In this paper, prismatic li-ion battery was analysed under the extreme atmospheric conditions, focusing on hot temperatures with an ambient temperature of 312.95 K (39.95°C).

Tab-P is the positive tab of the battery, typically made of copper, and Tab-N is the negative tab of the battery, typically made of aluminium, as mentioned in Table 1.

A comprehensive computational fluid dynamics (CFD) analysis, followed by a quantitative study, was conducted to examine the impact of mini-channel cooling on battery. Table 1. shows material specification used for the battery model.

Table 1. Material Specification

Specification	Tab–p	Tab–N	Cell	Coolant
Material	Aluminium	Copper	Cell Material	Water
Density $\left(\frac{kg}{m^3}\right)$	2719.1	8978.1	2092.2	998.20
Thermal capacity $(J/kg-K)$	871	381	678.1	4182.1
Thermal conductivity $\frac{(W)}{m-K}$	202	387.6	18.3	0.6

3.2 Geometry and Meshing

As shown in fig. 1, the dimensions for the battery model used is taken as (127×196×7) mm.

Figure 1. Battery model geometry with inclined mini-channel

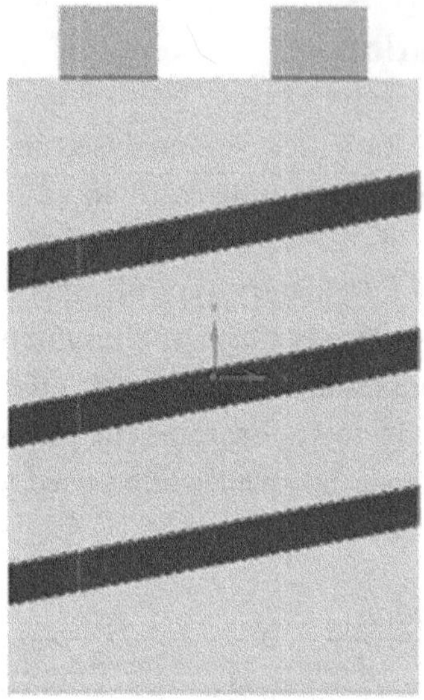

Aspect ratio: The aspect ratio represents the relationship between the mini-channel width (l) and its thickness (t). The thickness is fixed at 2 mm, while the channel width varies and can be adjusted as needed.

$$Aspect\ ratio\ (x) = \frac{width}{thickness} = \frac{l}{t} \tag{1}$$

Figure 2. Mini-channel

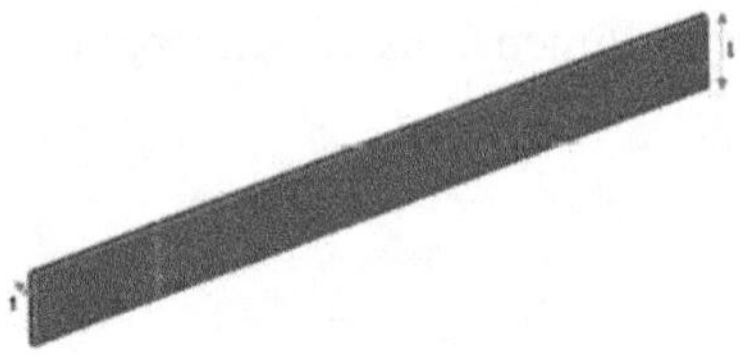

Figure 3. Meshing of the battery model

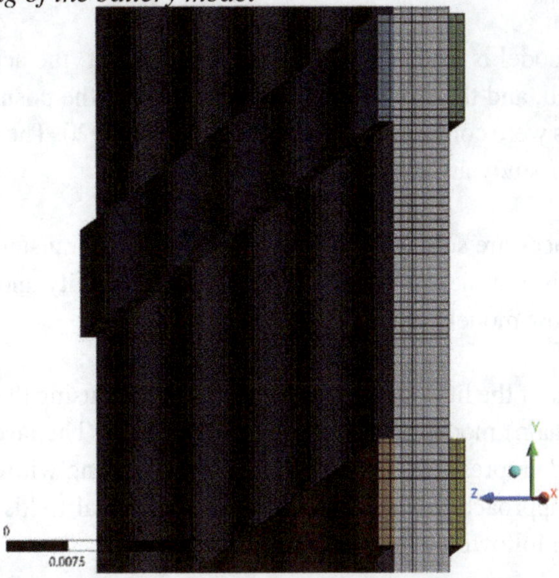

To ensure the mesh independence of the numerical results, simulations were conducted with four mesh sizes ranging from 125,120 to 2,050,580 elements. Table 2 summarizes the peak battery temperature for each mesh size. It is observed that beyond 1,764,566 elements, the variation in peak temperature becomes negligible (less than 0.1 K), confirming that the mesh with 1,764,566 elements is sufficient to achieve accurate results with optimized computational cost.

Table 2. Mesh Independence Study of Peak Battery Temperature for Different Mesh Sizes

Mesh Elements	Peak Temperature (K)
1,25,120	312.5
13,26,150	310.3
17,64,566	307.1
20,50,580	307

Power required to pump the liquid through channels is calculated by,

$$Wp = \Delta p \times Q \tag{2}$$

$$\Delta p = \frac{fl\rho v^2}{2D_h} \tag{3}$$

3.3 Model Validation

The battery model is primarily divided into two zones: the active zone, which represents the cell, and the passive zone, which includes the positive and negative tabs. Simulations were conducted using ANSYS Fluent 2020. The key assumptions considered in this study are as follows:

i. Battery surfaces are subjected to free convection heat transfer.
ii. The coolant is assumed to have constant thermal capacity and density.
iii. A laminar flow model is used.

The behaviour of the lithium-ion battery was analysed using the MSMD (Multi-Scale Multi-Domain) model available in ANSYS Fluent. The layer separating the anode and cathode represents the internal physics occurring within the cell.

The MSMD approach solves the electrical and thermal fields within the CFD domain using the following equations:

$$\partial \rho CpT - \nabla \cdot (k \ \nabla \ T) = q \tag{4}$$

$$\nabla \cdot (\sigma + \nabla \varphi +) = -J \tag{5}$$

$$\nabla \cdot (\sigma - \nabla \varphi -) = J \tag{6}$$

The NTGK semi-empirical electrochemical model, categorized under the MSMD framework, was utilized for battery simulations.

In this study, the battery model without a thermal management system (i.e., without mini-channels) was validated by comparing it with the numerical results from the literature by A. Verma et al. Figure 4 illustrates the battery's temperature variation over time, showing a comparison between the experimental model of A. Verma et al. and the present numerical model.

Figure 4. Temperature-Time plot without BTMS ((Verma et al., 2023) vs. present model.)

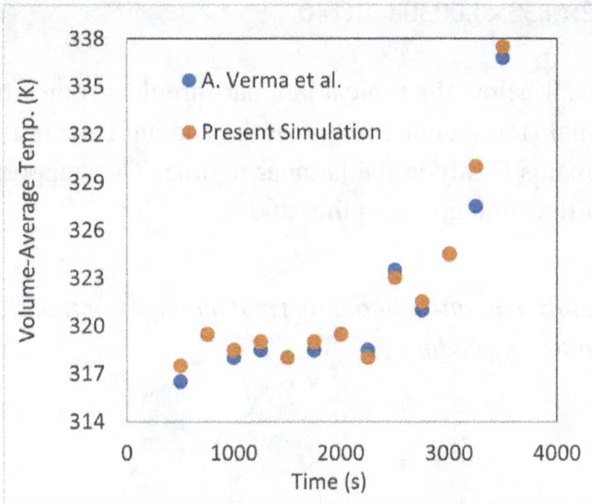

In the current simulation, the highest recorded temperature was 337.5*K*, compared to 336.8*K* in the reference case(Verma et al., 2023), resulting in a temperature difference of 0.2%. The depth of discharge was 0.45.

The results were then compared with the modified battery, which incorporated six mini-channels with different aspect ratios. Figure 5 presents the temperature contour of the simulated model.

To justify the assumption of laminar flow in the mini-channels, the Reynolds number (Re) was calculated using the channel dimensions and coolant properties. The Reynolds number is defined as

$$R_e = \frac{\rho v D_h}{\mu} \tag{7}$$

where ρ is the density of water (998.2Kg/m^3), v is the coolant velocity (0.33m/s), D_h is the hydraulic diameter, and μ is the dynamic viscosity (0.00089Pa•s) of water at room temperature. For a channel width of 10.6 mm and thickness of 2 mm (aspect ratio 5.3), the hydraulic diameter (D_h) for a rectangular cross-section is given by

$$D_h = \frac{2wt}{w+t} \tag{8}$$

Plugging in the values,

D_h= 0.0106+0.0022×0.0106×0.002≈0.00308m.

Thus,

$R_e = 0.00089998.2 \times 0.33 \times 0.00308 \approx 1.140$

This value is well below the typical laminar-turbulent transition threshold (Re ≈ 2,300) for internal flows, confirming that the coolant flow in all tested channel configurations remains firmly in the laminar regime. This supports the use of the laminar model for the simulations in this study.

Figure 5. Temperature contour without thermal management, indicating a hotspot, with peak temperature exceeding 337 K

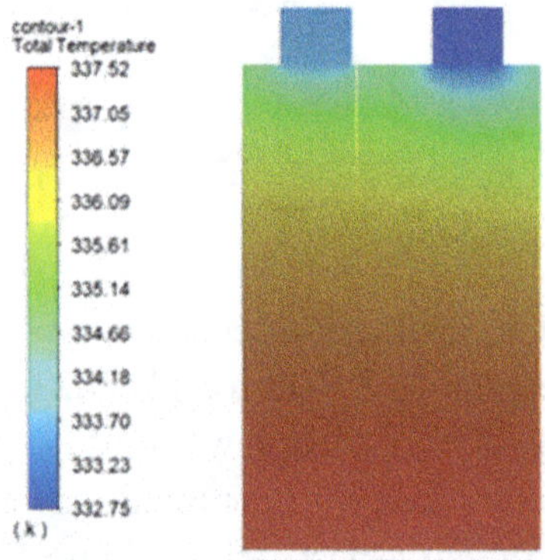

3.4 Boundary Conditions

The boundary conditions were defined as follows to ensure simulation reproducibility:

Heat Generation Rate: The volumetric heat generation (q) for the battery cell was computed using the NTGK semi-empirical electrochemical model in ANSYS Fluent, reflecting the operational C-rate of 4.25 (cell capacity 14.5 Ah) and depth of discharge 0.45.

Coolant Flow: Water was used as the coolant, with a constant inlet temperature of 299.98 K and velocity of 0.33 m/s. The flow regime was counter flow across six mini-channels, inclined at 11.3°, with all channel surfaces modelled as no-slip walls.

Thermal Boundaries: Battery external surfaces were subjected to free convection boundary conditions. Tab surfaces were assigned material properties as

listed in Table 1, with thermal insulation imposed at lateral faces to restrict heat transfer only through the intended channel paths.
Material Properties: Battery, tab, and coolant material properties were maintained constant as shown in Table 1. Fluid density and viscosity corresponded to standard water values at room temperature.
Simulation Specifics: All experiments used transient simulations with a time step of 15 seconds, structured mesh with 1,764,566 elements, and aspect ratios as detailed in Table 2.

All boundary condition values, fluid properties, and modelling choices are tabulated in Sections 3.1–3.2 and summarized in Table 1 (material) and Table 2 (fluid and channel parameters).

4. RESULT AND DISCUSSION

4.1 Battery With Thermal Control

After evaluating the battery's driving cycle without a Battery Thermal Management System (BTMS), six inclined mini-channels with five different aspect ratios were integrated into the battery, and a numerical analysis was conducted. The simulation was performed under identical working conditions with a cell capacity of 14.5 Ah and a discharge rate of 4.25 C. A laminar and transient flow model was used with a time step of 15 seconds. The inlet fluid temperature was maintained at 299.85 K, with a velocity of 0.33 m/s, and the mini-channels were inclined at an angle of 11.3°. The assumptions remained consistent with the no-BTMS case.

Table 2 presents the variation of the battery's maximum temperature for different mini-channel aspect ratios. It is clear from the table that increasing the aspect ratio (length/thickness) reduces the battery's maximum temperature. This trend occurs because increasing the aspect ratio increases the surface area of contact between the battery and the cooling mini-channels, which enhances convective heat transfer from the battery to the coolant. More surface area enables greater heat dissipation, thereby lowering the maximum temperature.

Table 2. Temperature value for different aspect ratio

Battery Maximum Temperature (K)	Mini-channel parameter (mm)		
	Width (l)	Thickness (t)	Aspect ratio (x=l/t)
311.11	10.6	2	5.3
309.13	11.6	2	5.8
307.95	12.6	2	6.3
307.11	13.6	2	6.8
307.00	14.6	2	7.3

However, this beneficial effect is significant only up to an optimal point. Beyond an aspect ratio of 6.8, further increases yield minimal reduction in temperature—only about a 0.035% difference between aspect ratios of 6.8 and 7.3—indicating negligible impact. This plateau arises because the surface-area-to-volume ratio of the mini-channels decreases as they become larger; the volume of coolant increases faster than the surface area, limiting additional improvements in heat transfer. At very high aspect ratios, the coolant becomes less effective at absorbing heat per unit volume, thus making thermal gains marginal.

Figure 6. Temperature contour with six inclined mini-channels (aspect ratio 6.8), showing reduced hotspot and maximum temperature below 308 K

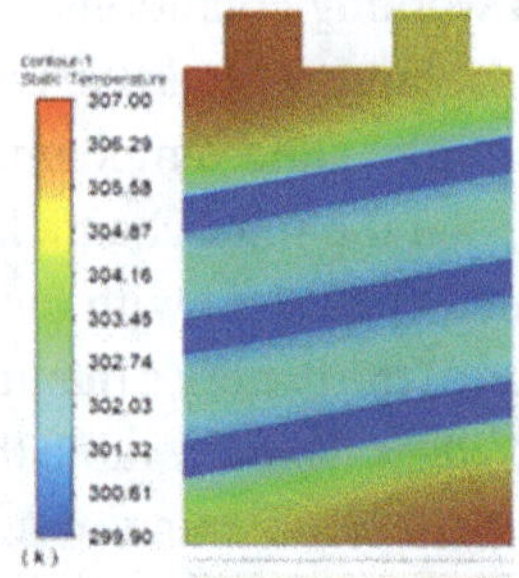

Moreover, increasing the aspect ratio beyond the optimum also leads to higher hydraulic resistance inside the mini-channels, which increases the pumping power required to maintain coolant flow. This higher pumping power translates into greater energy consumption, which may outweigh the slight cooling benefits. While enhanced cooling improves battery safety and lifespan, the increased operational energy cost reduces the overall efficiency of the battery thermal management system.

Our analysis indicates that beyond an aspect ratio of 6.8, the marginal decrease in battery temperature does not sufficiently justify the added pumping power demand, which rises significantly due to increased flow resistance. Therefore, selecting an optimal aspect ratio is crucial to balance effective thermal regulation with reasonable energy use, preserving the overall performance and driving range of the electric vehicle.

Figure 6 illustrates the temperature contour of the battery with six inclined mini-channels at the optimal aspect ratio of 6.8. The contour clearly shows a reduction in hotspot formation, with the maximum battery temperature kept below 308 K. The inclined channel orientation promotes localized cooling, guiding the coolant effectively along the battery surface, which helps in achieving a more uniform temperature distribution. The contour indicates that thermal gradients across the battery pack are minimized, which is crucial to prevent uneven aging and prolong battery life.

In addition to the reduction in maximum battery temperature achieved through optimization of the mini-channel aspect ratio, temperature uniformity within the battery cell was also critically evaluated. The maximum temperature difference (ΔT) across the battery cells was found to be approximately 3.5 K for the optimized design, which comfortably lies within the accepted range of 5 K for ensuring cell safety and minimizing uneven aging. This uniformity is facilitated by the inclined mini-channel geometry, which promotes more even coolant distribution and mitigates hotspot formation. Figure 6 visually confirms the reduced thermal gradients across the battery surface, highlighting the effectiveness of the cooling strategy in maintaining both low peak temperatures and high temperature homogeneity. Together, these thermal metrics affirm the practical viability of the proposed mini-channel design for electric vehicle battery thermal management.

It should be noted that the thermal performance of the proposed mini-channel system is sensitive to several operating conditions. Increasing coolant velocity can further reduce maximum battery temperature and enhance temperature uniformity, but may also proportionally increase pumping power and energy consumption. Higher ambient temperatures tend to degrade cooling effectiveness, requiring careful system calibration. Additionally, an increase in C-rate (discharge rate) causes greater internal heat generation, which may necessitate more aggressive thermal management strategies to maintain safe operating conditions. These factors highlight the importance of optimizing BTMS parameters for specific use-cases and driving scenarios, confirming the practical value and adaptability of the mini-channel approach for a wide range of electric vehicle applications.

While the thermal performance improvements demonstrated at the cell level are promising, practical implementation of the proposed mini-channel cooling system at the battery pack level presents further challenges. Effective manifold design becomes crucial to ensure uniform coolant distribution across numerous cells, preventing flow maldistribution and maintaining pressure balance critical for system stability. The pack-level integration must also consider structural reinforcements, electrical insulation, and packaging constraints, all without significantly increasing the overall weight and volume of the battery system. Furthermore, manufacturability considerations including the precision fabrication of mini-channels and manifolds at scale require advanced manufacturing techniques such as micro-machining or additive manufacturing. Addressing these scalability and practicality issues is imperative to transform the demonstrated lab-scale thermal management benefits into robust, commercially viable electric vehicle battery thermal management systems. Future work should focus on these integration challenges alongside continued optimization of cooling performance and energy consumption.

In summary, the observed thermal behaviour aligns with heat transfer fundamentals: enlarging the cooling surface area improves heat dissipation until geometric and fluid dynamic limitations restrain further gains. Considering pumping power and energy consumption is essential to optimize mini-channel design. A balanced approach ensures the battery temperature is maintained within safe limits while avoiding excessive energy use for coolant circulation, enabling efficient and practical battery thermal management.

Table 3. Comparative Performance of Battery Thermal Management Systems (BTMS)

BTMS Approach	Max Battery Temperature (K)	Temperature Uniformity (ΔT, K)	Key Benefits	Limitations
Air Cooling	~320 - 330	~7 - 10	Simple and low cost	Poor thermal performance at high load; large space requirement
PCM Cooling	~310 - 315	~3 - 5	Passive heat absorption	Saturation of latent heat reduces effectiveness in long high load
Cold Plate Cooling	~305 - 310	~3	Effective and uniform cooling	Added weight and complexity; higher cost
Mini-channel Cooling (This work)	307.11	3.5	Improved heat transfer area; localized cooling; good uniformity	Requires optimized aspect ratio to balance pumping power and efficiency

5. CONCLUSION

This study computationally analysed the thermal behaviour of lithium-ion batteries integrated with inclined liquid-cooled mini-channels, focusing on the influence of mini-channel aspect ratio on battery temperature control and pumping power requirements. Increasing the mini-channel aspect ratio from 5.3 to 6.8 reduced the maximum battery temperature from 311.11 K to 307.11 K, demonstrating a notable temperature reduction of approximately 1.3%. Beyond an aspect ratio of 6.8, further increases led to a marginal temperature decrease of only 0.035%, indicating diminishing thermal returns.

Correspondingly, pumping power demand increased with aspect ratio due to higher hydraulic resistance. Our analysis shows that excess pumping power required for aspect ratios above 6.8 outweighs the minor cooling benefits, underscoring the importance of optimizing aspect ratio to balance cooling efficiency and energy consumption.

Practically implementing this cooling strategy requires overcoming manufacturing challenges related to precise fabrication and system integration issues such as coolant sealing, electrical insulation, and compact design for EV battery packs. Cost, material durability, and real-world reliability under mechanical and thermal stresses also demand consideration.

Despite these challenges, the mini-channel cooling design significantly improves localized cooling and temperature uniformity, which are critical for battery safety and lifespan. Future work should address manufacturability, scalability, and cost optimization to facilitate integration into commercial EV battery systems.

Box 1. Nomenclature

BTMS: Battery Thermal Management System	***m*:** Mass flow rate (kg/s)
CFD: Computational Fluid Dynamics	***Q*:** Volumetric flow-rate (m^3/s)
PCM: Phase Change Material	***Ah*:** Cell capacity (Ah)
EVL: Electric Vehicle	***v*:** Coolant velocity at inlet (m/s)
MSMD: Multi Scale Multi-Dimensional	***t*:** Thickness of mini-channel (m)
NTGK: Newmann, Tiedeman, Gu and Kim	***ρ*:** Density (kg/m^3)
$\sigma-$: Effective electrical conductivity of the negative electrode	***φ+*:** Phase potential for the positive electrode
***p*:** Pressure drop (Pa)	***φ−*:** Phase potential for the negative electrode
C_p: Constant pressure specific heat (J/kg-K)	***J*:** volumetric transfer for current density
D_h: Hydraulic mean diameter (m)	**$\dot{q}$:** Heat generation rate
***k*:** Thermal conductivity (W/m-K)	**$\sigma+$:** Effective electrical conductivity of the positive electrode
***l*:** Width of mini-channel (m)	***F*:** Darcy friction factor

ACKNOWLEDGEMENTS

I extend my gratitude to NIT Rourkela for providing research facilities and expert guidance that greatly contributed to this work.

REFERENCES

Al-Hallaj, S., & Selman, J. R. (2002). Thermal modeling of secondary lithium batteries for electric vehicle/hybrid electric vehicle applications. *Journal of Power Sources, 110*(2), 341–348. DOI: 10.1016/S0378-7753(02)00196-9

Ding, Y., Ji, H., Wei, M., & Liu, R. (2022). Effect of liquid cooling system structure on lithium-ion battery pack temperature fields. *International Journal of Heat and Mass Transfer, 183*, 122178. DOI: 10.1016/j.ijheatmasstransfer.2021.122178

Ding, Y., Wei, M., & Liu, R. (2021). Channel parameters for the temperature distribution of a battery thermal management system with liquid cooling. *Applied Thermal Engineering, 186*(September 2020), 116494. DOI: 10.1016/j.applthermaleng.2020.116494

Guo, R., & Li, L. (2022). Heat dissipation analysis and optimization of lithium-ion batteries with a novel parallel-spiral serpentine channel liquid cooling plate. *International Journal of Heat and Mass Transfer, 189*. Advance online publication. DOI: 10.1016/j.ijheatmasstransfer.2022.122706

Jang, D. S., Yun, S., Hong, S. H., Cho, W., & Kim, Y. (2022). Performance characteristics of a novel heat pipe-assisted liquid cooling system for the thermal management of lithium-ion batteries. *Energy Conversion and Management, 251*(November 2021), 115001. DOI: 10.1016/j.enconman.2021.115001

Khateeb, S. A., Farid, M. M., Selman, J. R., & Al-Hallaj, S. (2004). Design and simulation of a lithium-ion battery with a phase change material thermal management system for an electric scooter. *Journal of Power Sources, 128*(2), 292–307. DOI: 10.1016/j.jpowsour.2003.09.070

Kim, J., Oh, J., & Lee, H. (2019). Review on battery thermal management system for electric vehicles. *Applied Thermal Engineering, 149*(September 2018), 192–212. DOI: 10.1016/j.applthermaleng.2018.12.020

Kong, W., Zhu, K., Lu, X., Jin, J., & Ni, M. (2021). Enhancement of lithium-ion battery thermal management with the divergent-shaped channel cold plate. *Journal of Energy Storage, 42*(February), 103027. DOI: 10.1016/j.est.2021.103027

Pesaran, A. A., Burch, S., & Keyser, M. (1999). An Approach for Designing Thermal Management Systems for Electric and Hybrid Vehicle Battery Packs Preprint. *The Fourth Vehicle Thermal Management Systems Conference and Exhibition 24-27, January*, 1–18.

Safdari, M., Ahmadi, R., & Sadeghzadeh, S. (2022). Numerical and experimental investigation on electric vehicles battery thermal management under New European Driving Cycle. *Applied Energy*, *315*(March), 119026. DOI: 10.1016/j.apenergy.2022.119026

Verma, A., Saikia, T., Saikia, P., Rakshit, D., & Ugalde-Loo, C. E. (2023). Thermal performance analysis and experimental verification of lithium-ion batteries for electric vehicle applications through optimized inclined mini-channels. *Applied Energy*, *335*(February), 120743. DOI: 10.1016/j.apenergy.2023.120743

Wright, D. R., Garcia-Araez, N., & Owen, J. R. (2018). Review on high temperature secondary Li-ion batteries. *Energy Procedia*, *151*, 174–181. DOI: 10.1016/j.egypro.2018.09.044

Xu, H., Zhang, X., Xiang, G., & Li, H. (2021). Optimization of liquid cooling and heat dissipation system of lithium-ion battery packs of automobile. *Case Studies in Thermal Engineering*, *26*(March), 101012. DOI: 10.1016/j.csite.2021.101012

Chapter 4

Computational Study on Natural and Mixed Convection Heat Transfer From a Rotating Solid Torus Suspended in Air: A Sustainable Thermal Management Approach

Kumud Ranjan
https://orcid.org/0009-0002-2771-8808
C.V. Raman Global University, India

Sachindra Kumar Rout
https://orcid.org/0000-0002-5471-1620
C.V. Raman Global University, India

Gloria Biswal
https://orcid.org/0000-0001-8078-2679
Indian Institute of Technology, Kharagpur, India

Jnana Ranjan Senapati
https://orcid.org/0000-0002-2672-3763
National Institute of Technology, Rourkela, India

ABSTRACT

This numerical investigation examines laminar mixed convection around a vertically oriented, rotating solid torus suspended in air, with emphasis on the interplay between

DOI: 10.4018/979-8-3373-3621-3.ch004

buoyancy and rotational effects. A detailed parametric analysis is performed over a range of Rayleigh numbers (10^3 to 10^7), rotational speeds (0 to 4), and aspect ratios (2.5 to 7.5) to quantify the heat transfer enhancement by the rotation of the torus. The temperature contours and the velocity vector plots are presented to elucidate the underlying flow physics and heat transfer mechanisms for both rotating as well as non-rotating cases. Utilizing natural convection to cool rotating components eliminates the need for energy-intensive active cooling systems, thereby enhancing overall energy efficiency and aligning with sustainable engineering principles

1. INTRODUCTION

Heat transfer from the rotating devices has attracted considerable research interest in the recent years, owing to its broad applications in the engineering sector. A deep insight into the heat transfer mechanisms of these devices is crucial to ensure efficient thermal regulation and reliable operation. In such systems, fluid motion and thermal transport are often influenced by a combination of buoyancy-driven and mechanically induced flow mechanisms. In natural convection, the fluid motion is driven by buoyancy forces resulting from temperature differences within the fluid, without any aid of external mechanical devices. However, if the device is rotated, it introduces a centrifugal force that induce significant fluid motion independent of thermal gradients. This forced convection component may strongly interact with the buoyancy-driven flows, resulting in complex flow structures and heat transfer patterns. When both buoyancy and mechanical forces contribute comparably to the overall flow behaviour and heat transfer, the regime is classified as mixed convection.

Numerous research investigations have been reported on heat transfer in a rotating geometry of different shapes through mixed convection. Anderson and Saunders (Anderson & Saunders, 1953) conducted an experiment to observe the behavioural change in heat transfer in a heated horizontal cylinder rotating about its axis. Different parameters such as rotational speed, temperature of the cylinder, and diameter of the cylinder were varied to analyse the outcome. They highlighted the significant role of the critical value of Reynolds number in affecting the Nusselt number. Kays et al. (Kays, W. M., I. S. Bjorklund, 1958) proposed a correlation to determine the Nusselt number for a rotating cylinder, considering the combined effects of rotation, natural convection, and crossflow, within a rotating Reynolds number ranging between 2000 to 45000.

Kurt M Becker (Becker, 1963) performed an experiment to examine the thermal characteristics of convective heat transport in a rotating heated horizontal cylinder immersed in water and proposed a correlation in connection with Nusselt number, Reynolds number and Prandtl number. Rajasekaran and Palekar (R. Rajasekaran and M. G. Palekar, 1985), theoretically analysed about the mixed convection past a rotating sphere and portrays how free stream conditions, rotation, and buoyancy influence the eruption of flow. Takhar et al. (H. S. Takhar, 2000) investigated mixed convection flow past a moving vertical slender cylinder, considering thermal and mass diffusion as the sources of buoyancy. Authors revealed that a decrease in the relative velocity between the surface and the free stream results in a corresponding decrement in surface skin friction. Bouakkaz et al. (Bouakkaz et al., 2014) numerically examined the thermo-fluid characteristics of a rotating cylinder for the laminar regime. The influence of Reynolds number and rate of rotation on the heat transfer were explored. Authors suggested that the value of average Nusselt number is directly proportional to the Reynolds number and is inversely proportional to the rate of rotation. Cheng et al. (Cheng et al., 2008) performed an experiment to examine the effects of thermal exchange in a radially rotating cylinder. Authors reported that heat transfer is enhanced by the effect of rotation to a greater extent with low Reynolds number. Ma et al. (Ma et al., 2015) experimentally investigated the characteristics of thermal exchange from a horizontal rotating cylinder surface and formulated a relation for determining the critical rotational Reynolds number which could be used for practical applications. This formula helps to ascertain if the effect of rotation on the Nusselt number is insignificant in its variation with rotational Reynolds number, across a broad range of Grashof numbers (10^4 to 10^8). Elghnam (Elghnam, 2014) examined, both experimentally and numerically, the heat transfer of a rotating heated horizontal cylinder kept in still air. He found that with increase in the Reynold number, the Nusselt number is less impacted by Grashof number and this dependency becomes negligible at higher Reynolds number. Thi et al. (Thi et al., 2025) performed a three-dimensional computational study for the experimental setup of Ma et al. (Ma et al., 2014) to analyse the thermal characteristics of a horizontally placed, heated cylinder rotating in air. Authors reported that, increased rotational Reynold number allows for a dense flow plume around the cylinder surface, thereby thickening the thermal boundary layer and thus reducing the rate of heat transfer. Sharma and Dhiman (Sharma and Dhiman, 2012), numerically investigated the influence of Prandtl number on thermal exchange in a rotating circular cylinder. They concluded that, higher Prandtl numbers lead to a significant decrease in the average Nusselt number as rate of rotation increases. Rana and Senapati (Rana and Senapati, 2020) numerically investigated the rate of heat transfer from a vertical hollow cylinder attributable to rotational effect. Authors concluded that, the percentage increase in the thermal exchange rate diminishes with rising Rayleigh number at a constant aspect ratio when transitioning from a non-rotating to a rotating cylinder. Khanafer and Aithal (Khanafer & Aithal, 2017) used the commercial software ADINA to numerically analyse mixed convection heat transfer in a lid-driven cavity with a rotating cylinder. The investigation examined the impact of the Richardson number, the non-dimensional angular velocity, and the direction of the cylinder's rotation on the average Nusselt number. In their other research work, Rana et al. (Rana et al., 2021) numerically studied the heat transfer from a vertically positioned hollow cylinder with stationary and rotating conditions by varying the Reynolds number in the range of 0 to 2100. Authors reported that a more significant reduction in mass flow rate of air is observed at larger L/D values with increasing Rayleigh number.

The existing literature indicates extensive research on natural and mixed convection involving various shaped geometries under both stationary and rotating conditions. However, limited studies appear to have focused on more complex geometries like the torus in similar contexts. Compared to cylinders and spheres, a torus introduces curvature and closed-loop topology, which significantly influence flow and heat transfer. Such features are directly relevant to applications in compact heat exchangers, rotating machinery thereby underscoring the novelty of the present study. In this study an attempt has been made to investigate the heat transfer characteristics of vertically placed solid torus suspended in air, for stationary and rotating conditions. Parameters such as Aspect ratio, Rayleigh number, and rotational speed were varied to examine their impact on the heat transfer rate. Temperature contours and Velocity vector plots are portrayed to have a clear visualization and understanding of the thermofluidic behaviour around the solid torus.

2. PROBLEM DESCRIPTION

Figure 1 depicts the isometric view of the rotating solid torus having constant surface temperature and suspended in ambient air whereas Figure 2 describes a two-dimensional physical domain with the applied boundary conditions. The three edges of the computational domain are provided with pressure outlet boundary condition and the right edge of the physical domain is subjected to axisymmetric boundary condition. The material chosen for the solid torus is Aluminium. This study comprises of the geometry with different aspect ratios (AR) of 2.5 $\left(D/d = \frac{50}{20}\right)$, 5 $\left(D/d = \frac{100}{20}\right)$ and 7.5 $\left(D/d = \frac{150}{20}\right)$. The working fluid in this study is air, with the surrounding environment held at a constant temperature of 300 K. The solid torus surface is maintained at a temperature of 326 K.

Figure 1. Three-dimensional view of the torus with angular rotation.

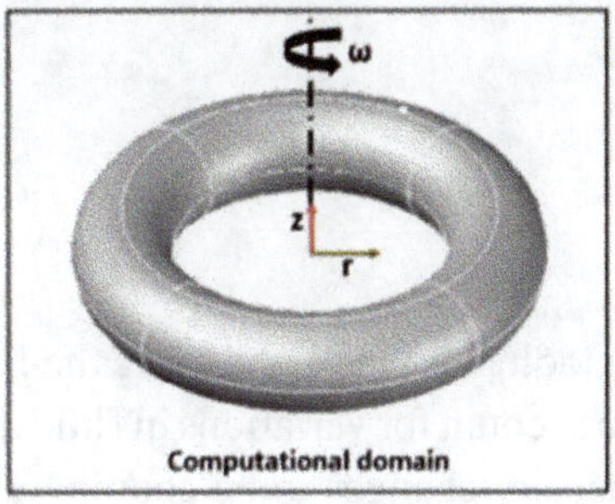

Figure 2. Two-dimensional axisymmetric computational domain with grid geometry.

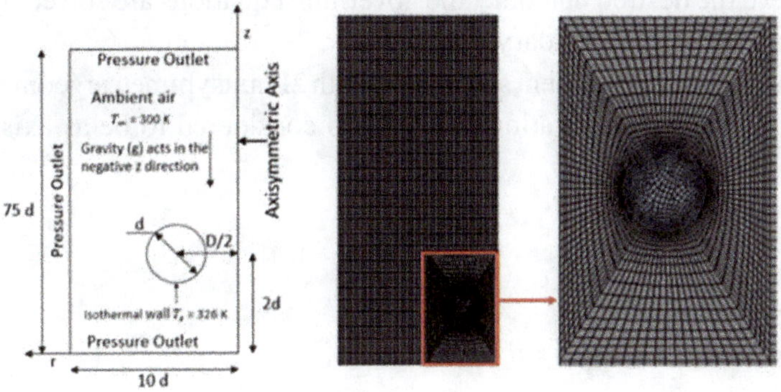

3. MATHEMATICAL METHODOLOGY

For the present investigation the fluid exhibits 2 dimensional, steady, incompressible and laminar flow. The surrounding ambient air is assumed to be as incompressible ideal gas.

Continuity equation:

$$\nabla.\vec{V} = 0 \tag{1}$$

Momentum equation:

$$\rho\left(\vec{V}.\nabla\right)\vec{V} = -\nabla p + \mu\nabla^2\vec{V} + \rho\vec{g} \tag{2}$$

Energy equation:

$$\vec{V}.\nabla T = \alpha\nabla^2 T \tag{3}$$

Since the temperature gradient is small, therefore the Boussinesq approximation method has been applied to account for variations in fluid density due to temperature changes, allowing these density changes to be coupled with the temperature field within the flow domain.

3.1 Boundary Conditions

To achieve the desired outcome, the governing equations are solved by applying the following specified boundary conditions.

At the axis: Since, the present study deals with 2D axisymmetric geometry, hence, the right side of the computational domain is considered to be as axisymmetric. Therefore, on the axis

$$\frac{\partial()}{\partial\theta} = 0 \tag{4}$$

At the torus wall: $T=T_s$, $u_r=u_z=0$ and

$$u_\theta = \frac{\omega D}{2} \tag{5}$$

The other three outer periphery of the domain is subjected to pressure outlet boundary condition:

$p=p_{atm}$ and $T=T_\infty$ (6)

3.2 Fundamental Relations Used for Determining Heat Transfer and Fluid Flow Characteristics

The calculation of thermo-physical parameters is carried out using the mean of the ambient temperature (T_∞) and the surface wall temperature (T_s).

Rayleigh number (*Ra*) based on the characteristic length ($\sqrt{A}$) can be expressed as

$$Ra = \frac{g\beta(\sqrt{A})^3(T_s - T_\infty)}{\alpha\nu} \tag{7}$$

The rotation of the torus is described through the Reynolds number (Re_D), which is derived based on the diameter of the torus and can be expressed mathematically as:

$$Re_D = \frac{D(\omega D/2)}{\upsilon} = \omega D^2 / 2\upsilon \quad (8)$$

The heat transfer through convection can be written as,

$$Q = hA(T_s - T_\infty) \quad (9)$$

The Nusselt number, considering the convective heat transfer can be formulated as,

$$Nu = \frac{hL_c}{k} = \frac{Q\sqrt{A}}{kA(T_s - T_\infty)} \quad (10)$$

The present numerical methodology has been validated with the existing literature of Acharya et al.(Acharya et al., 2018), along with the McAdams (McAdams, 1954) and Churchill (Churchill & Chu, 1975) corelations. From the plot, it can be clearly observed that the results of present numerical model are in good agreement with each other.

Figure 3. Validation Plot illustrating the variation of Nusselt number between the present model study and the published results

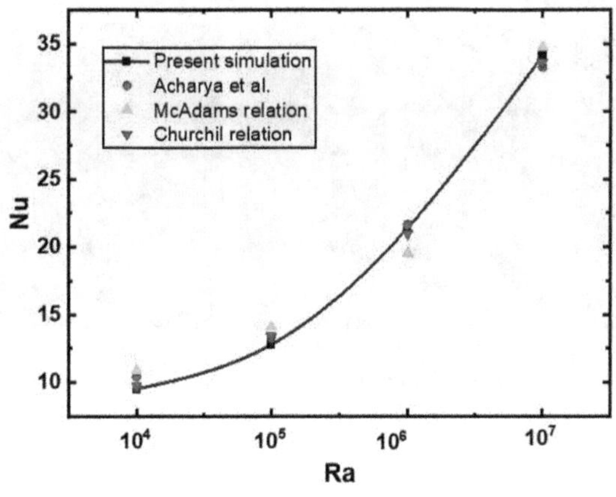

4. RESULTS AND DISCUSSION

4.1 Influence of Rotational Speed (ω) on the Temperature Plume

Figure 4 presents the temperature distribution around a vertically placed torus subjected to both rotating and non-rotating conditions for a range of rotational speed, while maintaining a constant Rayleigh number and aspect ratio. As the rotational speed increases, the influence of forced convection becomes more pronounced, superimposing on the natural convection effects. This results in the development of stronger and more extensive thermal plumes that spread further from the toroidal surface. The rotation intensifies the surrounding airflow, which reduces the thermal boundary layer thickness and consequently enhances the convective heat transfer rate from the cylinder. It can be observed that, in non-rotating condition ($\omega = 0$), the flow is primarily buoyancy-driven, resulting in symmetric and well-defined vertical plumes typical of natural convection. However, as rotation is introduced and increased, the symmetry breaks, and the flow behaviour transitions into a mixed convection regime where both buoyancy and inertial forces play significant roles. This change in flow structure is clearly reflected in the thermal plume shape and distribution, demonstrating how rotation alters the heat transfer mechanism in the system.

Figure 4. Pictorial view of the Thermal plume generated for various rotational speed for D/d = 2.5 and Ra = 10^3

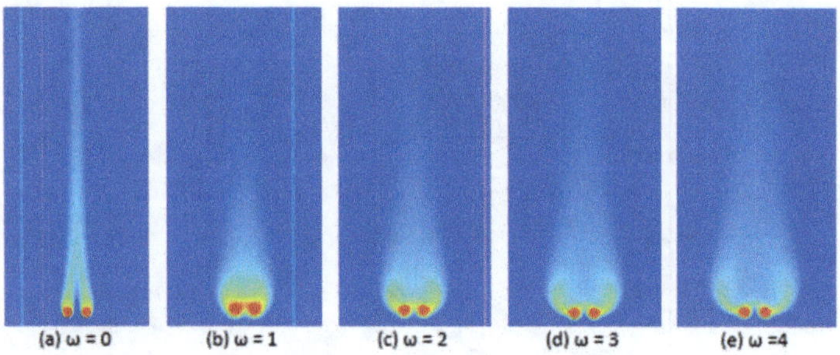

(a) $\omega = 0$ (b) $\omega = 1$ (c) $\omega = 2$ (d) $\omega = 3$ (e) $\omega = 4$

4.2 Influence of the rotational speed (ω) on the velocity vector field

Figure 5 illustrates the velocity vector fields around a vertical torus for both rotating and non-rotating cases at a constant Rayleigh number of Ra = 10^3 and aspect ratio of D/d = 2.5. Fig. 5a shows the behaviour of the fluid flow around the surface of the torus in the absence of rotation. Here, the flow is purely driven by natural convection, characterized by weak upward-moving currents that originate from buoyancy forces near the heated surface. From the figure it is very clear that the velocity near the torus wall remains low due to the dominance of the boundary layer effect. However, when the torus is rotated, a significant change in the flow pattern is observed. The rotation introduces strong forced convection, resulting in higher fluid velocities near the torus wall. This causes the boundary layer to get thinner and accelerates the fluid flow around the torus. As the rotational speed increases (at higher value of ω), the region of high velocity expands outward from the torus surface. This effect is especially evident when comparing the velocity plot for ω = 2, 3, and 4 (fig. 5c, 5d, and 5e respectively). With a gradual increase in rotational speed, the velocity distribution becomes broader and more intense, indicating a transition toward mixed convection dominated by rotational effects. The velocity increases by approximately 500% from the stationary to the maximum rotational speed of 4 rad/s.

Figure 5. Velocity vector plot illustrating the impact of rotational speed.

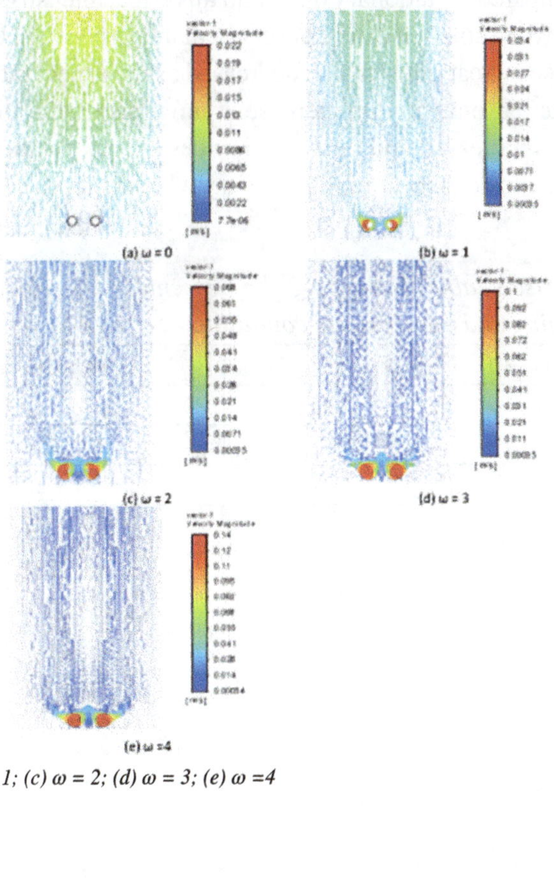

(a) ω = 0; (b) ω = 1; (c) ω = 2; (d) ω = 3; (e) ω =4

4.3 Effect of Rotational Speed (ω) on Heat Loss Rate at Different Aspect Ratio and Rayleigh Numbers

Figure 6 presents a comprehensive analysis of the effect of rotational speed on the heat loss rate from a solid torus surface for various aspect ratio (D/d) and Rayleigh number (Ra). This analysis is performed to understand how changes in geometry and flow parameters influence the combined modes of natural and forced convection heat transfer when considering rotational effect. A general trend observed across all cases is that the rate of heat loss increases as the D/d increases. This increase is evident for both stationary and rotating configurations of the torus. Physically, a greater D/d implies a larger surface area, resulting in a greater effective heat transfer surface area exposed to the surrounding fluid. This expanded surface area enhances thermal interaction, thereby promoting higher convective heat loss. More notably, the enhancement in heat transfer due to rotation becomes increasingly significant with the rise in rotational speed. For a fixed D/d and Ra, increasing rotational speed corresponds to higher rotational velocity of the torus. This leads to a stronger forced convection component, which supplements the natural convection mechanism already present in the system. The result is a greater percentage increase in heat transfer for rotating torus compared to stationary ones, and this effect intensifies with increase in ω. However, from the Table 1, it can be observed that for a constant Rayleigh number, the percentage rise in heat transfer gets decreased as the aspect ratio upsurges. This may be due to the reason that with increase in the aspect ratio, the surface area of the torus also gets increased and thus experiences a weak rotational effectiveness as compared to the lower aspect ratio geometry which is more compact.

Table 1. Comparison table illustrating the percentage increase in heat transfer considering rotating and non-rotating conditions at a fixed Rayleigh number of 10^5

D/d	Q (w) at ω = 0	Q (w) at ω = 3	Percentage increase in Q (W)
2.5	2.01	3.30	64.17
5	3.81	5.31	39.37
7.5	6.12	7.67	25.32

Figure 6. Plot illustrating the effect of rotational speed and Rayleigh number on heat transfer for various aspect ratio.

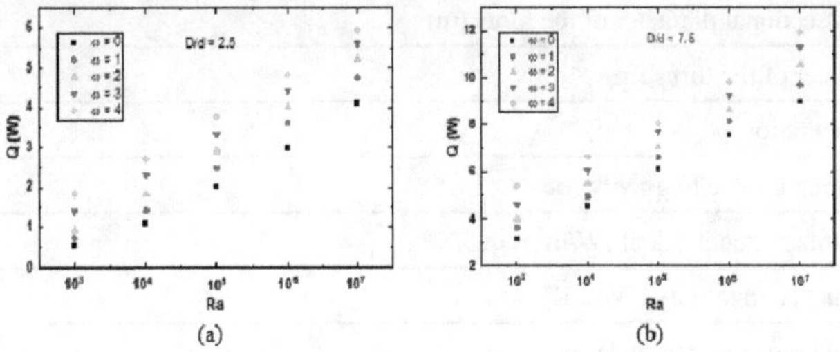

5. CONCLUSIONS

A comprehensive computational study has been conducted to examine the laminar mixed convection behaviour around a vertical torus under both rotating and non-rotating conditions, across a range of Rayleigh numbers (10^3 to 10^7), rotational speed ($0 \leq \omega \leq 4$), and aspect ratios (D/d = 2.5 to 7.5). The key findings of the present investigation are summarized as follows.

1) An increased Rayleigh number leads to enhanced heat transfer from the surface of the solid torus in both rotating and stationary conditions.
2) Compared to a stationary torus, a rotating torus exhibits a higher rate of heat transfer due to the additional influence of forced convection. This enhancement becomes more pronounced at higher rotational speeds. For an aspect ratio of 2.5, the percentage increase in heat transfer is approximately 64% from the stationary case to the maximum rotational speed considered in this study.
3) The velocity contour depicts that as compared to the stationary torus, the fluid nearby the surface of rotating torus experiences maximum velocity and gets more pronounced with the increase in rotational speed, thus, resulting in greater heat transfer efficiency. An increase of approximately 500% is observed between the stationary torus and the torus subjected to the maximum rotational speed of 4 rad/s.
4) As the aspect ratio rises (2.5 to 7.5), heat transfer improvement declines (approximately 60%), attributed to diminished rotational impact on the larger geometry.
5) Rotational speed significantly influences heat transfer in toroidal geometries by altering flow circulation and thermal boundary layers. These findings are relevant for engineering applications such as rotating machinery, where toroidal shaped objects are used.

Box 1. Nomenclature

A	surface area of the torus (m^2)
d	cross-sectional diameter of the torus (m)
D	diameter of the torus (m)
D/d	Aspect ratio
g	acceleration due to gravity (m/s^2)
h	heat transfer coefficient (W/m^2 K)
k	thermal conductivity (W/m K)
L_c	characteristic length scale (m)
Nu	Nusselt number
p	pressure (N/m^2)
Q	heat transfer rate (W)
Ra	Rayleigh number
Re_D	Reynolds number
r	radial coordinate
T_s	surface temperature of the torus (K)
$T\infty$	ambient temperature (K)
u_r	velocity component in radial direction (m/s)
u_z	velocity component in axial direction (m/s)
z	axial coordinate
Greek symbols	
α	thermal diffusivity (m^2/s)
μ	dynamic viscosity (kg/m s)
υ	kinematic viscosity (m^2/s)
ω	rotational speed (rad/s)

REFERENCES

Acharya, S., Agrawal, S., & Dash, S. K. (2018). Numerical Analysis of Natural Convection Heat Transfer from a Vertical Hollow Cylinder Suspended in Air. *Journal of Heat Transfer, 140*(5). Advance online publication. DOI: 10.1115/1.4038478

Anderson, J. T., & Saunders, O. A. (1953). Convection from an Isolated Heated Horizontal Cylinder Rotating about Its Axis. *Proceedings of the Royal Society of London. Series A. Mathematical and Physical Sciences*,555–562. https://doi.org/ DOI: 10.1098/rspa.1953.0080

Becker, K. M. (1963). Measurements of Convective Heat Transfer from a Horizontal Cylinder Rotating in a Tank of Water. *International Journal of Heat and Mass Transfer*, *6*(12).

Bouakkaz, R., Talbi, K., Khelil, Y., Salhi, F., Belghar, N., & Ouazizi, M. (2014). Numerical investigation of incompressible fluid flow and heat transfer around a rotating circular cylinder. *Thermophysics and Aeromechanics*, *21*(1), 87–97. DOI: 10.1134/S0869864314010089

Cheng, Y. J., Liao, Y. H., & Huang, C. K. (2008). Heat transfer on a radially rotating heated cylinder. *International Communications in Heat and Mass Transfer*, *35*(10), 1355–1359. DOI: 10.1016/j.icheatmasstransfer.2008.08.002

Churchill, S. W., & Chu, H. H. S. (1975). Correlating equations for laminar and turbulent free convection from a horizontal cylinder. *International Journal of Heat and Mass Transfer*, *18*(9), 1049–1053. DOI: 10.1016/0017-9310(75)90222-7

Elghnam, R. I. (2014). Experimental and numerical investigation of heat transfer from a heated horizontal cylinder rotating in still air around its axis. *Ain Shams Engineering Journal*, *5*(1), 177–185. DOI: 10.1016/j.asej.2013.09.008

Kays, W. M., & Bjorklund, I. S. (1958). Heat Transfer From a Rotating Cylinder With and Without Crossflow. *Transactions of the American Society of Mechanical Engineers*, *80*(1), 70–77.

Khanafer, K., & Aithal, S. M. (2017). Mixed convection heat transfer in a lid-driven cavity with a rotating circular cylinder. *International Communications in Heat and Mass Transfer*, *86*, 131–142. DOI: 10.1016/j.icheatmasstransfer.2017.05.025

Ma, H., Ding, Z., Cao, Y., Lv, X., Lu, W., Shen, X., & Yin, L. (2015). Characteristics of the heat transfer from a horizontal rotating cylinder surface. *Experimental Thermal and Fluid Science*, *66*, 235–242. DOI: 10.1016/j.expthermflusci.2015.03.031

Ma, H., Xia, W., Zhou, W., Liu, W., Guo, Z., & Zhang, C. (2014). Experimental investigation on the steady, external laminar mixed convection heat transfer characteristics around a large diameter horizontal rotating cylinder. *International Communications in Heat and Mass Transfer*, *57*, 239–246. DOI: 10.1016/j.icheatmasstransfer.2014.08.006

McAdams, W. H. (1954). *Heat Transmission*. McGraw-Hill.

Rajasekaran, R., & Palekar, M. G. (1985). Mixed convection about a rotating sphere. *International Journal of Heat and Mass Transfer*, *2*(5), 959–968. https://doi.org/https://doi.org/10.1016/0017-9310(85)90277-7

Rana, B. K., & Senapati, J. R. (2020). *Laminar Mixed Convection Over a Rotating Vertical Hollow Cylinder Exposed in the Air Medium*. Springer., DOI: 10.1007/978-981-15-7831-1

Rana, B. K., Singh, B., & Senapati, J. R. (2021). Thermofluid Characteristics on Natural and Mixed Convection Heat Transfer From a Vertical Rotating Hollow Cylinder Immersed in Air : A Numerical Exercise. *Journal of Heat Transfer*, *143*(2), 022601. DOI: 10.1115/1.4048830

Sharma, V., & Dhiman, K. A. (2012). Heat transfer from a rotating circular cylinder in the steady regime : Effects of prandtl number. *Thermal Science*, *16*(1), 79–91. DOI: 10.2298/TSCI1100914057S

Takhar, H. S. A. J. C. & G. N. (2000). Combined heat and mass transfer along a vertical moving cylinder with a free stream. *Heat and Mass Transfer, 36*(3), 237–246. https://doi.org/https://doi.org/10.1007/s002310050391

Thi, N., Quynh, N., Kuo, H., Liao, C., & Lin, K. C. (2025). 3D CFD analysis of convective heat transfer from a heated horizontal cylinder rotating in air: From laminar to turbulent flow. *Journal of Heat and Mass Transfer,* https://doi.org/DOI: 10.1115/1.4068377

Chapter 5
Design and Fabrication of Ornithopter

P. Srinivas Reddy
https://orcid.org/0000-0003-4632-1291
CVR College of Engineering, Hyderabad, India

Sk Mohammad Shareef
https://orcid.org/0000-0002-2634-9273
CVR College of Engineering, Hyderabad, India

Maturi Jeevan Kumar
CVR College of Engineering, Hyderabad, India

ABSTRACT

Ornithopters, or flying vehicles propelled by flapping wings, have been of great interest in recent years because of their application to micro aerial vehicles (MAVs). There are a host of new problems in vehicle dynamics and control that can be explored because of this renewed interest. An ornithopter is a robotic flight vehicle that generates lift and thrust by flapping its wings. An ornithopter equipped with a wireless camera will be developed to study flapping wing flight better. In this study, it is discussed the essential aspects of developing an ornithopter as well as its operation mechanism and various applications. This project begins with the design of the ornithopter mechanism, design of the mechanical, electrical, and control systems, followed by the manufacturing process. An ornithopter's data is obtained through kinematic and experimental analysis. Based on the ornithopter design, the wingspan is 126 cm, the weight is 370 grams, and the torsional length is 95 cm. This mechanism can achieve a maximum flapping frequency of 6-7 Hz based on the calculation results.

DOI: 10.4018/979-8-3373-3621-3.ch005

INTRODUCTION

1.1 Introduction to Ornithopter

Ornithopter flight, commonly referred to as bird flight (Mueller and DeLaurier, 2001), operates with only two degrees of freedom (DOF): the primary flapping motion and a minor deviation from the stroke plane, known as out-of-plane flapping. Unlike insect flight, which involves three degrees of freedom, ornithopter flight replaces the active wing rotation with passive wing rotation, a phenomenon termed 'feathering.' This passive rotation arises from the inertial forces generated by the wings during flapping, effectively simplifying the kinematics as the wing rotation occurs naturally.

As micro air vehicles (MAVs) (Buchner et al., 2017) shrink in size, challenges in fabrication and the complexity of airflow dynamics within their operational Reynolds number range become more pronounced. Drawing inspiration from a red-tailed hawk, researchers have developed a small-scale ornithopter with a single flapping frequency to replicate the straightforward and dominant flapping-wing motion of birds. However, the design process for ornithopters is hindered by a lack of comprehensive understanding of the key design parameters and their interdependencies in relation to flapping-wing flight dynamics (Warrick et al., 2010). Studies indicate that incorporating a tail wing can enhance the longitudinal stability of ornithopters, even under the continuous flapping motion of their wings.

1.2 Definition of Ornithopter

An ornithopter is a type of aircraft that achieves flight by mimicking the flapping motion of wings (Liger et al., 2002), inspired by the natural flight of birds, bats, and insects. While the designs of these machines can vary, they are typically constructed to be comparable in size to the flying creatures they emulate

1.3 Working of Ornithopter

The operation of an ornithopter is inspired by the natural flapping-wing mechanism of biological birds is shown in Figure.1.In this process, the rotational power generated by a BLDC motor (DeLaurier and Chronister, 1994) is transformed into a flapping mechanism through various components, including gears, a gearbox, and pistons. As the motor rotates, the power and torque are converted from rotational motion into the linear motion of the wings, enabling unidirectional wing movement.

Figure 1. Working model of ornithopter

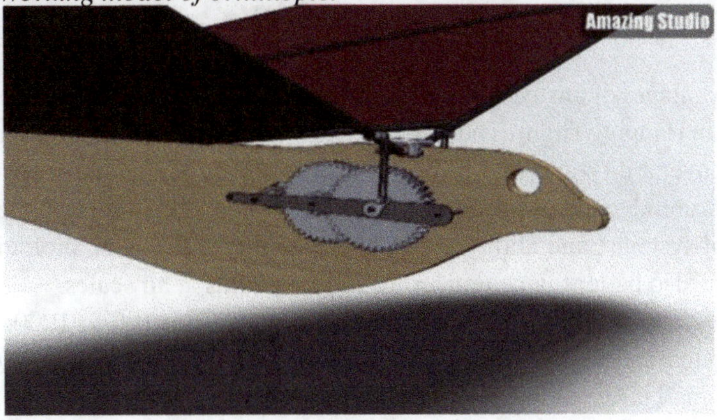

1.4 Application of Ornithopter

- In order to conserve the Gunnison sage grouse, Colorado Parks and Wildlife has used ornithopters. With the use of a remotely controlled artificial hawk, they can keep the grouse grounded, making it easier to capture them.
- Ornithopters can be used for military purposes, such as aerial reconnaissance, because they are designed to look like birds or insects. By blending into the natural surroundings, they can monitor potential adversaries without alerting them.
- The ornithopters are designed to be capable of hovering and maneuvering in tight spaces using video cameras. A remotely piloted ornithopter, modeled after a large hummingbird, was demonstrated in 2011 by AeroVironment, Inc.
- Future developments will aim to eliminate motors and gears from current ornithopter designs, enabling more efficient and natural motion.

1.5 Scope of the Project

- In military applications, such as aerial reconnaissance, ornithopters can be used without alerting enemies that they are being watched.
- A number of ornithopters have been flown with video cameras onboard, some of which are capable of hovering and maneuvering in narrow spaces. Ornithopters are primarily used by spies for gathering information.

1.6 A Study of Low Reynolds Number Aerodynamics Applied to Flapping MAVs

The advantages of unsteady aerodynamics are evident at low Reynolds numbers, especially in flapping flight. The lapping wings of insects and hummingbirds have high maneuverability and agility because of unsteady aerodynamics rather than steady-state aerodynamics. In addition to their superior power supply, better stability and control systems, and ability to fly in fluctuating conditions, biological flight systems are also better than engineering flight systems at all scales.

An insect's wing chord Reynolds number ranges from 100 to 1000; its wings are corrugated and curved and it uses unsteady effects to stay aloft. From large insects to small birds, the wing chord Reynolds number ranges between 1000 and 15000, and they use conventional airfoil circulation and are sensitive to transitions and separations. Increasing wing beat frequency reduces the size of MAVs, making unsteady flight regimes a more critical feature. Adapting the flight mechanisms and wing structures of insects for our flapping vehicle enabled us to achieve those characteristics. As shown in Figure 1. Different Reynolds number regimes have different flight mechanisms and aerodynamic characteristics. According to Pennycuick, flight speed and bird mass are related given by

$$U=4.77\ m^{1/6} \quad (1)$$

where U is the flight speed in $m \cdot s^{-1}$ and m is the mass in g. Greenawalt computed from statistical data the correlation between wing flapping frequency f (Hz), vs. wing length l (cm), to be

$$f l^{1/6} = 3.45 \quad (2)$$

While Azuma showed that the correlations between wing flapping frequency f (Hz) and mass, m (g), for large birds and small insects are

$$f(\textit{large birds}) = 116.3\ m^{-1/6} \quad (3)$$

$$f(\textit{small insects}) = 28.7 m^{-1/3} \quad (4)$$

From Equation. (1)–(4), relationships between wingtip speed and mass can be derived. These relations are

$$\textit{Wingtip speed (large birds)} = 11.7 m^{-0.065} \quad (5)$$

Wingtip speed (small insets) =9.7m^-0.043 (6)

1.7 Aerodynamics of Flying Bird

Birds fly with flapping wings, which has several advantages over fixed-wing aircraft:

- When flying at low speeds and small sizes, flapping wings provide greater maneuverability and potential energy savings. Due to unsteady separating flows' complex, nonlinear nature, developing an aerodynamic theory for flapping flight remains challenging.
- The ornithopter uses flapping or oscillating airfoils instead of rotary motion, unlike airplanes and helicopters. In the same way that helicopter wings provide lift and thrust, ornithopter wings serve a dual purpose.
- An advantage of flapping wings is that they can be set to zero angle of attack on the upstroke, leading to a minimal amount of resistance while moving through the air. In addition to generating lift and thrust, flapping airfoils also minimize drag, making flight more efficient.

2. THEORY ON BIRD FLIGHT

2.1 Bird Flight Theory

Despite the body and tail also contributing to flight control and lift enhancement, bird wings generate most of the aerodynamic forces that are needed for flight. Unlike fixed-wing aircraft, which integrate and separate lift and thrust functions, this is different from an aircraft with an angled wing. A bird's aerodynamic force is determined by both its wing shape and motion, the latter of which is actively controlled by its neuromuscular system. Sensors provide information to this system. A bird's wings can be oriented to support both body weight and thrust because they move relative to their bodies. The arm and hand wings can change shape (morph) in response to active joint movements around the shoulder.

Within the body's frame of reference, aerodynamic force can be subdivided into lift, thrust, and drag.

2.2 Definition of Lift, Thrust and Drag

Avian flight involves lift, which is perpendicular to oncoming air, and thrust, which is parallel to oncoming air. Birds are propelled forward by thrust, while they are slowed down by drag. A forward-flying bird relies on lift to support its weight, while thrust counteracts drag. An aerodynamic analysis of a wing looks at lift, thrust, and drag in the frame of reference of the segment. The force components are defined in relationship to local air velocity to determine the lift, thrust, and drag of a wing segment.

When the bird flaps, its forward velocity is combined with the wing's motion, which is influenced by the stroke angle, spanwise position, and angular velocity of its wings. Except for hummingbirds, the downstroke is crucial to weight support and thrust generation. A downward wing velocity contributes to thrust and weight support during a downstroke in forward flight.

In addition to air density, which varies with altitude, aerodynamic forces are also influenced by velocity at the wing's radius of gyration, where pressure acts. When air density is low, lift, drag, and thrust are decreased, but as wing surface area increases, lift, drag, and thrust increase. A wing's shape and angle of attack (the angle at which the wing meets the oncoming air) influence lift, drag, and thrust coefficients. By measuring forces at an angle of attack, air density, wing area, and flight velocity, lift and drag coefficients can be calculated.

As measured by the Reynolds number (Re), the ratio of inertial to viscous forces strongly influences force coefficients for birds. A diving falcon's Re is around a million, while hummingbirds' Re is about 5,000-15,000, and it ranges from 5,000-15,000 for birds flying near sea level. For birds' morphing wings with low Reynolds numbers, determining these coefficients is more difficult than for fixed-wing aircraft with high Reynolds numbers. Hovering and gliding flights using prepared wings have been studied successfully using force coefficient calculations.

2.3 Types of Bird Wing Structures

1. Passive soaring wings are characterized by long primary feathers that spread out, forming "slots" that enable the bird to catch vertical columns of hot air, known as "thermals," and rise higher in the air. Birds with this type of wing include eagles, most hawks, and storks. A notable example of a bird with passive soaring wings is the Bald Eagle.
2. Active soaring wings are long and narrow, enabling birds to soar without flapping their wings for extended periods. However, these birds rely more on wind currents compared to passive soaring birds. Examples include albatrosses, gulls, and gannets.
3. Elliptical wings are designed for short bursts of high speed and allow for fast take-offs and tight maneuvering. Birds with this wing type can maintain their speed for a while. Examples include crows, ravens, blackbirds, sparrows, and thrushes (such as the American Robin).

4. Hovering wings are small and quick, with the bird's nerves and muscles specially adapted for rapid movement. This allows birds to hover in place, as seen in the Ruby-throated Hummingbird, which can appear to float in space while feeding on nectar.
5. High-speed wings are long and thin, though not as long active soaring wings. Birds with this wing type are incredibly fast and can maintain their speed for longer periods. Examples include swifts,
6. ducks, falcons, terns, and sandpipers.

The various types of bird wings and structure which are explained above are shown in Figure. 2 & 3 .

Figure 2. Various shapes of bird wings.

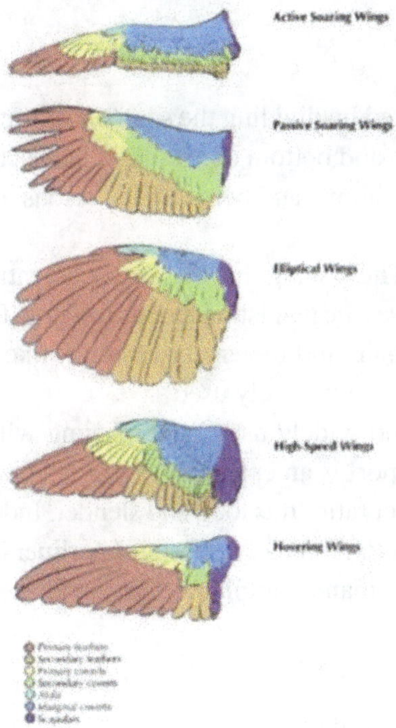

Figure 3. Structure of the bird wing

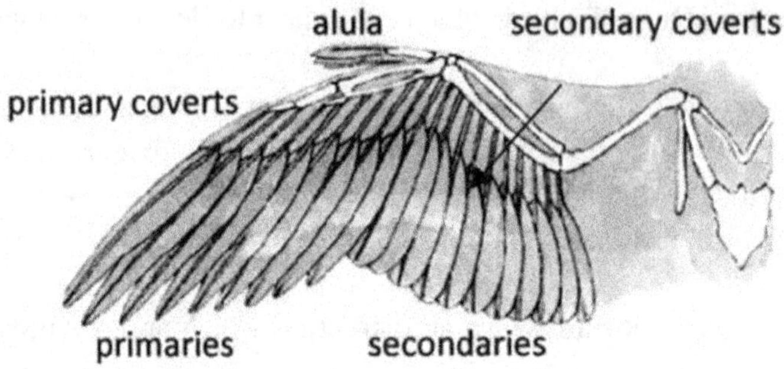

2.4 Wing Aspect Ratio

An aspect ratio is calculated by dividing the span of a wing by its average chord. The distance between the top and bottom of a wing indicates how long and slender it appears when viewed from above and below. The various aspect ratios of wings is shown in Figure.4 Below

Short and stubby wings: These wings have a low aspect ratio. In addition to their structural efficiency, they have a faster instantaneous roll rate. In fighter aircraft, like the Lockheed F-104 Starfighter, and in very fast aircraft like the North American X-15, low aspect ratio wings are typically used.

It is a general-purpose and widely used type of wing with a moderate aspect ratio. A Douglas DC-3 transport is an example.

This wing has a high aspect ratio: It is long and slender. Induced drag is reduced, which makes them more aerodynamically efficient. Airliners like the Bombardier Dash 8, as well as high-performance sailplanes like the Glaser-Dirks DG-500, use wings with high aspect ratios.

Figure 4. Various aspect ratios of wings

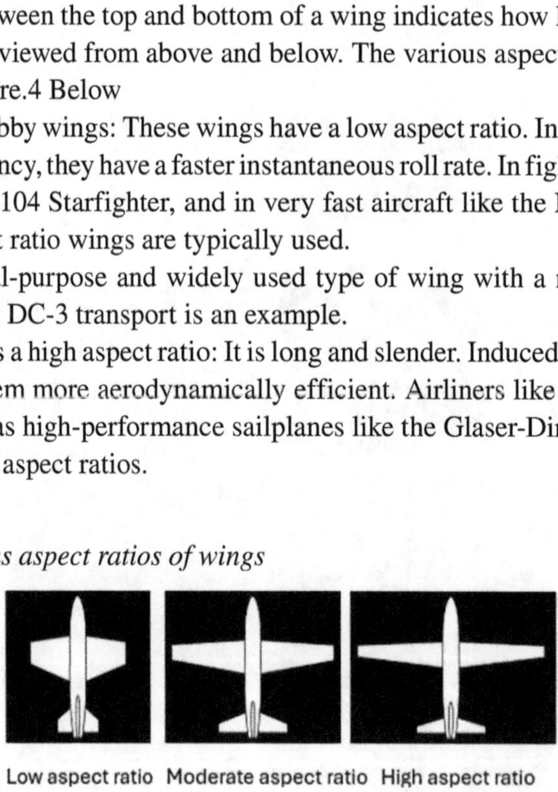

2.5 Types of Bird Flight

Understanding the main types of bird flight is helpful before discussing how a bird's wings generate forces. It can be divided into two main categories and they are shown in Figure.5 below.

Flapping Flight

During power-on flight, the bird continuously flaps its wings to generate thrust and lift forces for propulsion and to counter gravity. Although all birds flap their wings in a similar manner, there are subtle differences between small and large birds. A flexible wing's motion goes beyond just up and down; it also moves forward and backward, and parts twist as it flaps.

Gliding and Soaring

Birds that glide are not flapping their wings, but rather using gravity to fly. When the bird flies in still air, aerodynamic forces and gravity combine to allow it to glide steadily. Gliding birds use air currents with vertical velocity components, such as thermals, slope currents near mountains, or cliffs, to soar. As the wind speed increases in the upward direction, dynamic soaring occurs. This particular form of soaring occurs when there is gradient velocity in the wind.

Figure 5. Flapping and gliding motion of bird wings.

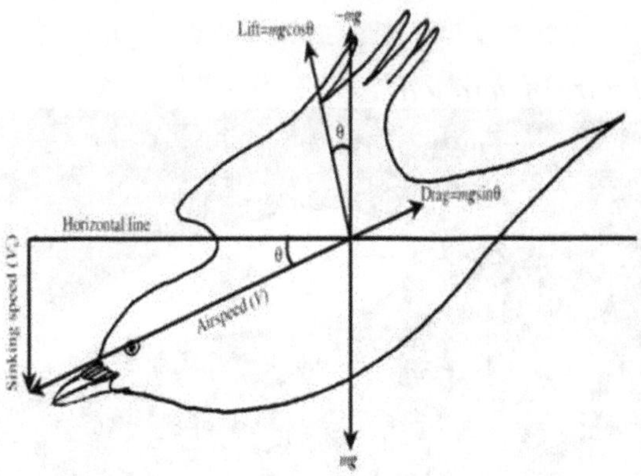

When a bird is gliding, it doesn't need to exert any effort, but it can't remain airborne indefinitely. The wings are extended out to the sides of the body without flapping. As the wings slice through the air, they are angled slightly, which pushes the air gently downward. This downward motion of air generates a reaction force in the opposite direction, known as lift. Lift is a force that acts roughly perpendicular to the wings and prevents the bird from falling.

Soaring flight is only possible under certain conditions and in specific locations. For instance, warm air heated by the sun rises from the ground into the atmosphere, creating a rising air current known as a thermal. Thermals typically form along the slopes of hills but can also occur over flat terrain. As the air ascends, it expands and cools due to the lower pressure at higher altitudes. Eventually, the water vapor within the thermal may cool enough to condense, forming the tiny droplets of liquid water that constitute a cloud. On sunny days with puffy clouds, it's a good opportunity for thermals to form, and you might spot hawks or vultures soaring in the sky.

3. THRORY OF AEROFOIL

3.1 Aero Foil Theory

Aerodynamic forces are generated when an airfoil-shaped body moves through a fluid. Among this force's components is lift, which acts perpendicular to motion, and drag, which acts parallel to motion. Airfoils used in subsonic flight typically feature rounded leading edges, sharp trailing edges, and symmetrically curving upper and lower surfaces. Hydrofoils, like airfoils, serve a similar function in water, similar to airfoils. The nomenclature and forces acting on aero foil are shown below in Figure.6 &7.

Figure 6. Nomenclature of aero foil

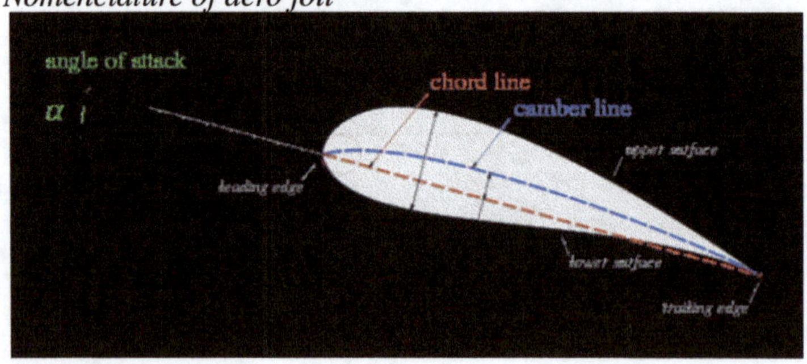

Figure 7. Forces acting on aero foil

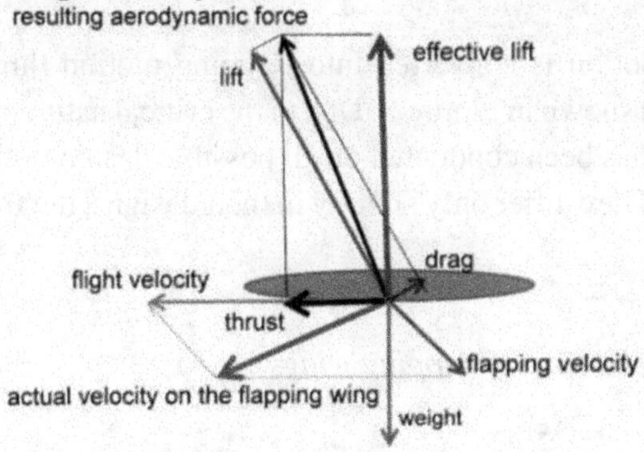

3.2 How Do Birds Generate a Lift?

Flying requires wings to generate downward momentum in the surrounding air, which is explained by Newton's Third Law: every action has an equal and opposite reaction. A bird's weight is counterbalanced by a lift force derived from Newton's Second Law. It is necessary to consider the downward velocity component deflected by the wing and the mass flow, which is proportional to the air density, the wing area, and the downward velocity component, in order to achieve this downward momentum. A flight's velocity is determined by its speed and angle of attack. Mass flow generated by a deflected wing is influenced by the area of the wing, the curvature of the wing, and, in birds, by feather manipulations that increase lift. Lift, however, is fundamentally a function of angle of attack.

Exceptionally, hummingbirds don't flap their wings to generate lift or thrust. It is a complex, unsteady, three-dimensional flapping motion, which changes continuously as the wings move. A traditional approach to studying bird flight is to assume that the forces on a flapping wing are the same as those on a wing in steady motion. As a result, it is more accurate to view bird flight aerodynamics as unsteady, especially when dealing with insect flight.

In addition to generating lift and thrust, wings also cause drag. An optimal wing design minimizes momentum loss (drag) while maximizing lift. A high-aspect-ratio wing generates two counter-rotating vortices during its downstroke phase, one on its leading edge and one on its trailing edge. Lift is generated as a result of these vortices, which force air downward. As opposed to high-aspect-ratio wings, low-aspect-ratio wings generate lift primarily through tip vortices that form a couple.

4. FLAPPING WING MECHANISM

Motor rotary motion is converted into flapping motion through the flapping wing mechanism as shown in Figure.8. Due to the critical nature of this mechanism, extensive research has been conducted on all possible design options for Micro Air Vehicles (MAVs). They differ only slightly in their designs, but their general design is quite similar.

Figure 8. Basic mechanism of flapping wings

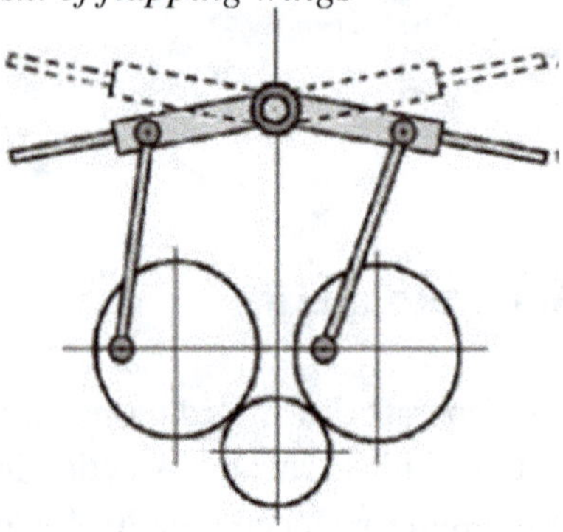

4.1 Staggered Crank Design

Flapping wing mechanisms are based on staggered crank designs as shown in Figure.9. Specifically positioned connector rods ensure that the left and right wings flap in a symmetrical manner. Hobbyists who want to make their own ornithopter using household items prefer this design. Rubber bands must be modified to replace motors as power sources.

Figure 9. Staggered crank.

4.2 Single Gear Crank Design

This crank design shown below in Figure.10 is based on the Biomimetic Milli Systems developed by the University of California, but it is more complex than it appears at first glance. There is no difference in height between the wings in this design. Connector rods and wing hinges must expand and contract as the mechanism flaps, making the system more complex.

Figure 10. Single gear crank.

4.3 Dual Gear Crank Design

Each wing hinge is independently controlled by two gears. It is possible to design a drivetrain in different ways. According to Figure.11, one version drives both secondary gears with a pinion wheel, causing them to rotate in unison. A second design uses a pinion gear to drive a secondary gear, which then drives a third gear. This configuration rotates the secondary gears counterclockwise. In addition to being simpler to implement, this design reduces wing symmetry misalignment.

Figure 11. Dual gear crank.

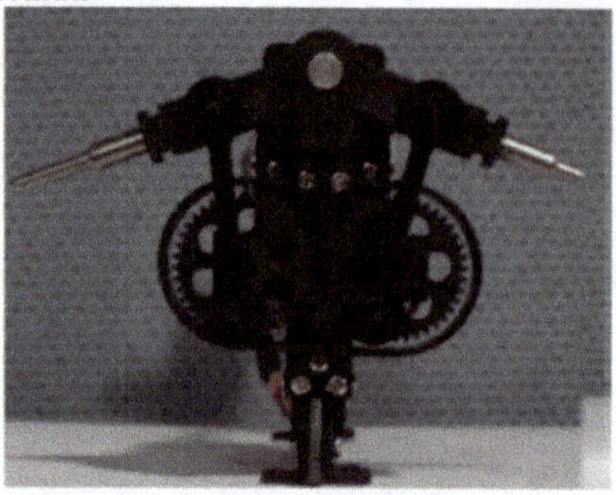

4.4 Transverse Shaft

This Figure.12 illustrates another variation on the flapping mechanism: the transverse shaft. A symmetrical flapping motion can be achieved with this design. Despite its heaviness and complexity, it is also one of the most complex designs. Due to the non-coplanar position of the flappers and the gears, a connector rod that can rotate is required for this configuration.

Connector rods in this design are equipped with ball bearings, adding weight. Furthermore, this mechanism uses more gears than any other flapping mechanism. Typically, transverse shafts are used in larger Micro Air Vehicles (MAVs), where their larger wings can compensate for their increased weight.

Figure 12. Transverse Shaft.

5. SELECTION OF ELECTRONICS

The selection of electronics for an ornithopter depends on the power required to generate lift and control the motion and direction of flight. These electronics must be capable of providing the necessary energy and control mechanisms to ensure the ornithopter can achieve and maintain flight effectively.

5.1 Selection of Motor

To achieve a specific flapping rate of the wings, a brushless direct current (BLDC) motor is used in conjunction with specially designed gears as shown in Figure.13 &14 below. These gears convert the rotational power of the motor into the reciprocating motion required by the piston. A BLDC motor, also known as BL motor, is an electronically commutated DC motor without brushes. The motor's controller delivers pulses of current to the motor windings, which regulate the speed and torque of the synchronous motor.

Figure 13. BLDC motor

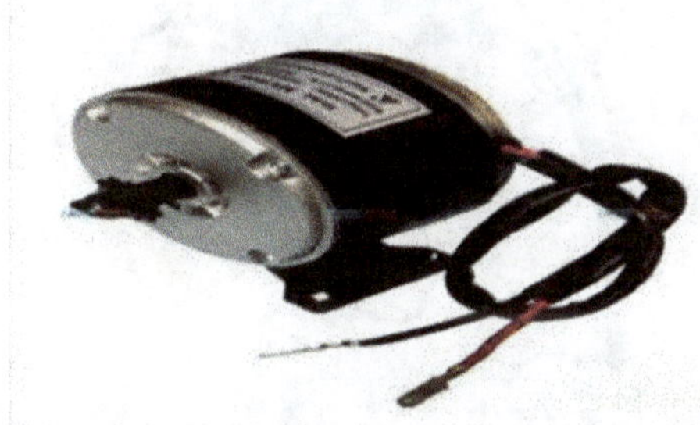

Specifications of the motor

- Power: 1800 kv
- Rated voltage: 7.4 volts
- Current: 0.8 A
- Speed:3500 rpm (no load)
- Motor size: 27.7×26.3 mm.

Circuit diagram of BLDC motor

Figure 14. Circuit diagram of BLDC motor

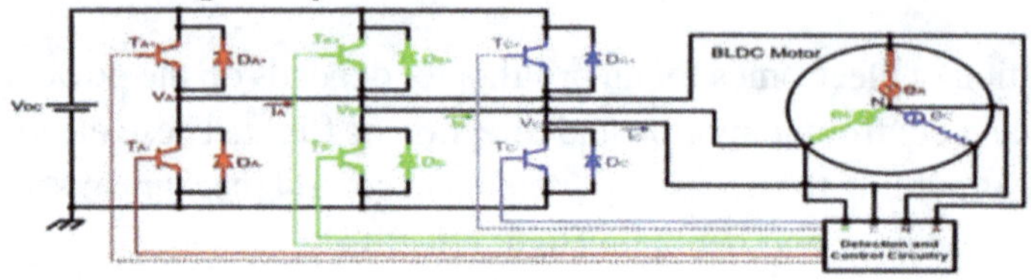

5.2 Selection of Speed Controllers

An Electronic Speed Controller (ESC) shown in Figure.15 below regulates the movement or speed of a brushless motor by activating the appropriate MOSFETs. This action creates the rotating magnetic field necessary for the motor to rotate.

Figure 15. Electronic speed controller (ESC)

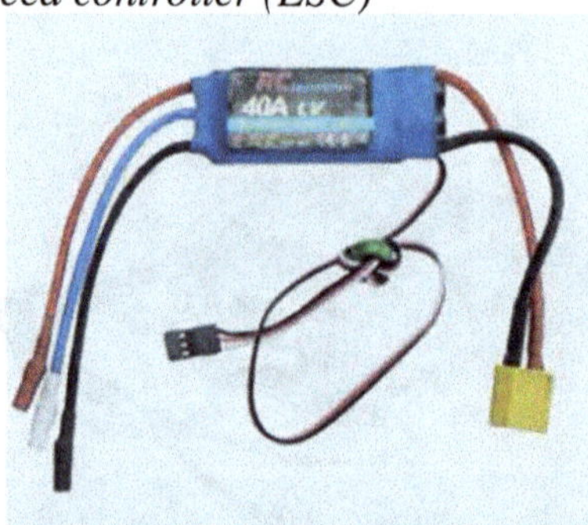

The speed of the motor increases as the frequency of the ESC (Electronic Speed Controller) rises, meaning the ESC cycles through its six intervals more quickly. An ESC is an electronic circuit as shown in Figure.16 below that controls and regulates the speed of an electric motor. It can also provide functionalities like reversing the motor's direction and dynamic braking. Miniature ESCs are commonly used in electrically powered radio-controlled models, while full-size electric vehicles use similar systems to control the speed of their drive motors

Circuit diagram of electronic speed controller

Figure 16. Circuit diagram of electronic speed controller

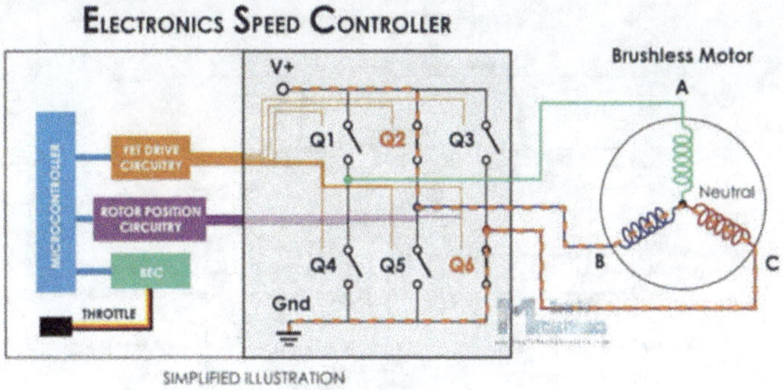

5.3 Selection of Transmitter and Receiver

Figure. 17 shows Radio-controlled (RC) toys, also known as remote-controlled toys, are self-powered and can be operated from a distance using a remote control that communicates via radio waves.

Figure 17. Remote or radio controller and receiver

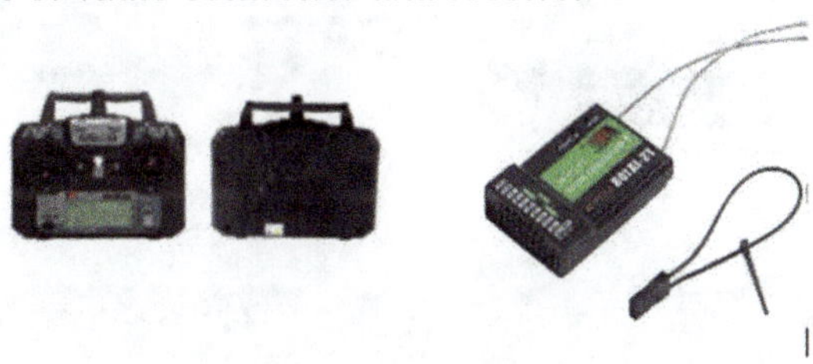

Figure 18. Transmitter diagram

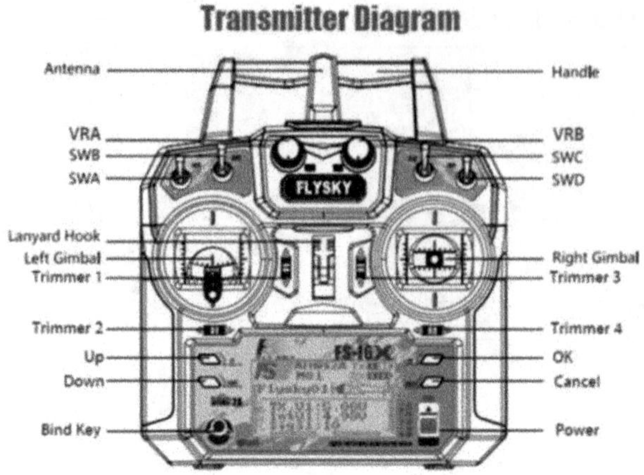

Figure 19. Binding of transmitter and receiver

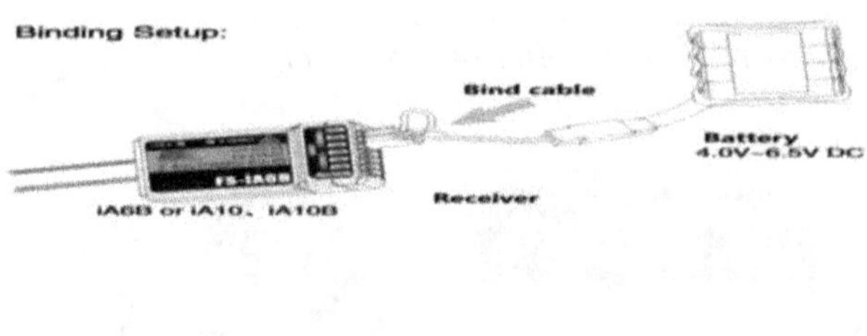

When a control on a radio-controlled (RC) toy is activated, the transmitter sends a specific number of electrical pulses corresponding to that action via radio waves. The transmitter requires its own power source, typically a 7.5-volt battery, to send these signals. Without the battery, the transmitter cannot communicate with the receiver.

Upon receiving the radio waves, the motors in the RC toy are activated to perform the desired action. The power source supplies energy to all working parts, including the motor. The transmitter shown in Figure.18 & 19 above enables control through radio waves, while the receiver interprets these signals and activates the motors accordingly. When a button is pressed on the transmitter to move the RC toy forward or backward, a pair of electrical contacts touch, and the receiver identifies these signals and sends them to the circuit.

Specifications and Features

Specification for FS-i6X RC Transmitter:

1. Item: FS-i6X RC Transmitter
2. Tx Channel: 6
3. Model Type: Fixed-Wing/Glider/Helicopter
4. RF Range: 2.408-2.475GHz
5. RF Power: < 20dBm
6. RF Channel: 135
7. Bandwidth: 500KHz
8. 2.4GHz System: AFHDS 2A / AFDHS
9. Modulation Type: GFSK
10. Stick Resolution: 4096
11. Low Voltage Warning: < 4.2V
12. DSC port: PS/2 Port PPM
13. Chargeable: No
14. Antenna Length: 26mm (Dual Antenna)
15. Weight: 392g
16. Power: 6V DC 1.5A
17. Display: STN Translative Display, LCD 128×64 Lattice, VA 73x39mm, LCD with white backlight

Specifications for FS-iA10B RC Receiver:

1. Rx Channels: 10
2. Frequency range:2.4-2 .48 GHz
3. Transmitting power: not more than 20dBm
4. Antenna length: 26 mm * 2 (dual antenna)
5. Input power:4.0-6 .5 V DC
6. Dimensions: 47 x 33.1 x 14.7 mm
7. Data acquisition interface: Yes
8. Model Type: Airplane / Glider / Helicopter
9. Compatible Transmitter: Compatible with FS-i4, FS-i6, FS-i10, FS-GT2E, FS-GT2G

5.4 Selection of Battery

The selection of the battery for an ornithopter is based on the power required by all electronic components and also considers the weight of the battery. If the battery is too heavy, it will affect the overall weight of the ornithopter. For the design of the ornithopter, Figure.20 a battery with the following specifications is chosen.

- Current: 900 mah
- No.of cells: 2s
- Voltage:7.4 volts

Figure 20. 2s cell battery

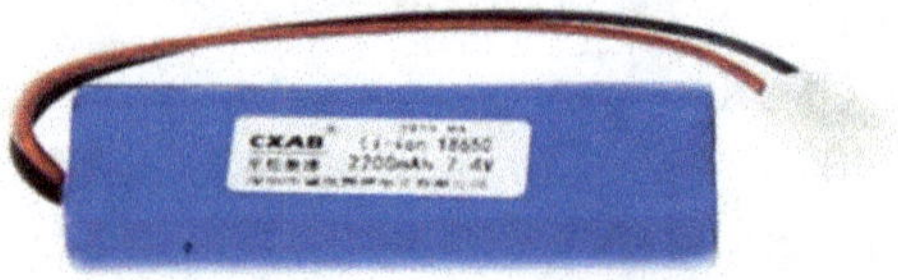

5.5Selection of Servo Motor

A servo motor is an electrical device that can precisely push or rotate an object. It is used when you need to rotate or move an object to specific angles or distances. A servo motor is essentially a simple motor that operates through a servo mechanism as shown in Figure.21 below.

Figure 21. Servo motor

A servo motor is used to change the direction of the ornithopter, and it is connected to the tail of the ornithopter. Two servo motors are installed on the right and left sides of the V-shaped tail. Specific signals from the remote control are sent to these servo motors to adjust the direction of the ornithopter's flight. These electronic components control the flight and enable the ornithopter to move at specific speeds and directions.

Table 1. Specification of 2 servo motors

Modulation:	**Analog**
Torque:	4.8V: 25.00 oz-in (1.80 kg-cm)
Speed:	4.8V: 0.12 sec/60°
Weight:	0.32 oz (9.0 g)
Dimensions:	Length:0.91 in (23.0 mm) Width:0.48 in (12.2 mm) Height:1.14 in (29.0 mm)
Motor Type: 3-pole	3-pole

6. DESIGN AND FABRICATION OF ORNITHOPTER

Design of ornithopter is distinguished into different part as of following:

1) Design of wings
2) Design of gears
3) Design of gear box
4) Design of fuselage
5) Design of tail
6) Design of frame

6.1DESIGN OF WINGS

The wing is the component responsible for generating lift. The design of the wing is inspired by the biomimicry of a falcon wing. A bird wing is divided into three sections as shown in Figure.22 below.

Primary Section: This section of the wing has a pure membrane structure and is the largest area. It generates the maximum amount of lift and thrusts, with the formation of a vortex.

Secondary Section: This section is a combination of an aero foil and a membrane. The thick aero foil gradually transitions into a membrane as it moves away from the ornithopter body.

Tertiary Section: This section is located near the ornithopter body. It is an aero foil section with high camber and a small area. It provides a small amount of lift during gliding and upstrokes.

Figure 22. Representation of bird wing

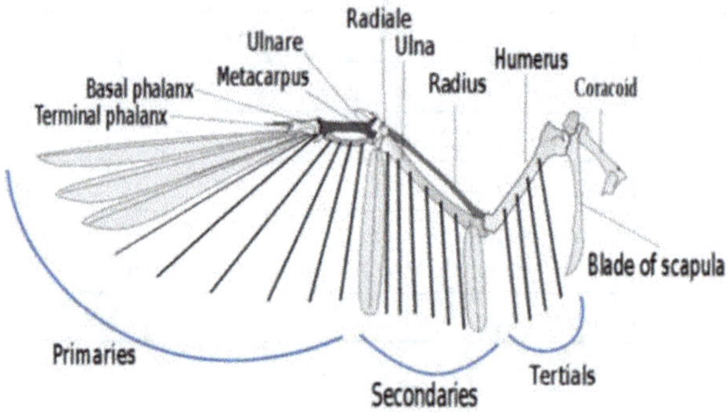

Calculation of wingspan

- The wingspan of the ornithopter is determined based on biological bird references. The references are made using the Red Hawk bird.
- The wing width is calculated by taking 1/8 part of the fuselage length – total length of the fuselage

Fuselage length= 400*mm*

W= *F* – 1/8 * *fuselage length*= 1/8*400

W= 400 – 50= 350*mm*

W= 35*cm*

- The wing width is also known as chord length.

Figure 23. Design of wings using SolidWorks software

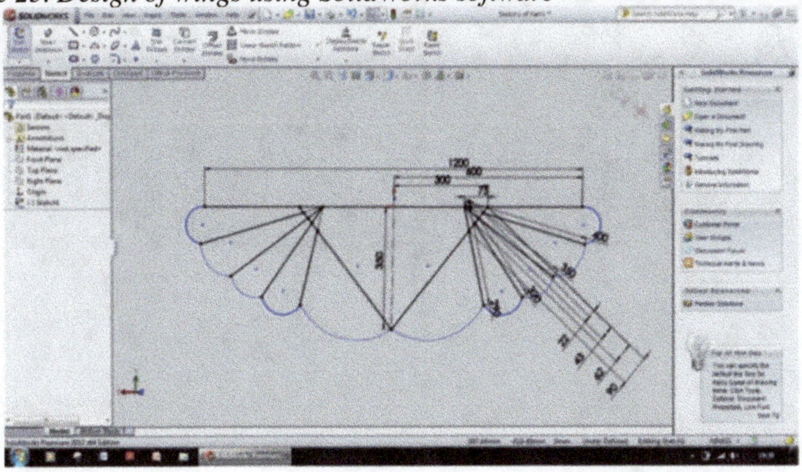

Fabrication of Wings

As a key component in ornithopter design and construction, the wings are crucial to its performance. Wings play a crucial role in flight performance aspects such as endurance, speed, and maneuverability. It is necessary for an ornithopter's wings to be capable of both thrust, which propels it forward, and lift, which keeps it aloft. A strong enough force is needed to counteract drag and the weight of the ornithopter.

When the wings flap, they change shape, generating lift and thrust. The chamber undergoes a significant change as the wing accelerates due to aerodynamic loading. As a result of this deformation, the wing takes on an airfoil shape instead of its flat plate shape. Furthermore, aerodynamic loading generates thrust through large angles of attack. A flight-sustaining force results from the combination of these two effects. The fabricated wings of Ornithopter are shown in Figure. 24 below.

Ornithopters with thin wings have a limited angle of attack, but offer optimal minimum-drag performance. Hawk weighs 181.3 grams, compared to the prototype ornithopter's 248.8 grams, so its wings are smaller.

Figure 24. Fabricated wings of ornithopter

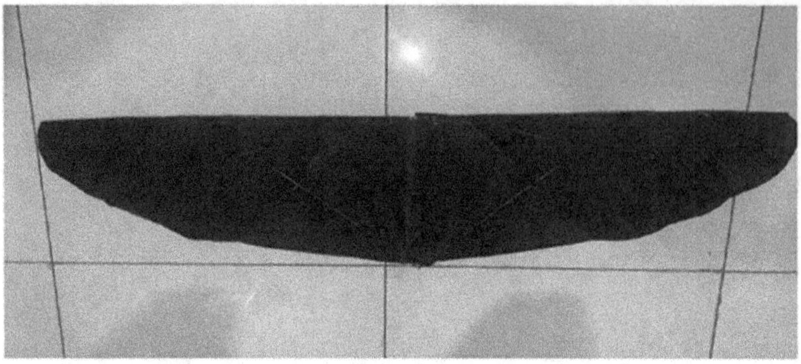

Materials used for construction

1) Carbon fiber rods for primary spar: 3mm carbon fiber rods
2) Carbon fiber rods for secondary spar: 2mm cabon fiber rods
3) Carbon fiber rods for ribs spar: 1.5 mm cabon fiber rods
4) Wings aerofoil membrane: Epp (expanded poly propylene)
5) Length of the wing span = 1200 mm
6) Chord length = 350 mm

6.2 Design of Gears

Gears and cogs are rotating parts of machines that have teeth cut into them or, in the case of cogwheels, teeth inserted into the wheel to transmit torque. The direction, speed, and torque of a power source can be changed by gearing devices. The gear ratio of a machine usually creates a mechanical advantage, which is why gears are considered simple machines.

There is no difference in the shape of the teeth on the two meshing gears. Generally, a gear train or transmission is composed of two or more meshing gears working together. Translation can be produced instead of rotation by meshing a gear with a linear toothed part called a rack.

Two gears meshing, if one gear is bigger than the other, produce a mechanical advantage, with the rotational speeds and torque differences based on their diameters..

Calculation of Gear Reduction Ratio for Obtaining Flapping Rate

For the generate of lift by the wing, the flapping rate of the wings should have between '6-8' for the Selected design of the ornithopter. Where the minimum flapping frequency per second is ‘6’ and the maximum frequency should be ‘8’. By selecting various gears, we can obtain such flapping frequency for the wings from the motor power and Rotational speed of the motor.

Gear reduction is made to calculate the flapping wing rate for required flight.

A. Main gear: 50 teeth
    C. Secondary gear: 90 teeth
B. Pinion gear: 10 teeth
    D. Driver gear: 12 teeth

Appling gear reduction ratios:
A: B: C: D
(50÷10)×(90÷12)=1: 37.5=1/37.5
Rotational speed calculation:
7.4*v**1870*rpm/v*= 13,690*rpm* (from motor specifications)
Rotation speed per second:
13,690/60=228 (rotation per sec)
Calculation of flapping rate of wings:
Rotational speed of gears * gear ratio
226*1/37.5=6.09 (flaps per second)

By the above calculation the flapping frequency per second is obtained has ‘6.08’ And the required frequency should have between ‘6 – 8’. So the design of gears can obtain by calculating The ration reduction rate of the selected gears. The Figure. 25 shows the Fabricated gears.

Figure 25. Fabricated gears for the gear box

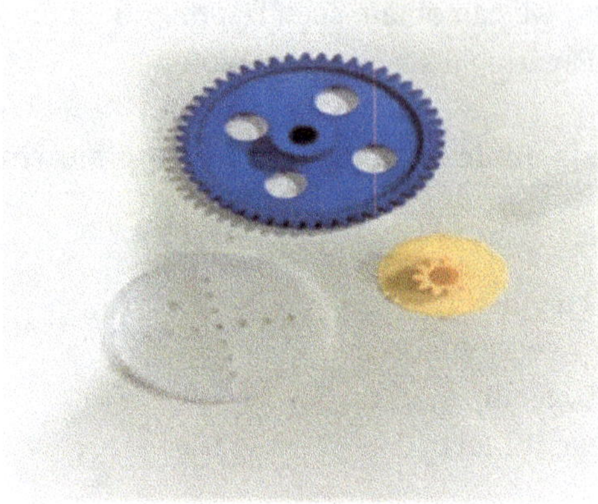

6.3 Design of Gear Box

Gear box is a mechanical device where the gears are used for transmitting power from one part of a machine to another.

A crucial part of an ornithopter is its drive mechanism, which converts electric power to flapping wings with the help of a battery. In order for this system to function effectively, it must be capable of surviving extremely large forces that reverse direction dozens of times every second, while also being extremely light and durable.

A drive system can be divided further into four components: the electric motor, gear reduction stages, linkages to convert high torque rotations into reciprocating motions, and the links that connect to the spars of the wings. Power-to-weight ratio is maximized by integrating these components closely, but the analysis shows they are easily separated.

1) To increase speed and decrease torque
2) To decrease speed and increase torque

In designing the ornithopter gear box, the goal is to reduce speed and increase the power from the motor, converting the rotational speed into the flapping wing mechanism.

The gear box is located at the front of the fuselage, where the gears are mounted in their respective positions. A dual position made of carbon fiber is attached on both sides of the driven gear, transferring the rotational motion of the gears into oscillating motion.

This oscillating motion is achieved through a single crank and slider mechanism as shown in Figure.26 & 27 below, with the connecting piston connected at one end to the driven gear and at the other end to the wings' main spars, unstoppers.

Figure 26. Single slider mechanism of piston connection rod

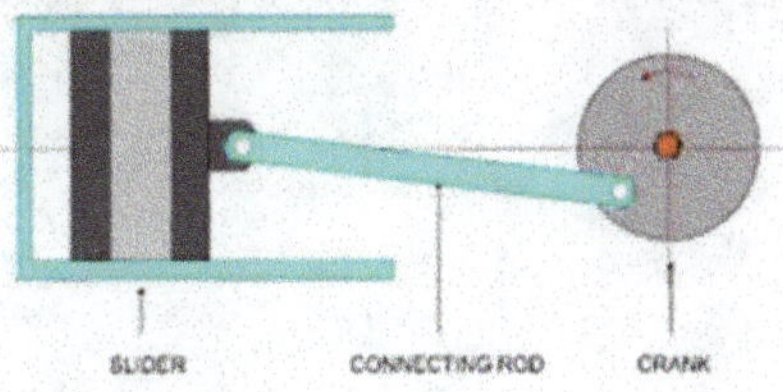

Figure 27. Crank and slider mechanism of ornithopter wings

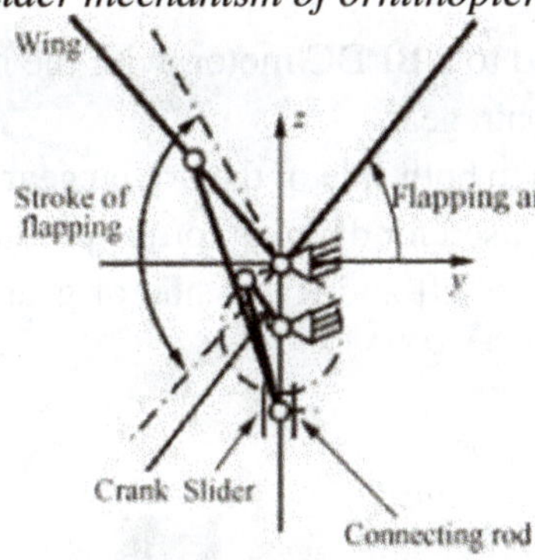

The gear box is connected to a BLDC motor, where the motor's shaft is linked to a pinion gear that drives the main gear.

Figure 28. Design of gear box using Creo parametric software

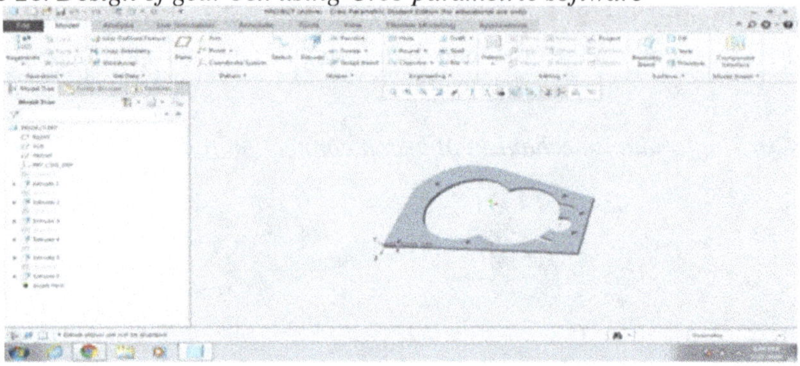

Fabrication of Gear Box

The gear box is connected to a BLDC motor, with the motor's shaft linked to a pinion gear that drives the main gear.

A gliding stopper is fixed to both side of the driven gear which is connected to a lever. This gliding stopper is used for dapping torque produced by the wing during gliding of the ornithopter. The left and Right side of gear box is shown in above Figure.29 & 30 above.

Figure 29. Left side view of gear box

Figure 30. Right side view of gear box

- Total length of gear box = 118 mm
- Total width of gear box = 65 mm

6.4 Design of Fuselage

The fuselage, or body of the air vehicle, is a long, hollow tube that holds all the components of the aircraft together. It is designed to be hollow to reduce weight. The shape of the fuselage is typically determined by the mission of the aircraft. For example, a supersonic fighter plane has a very slender, streamlined fuselage to minimize drag associated with high-speed flight.

In the case of the ornithopter, the fuselage holds all the mechanical and electrical components. It is divided into three sections as shown in Figure.31 below.

1. Front Section (Nose): This section contains mechanical devices such as gears, the gear box, BLDC motor, bearings, and connecting piston rods. The front section is constructed aerodynamically to reduce drag caused by the trusses of the fuselage.
2. Middle Section (Body): The middle section houses all the electrical and electronic devices, including the battery, electronic speed controller (ESC), and receiver. These components are strategically placed to balance the overall weight of the fuselage.

3. Back End Section: The tail of the ornithopter is attached to the back end section using threading and super glue. During flight, the major weight of the fuselage is concentrated in this section.

The material used for the fuselage consists of 3mm diameter hollow rods with a 4mm hollow diameter.

Figure 31. Fuselage of the ornithopter

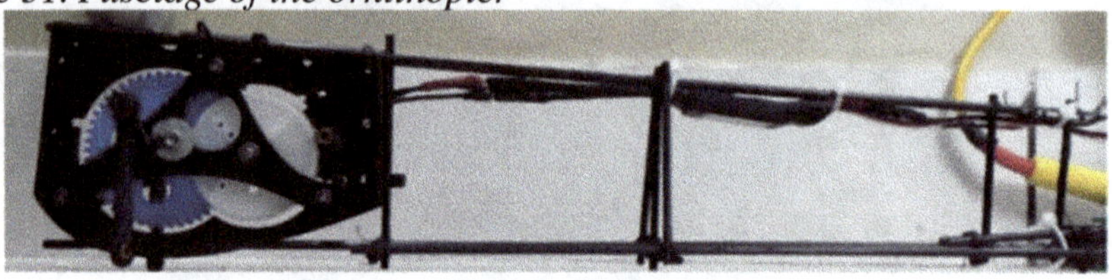

Total length of the fuselage:38.5 cm Total width of the fuselage: 6.8 cm

6.5 Design of Tail

Fixed-wing aircraft, as well as helicopters, gyroplanes, and ornithopters, have a tailplane, or horizontal stabilizer. Additionally, ornithopter tail assemblies have two degrees of freedom that can be controlled.

An angle-attached servo is located directly on the tail, and another is located further along the body, which is connected by a linkage. Due to the inherently co-dependent relationship between elevator and rudder, angling the rudder servo is crucial. In this coupling, the tail moves vertically when the rudder servo is activated, unless it is at a zero-angle position. When the tail is angled relative to the rudder servo, it is possible to maintain a zero-angle position regarding the ornithopter body, ensuring that the tail remains near its trimmed position when flying horizontally.

While this arrangement doesn't resolve issues caused by the elevator servo displacing the tail from its horizontal trim, it offers a significant advantage when utilized effectively. Another notable feature of the tail design is the elevator servo linkage, which, though it may seem redundant, fulfills two critical roles. Firstly, it reduces the servo's range of motion, which is otherwise excessive for the tail's requirements. This reduction, combined with the mechanical advantage provided by the linkage, allows for the use of a smaller servo. Secondly, it shifts the servo, a substantial weight component, toward the front of the frame. This adjustment is vital for achieving proper balance, as the ornithopter's frame often becomes rear-heavy when the center of gravity is considered.

Configuration of Tail

When an airfoil-shaped body moves through a fluid, it generates an aerodynamic force. An element of this force that acts perpendicularly to the direction of motion is called lift. Drag refers to the component that runs parallel to the motion direction. A subsonic airfoil typically features a rounded leading edge, a sharp trailing edge, and a symmetrical curvature on both surfaces during subsonic flight. A hydrofoil is a foil designed for use in water and serves the same purpose as an airfoil. Fixed-wing aircraft, as well as helicopters, gyroplanes, and ornithopters, are powered by tail planes (also referred to as horizontal stabilizers) located behind their main lifting surfaces.

Other than throttling the drive motor, the tail section of an ornithopter provides both controllable degrees of freedom.

In the tail of the ornithopter, one servo is directly attached at an angle to the tail. In addition, there is another linkage higher up the body that gives it a rocking motion. It is important to mount the rudder servo at an angle since an elevator and a rudder function as a single unit. It is only at zero angle where it does not have any control authority that the tail moves vertically along with the rudder servo. When the tail is angled to the rudder servo, the servo can be held at zero-angle with reference to the ornithopter body, and the tail will remain near its trimmed position during horizontal flight.

Although it does not solve the problem when the elevator servo moves the tail out of trim horizontal, exploiting it is a big advantage where it applies. In addition, the elevator servo linkage appears unnecessary but serves two important purposes. There is first a large servo motion range for the tail that is not necessary. With a linkage, not only is the range of motion reduced, but a smaller servo can be used due to the mechanical advantage. The servo is also moved forward in the frame, which is one of the major points of weight on the ornithopter. In spite of their apparent lack of importance, the frame becomes rear-heavy once the centre of gravity is placed. The Design and construction of V-Tail is shown in Figure. 32 & 33 below.

Design of V- Tail

Figure 32. Design of V-tail

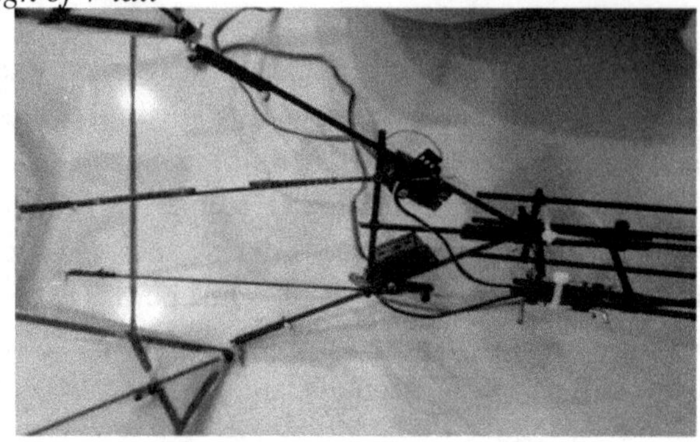

Construction of V- tail

Figure 33. Construction of V-tail

6.6 Assembly of Ornithopter

The final assembly as shown in Figure.34 below involves the thorough fixing and binding of mechanical and electrical parts. Fasteners, screws, stoppers, and other binding materials are used to assemble the body parts together. An inspection is carried out to ensure there are no loose connections or faulty assembled parts. The body is then painted with a solid color, and an initial test is conducted before the actual flight of the ornithopter. The outer frame protects the inner parts of the ornithopter from impacts in case of any failure during flight. Further balancing is carried out by testing the actual flight, observing flight paths, and analyzing the loads acting on the ornithopter.

Figure 34. Assembly of ornithopter

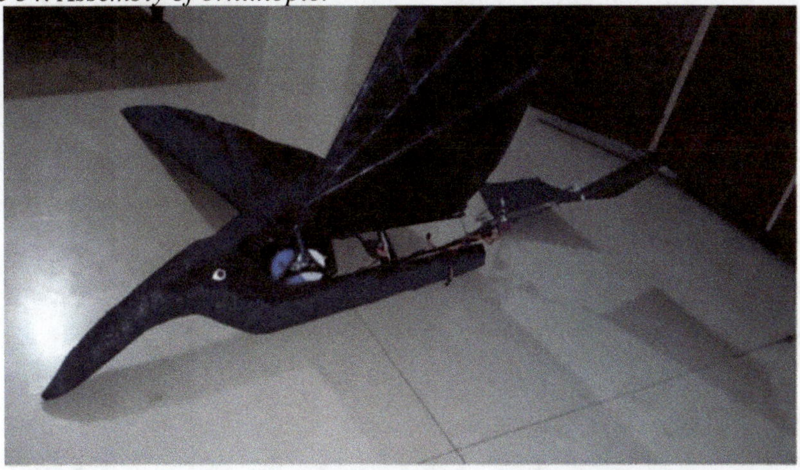

7. FLIGHT MECHANISM OF ORNITHOPTER

Taking off an ornithopter from the ground requires support, which is why they cannot take off on their own. An ornithopter's motion is triggered when hand launching is used to provide support. An ornithopter simulator controls the dynamics of the ornithopter by switching on the motors and servos.

It is necessary to adjust the throttle or directions in order to achieve a smooth, stable flight. Depending on the position of the rudders, the plane's direction of flight can be changed:

1. An ornithopter will deflect downward if both rudders are pointed upward.
2. Ornithopters deflect upward when both rudders are in the downward direction.
3. A leftward rudder directed upward deflects an ornithopter leftward if a rightward rudder pointed upward.
4. If the right rudder is pointing downward and the left rudder is pointing upward, the ornithopter will deflect to the right.

7.1 Ornithopter Stability and Control

At the rear of an ornithopter, there is usually a horizontal stabilizer. To keep the nose elevated, the tail generates downforce, which is usually around 15 degrees perpendicular to the wing, though it can vary by center of gravity and other factors.

Ornithopters can turn left or right because of small differences between their wings. As an example, wooden wing spars may have different weights, or carbon rods or membranes may have different mechanical properties.

Weight can be added to the wingtip on the outside of a turn to correct it. By adding weight, the weighted wing is inhibited from flapping while the unweighted wing is increased in flapping amplitude through inertia.

8. RESULTS

Table 2. Component and sub-assembly weights.

Component and Sub-assembly	Weight (g)
Lithium–Polymer Battery	30
Brushless Motor	20
Receiver	15
Electronic Speed Controller	36
2 x Servo Motor	18
2× Driving Arm and Linkage	8
Tail Structure	12
Tail membrane	7
Gear box	20
Fuselage	80
Fasteners, Screws, Nuts and Wheel-Locks	5
Wings Barings Outer form frame	85 10 25
Total Weighr	370

1) The total weight of the ornithopter under ideal conditions is the sum of the weights of all its components, which is 370 grams. To achieve lift, the ornitho-

pter must generate a lift force greater than its weight (L > W). Using a flapping frequency of 6.02 Hz, the required lift force and thrust are achieved.

2) The velocity or forward thrust of the ornithopter is controlled using a speed controller connected to a BLDC motor, which provides an output power of 6.7 Watts. The maximum velocity of ornithopter is 20 km/h.
3) The total output power delivered by the battery is 7.5V and 0.900 amps, which equals 405 watts per hour. The input power of the electrical devices, such as the speed controller, receiver, and two servo motors, is approximately 3500 watts per hour. So,
4) the total time of flight of the ornithopter is given by

   = **Total power consumed by the electronics / total power delivered by the battery**

   = 3500/405

   = 8.6 mins
5) The total flight time for the ornithopter ranges between 7 to 9 minutes. However, the flight time can vary by 2 to 3 minutes due to environmental factors such as wind, temperature, and humidity, which can impact the electrical and mechanical devices of the ornithopter.
6) The flight range of the ornithopter depends on the receiver and transmitter. The Flysky F6 i6 transmitter has a range that varies between 500 meters to 800 meters.

9 CONCLUSION

In this project, a fully functional and operational ornithopter is designed and fabricated using mechanical and electrical components such as a BLDC motor, electronic speed controller, transmitter and receiver, 2S battery, 2 servo motors, gears, and a gearbox.

The ornithopter flies at a height of about 20 meters within a radius of 1 km from the remote system, mimicking a bird as seen from the ground. Future editions may include GPS or other advanced sensors. The ornithopter is inexpensive to fabricate and assemble. Repairs can be made quickly in the field without the need for specialized equipment.

Further research can be conducted to develop this ornithopter on a larger scale, increasing its thrust and overall performance.

This work aligns closely with the scope and objectives of the book titled "Advancing Sustainable Engineering Through Next-Generation Thermo-Fluid Systems." The study contributes to sustainable engineering by integrating lightweight materials, efficient power transmission, and bio-inspired aerodynamic mechanisms that optimize energy utilization. The ornithopter's design reflects advancements in next-generation thermo-fluid systems through the application of low-power propulsion and enhanced aerodynamic efficiency. Thus, the presented work not only demonstrates mechanical innovation but also supports the broader goal of developing energy-efficient and environmentally conscious flight systems.

Future Scope of the Project

- This project can be further developed by increasing the overall flight time of the ornithopter by increasing battery capacity and reducing the drag produced by the ornithopter's body.
- Research and analysis should also be conducted on this ornithopter to measure the airflow through the wing's surface, which affects the thrust developed by the wings.

REFERENCES

Buchner, A. J., Honnery, D., & Soria, J. (2017). Stability and three-dimensional evolution of a transitional dynamic stall vortex. *Journal of Fluid Mechanics*, *823*, 166–197. DOI: 10.1017/jfm.2017.305

DeLaurier, J. D., & Chronister, N. (1994). An ornithopter wing design. *AIAA Journal*, *40*(1), 10–18. DOI: 10.2514/3.13973

Liger, M., Pornsin-Sirirak, N., Tai, Y.-C., Ho, S., & Ho, C.-M. (2002). Large-area electrostatic-valved skins for adaptive flow control on ornithopter wings. In *Proceedings of the 15th IEEE International Conference on Micro Electro Mechanical Systems* (pp. 247–250). IEEE. DOI: 10.1109/MEMSYS.2002.984274

Mueller, T. J., & DeLaurier, J. D. (2001). An overview of micro air vehicle aerodynamics. In P. Zarchan (Ed.), Fixed and flapping wing aerodynamics for micro air vehicle applications (Vol. 195). American Institute of Aeronautics and Astronautics (AIAA).

Warrick, D. R., Tobalske, B. W., Powers, D. R., & Dickinson, M. H. (2010). The aerodynamics of hummingbird flight. American Institute of Aeronautics and Astronautics Conference Proceedings, 1–5.

Chapter 6
Evaluation of Drying Characteristics of Orthodox Fermented Tea Particle Under Convective Drying

Shantanu Kumar
https://orcid.org/0000-0002-6203-9766
Indian Institute of Technology, Delhi, India

K. Ravi Kumar
https://orcid.org/0000-0002-5739-3744
Indian Institute of Technology, Delhi, India

Kavya Dashora
Indian Institute of Technology, Delhi, India

ABSTRACT

Tea being a non-alcoholic drink places second among all the drinks. Black tea is the highest consumed tea worldwide. Black tea is of two types: Orthodox and Crush-Tea-Curl. Drying is a crucial step in black tea processing. It reduces moisture content. Tea drying especially arrests the oxidation reaction started during oxidation. In the current study, cylindrical tea particle is considered under a natural hot air-drying environment. Orthodox tea particle can be considered cylindrical in shape. A finite difference method-based explicit scheme is used to discretize the governing and boundary conditions. Drying temperatures of 160°C, and 170°C give acceptable moisture content in the drying duration of 1800 s (approx.) and 1500 s – 1800 s,

DOI: 10.4018/979-8-3373-3621-3.ch006

respectively. Drying at 160°C and 170°C gives mean moisture contents of 0.063 g/g (d.b.) and 0.038 g/g (d.b.), respectively in 1800 s. Drying at higher temperatures shows faster moisture ratio reduction as compared to lower drying temperatures. Moisture ratios corresponding to 160°C and 170°C are found to be 0.017 and 0.003, respectively for 1800 s.

1. INTRODUCTION

Tea is among the top drinks in the world. It is a non-alcoholic drink. It has been a significant commodity in shaping human culture, history, and economy. As per data published by Tea Board India, the annual global tea production was recorded as 6422.66 million kg in 2022, and the global tea producers were China, India, Kenya, Turkey, Sri Lanka, Vietnam, and Indonesia. In addition, the world's top six exporters of tea were Kenya, China, Sri Lanka, India, Vietnam, and Argentina. The total annual global tea export was 1830.97 million kg in 2022 (Tea Board India, 2022). Tea is classified as green tea, white tea, black tea, oolong tea, etc. All kinds of tea are produced from young tea shoots of the same tea plant (Camellia sinensis). The differences in color, taste, aroma, fragrance, and shape in these tea come from the processing steps involved in the production. Production of tea is started with the plucking of tea shoots either manually or mechanically (machine). Manual plucking is preferred because of having added advantages of selective plucking to get good quality of tea. However, manual plucking is a labor intensive and gives lower productivity. Among all the tea, black tea is largely produced and consumed in the world. It accounts for approximately 74% of total tea production globally and 75% of total tea consumption globally. Various processing steps are involved in the production of black tea such as withering, maceration, oxidation, drying, sorting and grading, and packaging. Withering helps in partial reduction of moisture content in harvested tea shoots on storing about 12-18 hours. Withering changes the texture of leaves and makes leaves soft. Maceration process gives a certain shape and size to the withered leaves to meet the required market trends. During maceration, rupturing of tea leaves happens, that exposes the chemicals within leaves. Black tea is of two types; (a) CTC (Crush-Tear-Curl) tea and (b) Orthodox (Rolled) tea. The main difference between CTC and orthodox tea is in their maceration techniques. CTC tea is crushed, torn, and curled into small, granulated particles; however, orthodox tea is in the twisted form of leaves through rolling. A gentle rolling is performed manually (traditionally) or using a rolling table machine on the selectively plucked (manually plucked one bud and two youngest leaves) tea leaves. During maceration (rolling) tea leaves are ruptured and twisted. The content (liquid) inside leaves comes out during rolling and interacts with the air and thus oxidation starts. The quality attributes develop during oxidation process (Sanyal, 2016). The final shape of oxidized twisted leaves can be considered cylindrical in shape based on their appearance.

Drying is a crucial processing step. It not only reduces the moisture content to a desired level (less than 5% w.b.) but also stops the oxidation reaction (fermentation) to avoid quality loss. The oxidation of tea particles happens because of enzymes (polyphenol oxidase and peroxidase) in the presence of oxygen. The inactivation of these enzymes is crucial to stop oxidation reactions to avoid quality loss. Further, two conditions during drying, namely casehardening and stewing, are necessary to avoid quality loss of tea. Casehardening happens because of high drying air temperature initially, and stewing happens on drying at a low drying temperature which makes tea dull and soft. Both hamper the quality of the tea leaves (K. R. Kumar et al., 2023; Sanyal, 2016). Thus, the selection of drying conditions is important to avoid quality loss because of over and under-drying of tea particles. Further, both electrical and thermal energy are consumed in tea factories. However, the main energy-consuming processes in tea factories are drying and withering which require thermal energy; thus, thermal energy is mainly consumed (>80%) in the tea factories. Tea factories typically consume electrical energy in the range of 0.21–0.5 kWh per kg of produced tea while thermal energy in the range of 3.5–6 kWh per kg of produced tea. The thermal energy requirement for drying mainly comes from the burning of fossil fuel in the furnace of a tea factory (K. R. Kumar et al., 2021; Kumari et al., 2021).

During drying, heat and mass interactions take place between the hot air and moist tea material. Hot air provides heat energy to the tea material and evaporates the water on the surface of the material and carries the water vapor with it. Further, created moisture concentration difference between the surface and interior of the material induces moisture diffusion from the interior to the surface of the material. Numerical studies on drying consider heat and mass interaction phenomenon and predict the moisture and temperature distribution inside the tea material at any point of time. In one such numerical study on drying of green tea leaves in a multi-cabinet dryer, trays were modeled as fixed beds of leaves. Finite difference method was used to solve the equation of heat and mass transfer. Further, changes in the moisture content and temperature of tea leaves and drying air were determined. The modeling results were compared with experimental results at drying air temperatures of 60°C and 70°C and showed good agreement between two (Shomali & Abbasi Souraki, 2019). Further, a transient study of heat and mass transfer of drying of black tea in an industrial tea dryer was conducted to assess the distribution of temperature and moisture content in the dryer during drying. The descending and ascending trends for moisture content and drying air temperature, respectively were noticed in flow direction of tea leaves inside dryer. Further, good agreement was recorded on comparing experimental and numerical results (Zeng, Han, Li, et al., 2024). Furthermore, in one study, the modeling of mass transfer in drying of CTC tea particle was conducted considering drying air temperatures of 80°C, 90°C, 100°C, 110°C, 120°C, and 130°C. The impact of drying air temperatures on moisture content distribution, moisture content, and moisture ratio was analyzed; and it was determined that drying air temperatures of 110°C, 120°C, and 130°C were preferred temperatures for drying in the drying durations of 1500 s, 1200 s - 1500 s, and 1200 s. However, temperatures of 80°C, 90°C, and 100°C were not shown required amount of drying of tea particle. Further, the effect of size of particle was also observed on drying and particles of size less than 1 mm were recommended to get proper drying performance. In addition, it was suggested that the model considered could be used for a single-tray solar tea dryer (S. Kumar et al., 2024). An experimental study on drying orthodox tea using blast type hot air-drying oven was conducted. Evaluation of effect of temperature and thickness of tea layer on drying, effective moisture diffusivity, heat and mass transfer coefficients, and thermodynamic performance parameters were performed. The drying temperature of 70°C, 80°C, and 90°C and thickness of layer of 10 mm, 15 mm, and 20 mm were considered for the study. Results showed that increasing the thickness of layer increases the drying time, however, for same thickness of layer increasing the drying temperatures provided higher drying rate. The values of effective moisture diffusivity for drying at 70°C were found to be 4.4969×10^{-7} m^2/s, 6.8875×10^{-7} m^2/s, and 1.0222×10^{-6} m^2/s corresponding to the thickness of 10 mm, 15 mm, and 20 mm, respectively (Zeng, Han, Wang, et al., 2024). Further, a recently published experimental study on drying Yunnan Congou black tea was conducted to assess the effect of drying techniques on quality attributes. Drying techniques such as pan-fired drying, sun drying, and hot air drying were adopted with non-targeted metabolomics, electronic tongue, computer vision technology, and targeted quantitative analysis. The results showed that hot air drying had provided higher redness, sweetness, yellowness, and brightness to tea; however, sun drying had provided higher bitterness and tea pigment content, and pan-fired drying had demonstrated astringency in the tea (Lu et al., 2025). Further, in onc study effects of drying techniques such as hot air-based drying, roasting, and combined effect of hot air and roasting on sensory quality and flavor compounds of summer Congou black tea. The results of combined effect of drying showed the improved sensory and flavor quality of tea (Liu et al., 2025).

Orthodox tea is a premier tea as it requires significant human intervention and time in production to inculcate quality attributes in the tea. During production of orthodox tea, objective is to preserve the integrity and subtleness of whole leaf. Drying needs to be performed on oxidized rolled tea leaves to stop enzymatic reaction to achieve required quality attributes. Numerical studies available particularly on drying of orthodox tea leaves are limited; that necessitates further exploration on studying of drying behavior of orthodox tea. Thus, in the present study, mass transfer modeling of orthodox tea leaves is conducted to find out the suitable drying conditions for orthodox tea under various operating conditions in a single tray dryer. Orthodox tea needs to be dried from a moisture content of 65% (w.b.) to 3% (w.b.) (or preferably < 5% (w.b.)). The orthodox tea leaves are considered as kept in a single-tray dryer and facing convective heat and mass interaction with hot air coming from bottom side of dryer.

2. METHODOLOGY

The schematic of the drying of orthodox tea particles in a single-tray dryer is shown in Figure (1). Orthodox tea particles in the form of small cylindrical particles are kept over the tray of the dryer and hot drying air is supplied from the bottom side of the dryer. During drying hot air comes in contact with tea particles and performs heat and mass interactions.

Figure 1. Schematic of drying of Orthodox tea particles in a single-tray dryer.

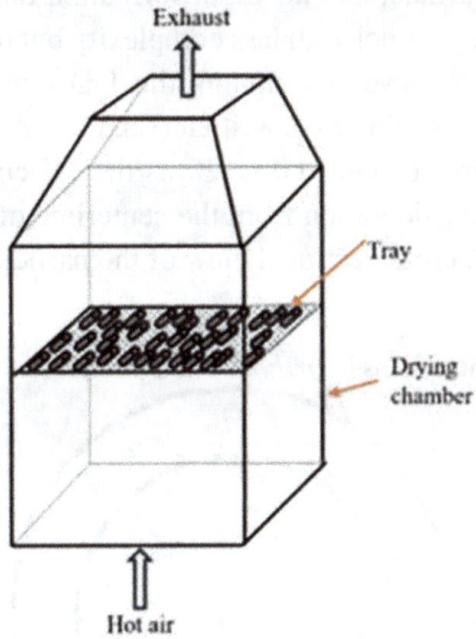

A single orthodox tea particle as a representative of whole drying bed is considered to perform the drying study, and the following assumptions are taken into consideration:

1. Orthodox tea particles are assumed to be cylindrical in shape based on their method of maceration technique i.e., rolling.

2. One-dimensional moisture transfer in cylindrical coordinates is considered for drying because of particle's smaller diameter. Thus, convective heat and mass interactions can be neglected on the circular faces of cylindrical particle.
3. The particle is facing a uniform hot air environment around its surface. This is because the diameter of particle is small, and continuously coming air crosses the particle fast. Thus, the uniform hot air environment is available all the time around the particle.
4. Hot air passes over the particle naturally and convective mass transfer of moisture happens from the surface of the particle to the air. Since there is no forced mode flow consideration involved for air, air moves over particle's surface through natural mode of flow.
5. Tea particle is considered uniform and isotropic in nature.
6. No shrinkage is considered in the material during drying.
7. No internal vapor formation is happening in the particle.

The above assumptions of not considering shrinkage, considering uniform and isotropic property of particle, and no vapor formation during drying are taken to create an efficient drying model with less complexity but to get important practical results. Considering the above assumption, the 1-D moisture transfer governing equation in cylindrical coordinates is written (Eq. (1)) (Arunsandeep et al., 2018), where "*MC*" is moisture content (d.b.), "*D*" is the effective moisture diffusivity, "*r*" is any distance of consideration from the centerline of particle, and "*R*" is the radius of the particle. A cross-sectional view of the particle is shown in Figure (2).

Figure 2. Sectional view of single orthodox tea particle.

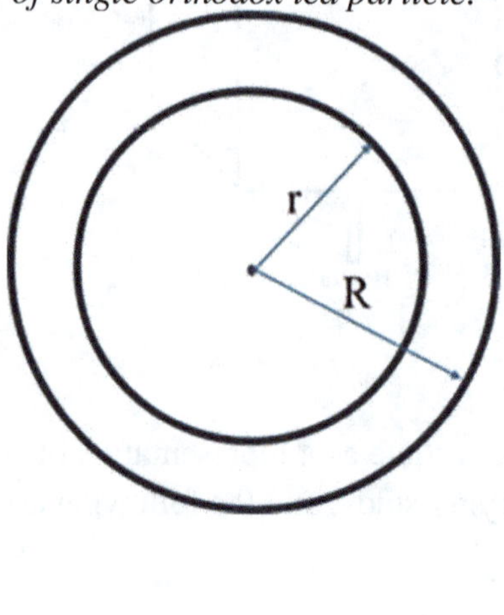

$$\frac{\partial}{\partial t}(MC(r,t)) = \left(\frac{D}{r}\right)\frac{\partial}{\partial r}\left(r\frac{\partial}{\partial r}(MC(r,t))\right) \qquad (1)$$

In addition, equations for initial and boundary conditions are given in Eq. ((2) - (4)) (Kaya et al., 2007). Eq. (2) shows the initial condition of moisture content (MC_0) of particle.

$$\text{At } t = 0;\ MC = MC_0 \qquad (2)$$

Eq. (3) shows that there is no mass interaction at the center of the particle. Since hot air comes in contact only over the surface of particle.

$$\text{At } r = 0\ ;\ \frac{\partial}{\partial r}(MC(0,t)) = 0 \qquad (3)$$

However, Eq. (4) shows convective mass interaction at the outer surface of the particle with air.

$$\text{At } r = R\ ;\ \frac{\partial}{\partial r}(MC(R,t)) = -\frac{h_m}{D}\left(MC - MC_{air}\right) \qquad (4)$$

Furthermore, the correlation mentioned in Eq. (5) is used to model the airflow around the cylindrical particle in natural convection to calculate the value of the heat transfer coefficient (h) (Çengel & Ghajar, 2015).

$$\overline{\mathrm{Nu}} = \frac{\mathrm{hd}}{\mathrm{k_{air}}} = \left[0.6 + \frac{0.387Ra^{1/6}}{\left[1 + \left(\frac{0.559}{Pr}\right)^{9/16}\right]^{8/27}}\right]^2 \qquad (5)$$

Eq. (6) is the relation predicting analogy between heat and mass transfer and used to calculate the value of the mass transfer coefficient (h_m) from the heat transfer coefficient (h).

$$\mathrm{h_m} = \mathrm{h}\left[\frac{\mathrm{D_a Le^{1/3}}}{\mathrm{k_{air}}}\right] \qquad (6)$$

In Eq. (4), MC_{air} is the moisture content present in the air and MC is the moisture content present on the surface of the particle (assuming equal to moisture content present in the air over the surface). Eq. (5) and Eq. (6) contain parameters such as Prandtle number (Pr), Rayleigh number (Ra), the thermal conductivity of air (k_{air}), Lewis number (Le), and water vapor diffusivity in the air (D_a).

The finite difference method (FDM) with an explicit scheme (forward in time and central in space (FTCS)) is used for discretization of governing equation (Eq. (1)) and boundary conditions (Eq. (3) & Eq. (4)) into simple algebraic equations

(Eq. (7), Eq. (8), and Eq. (9)). In these equations, "*i*" and "*n*" are space and time marching parameters, respectively. Where $i = 1, 2, 3, \ldots\ldots, m$ and $n = 1, 2, 3, \ldots\ldots, tr$; "*m*" is the total number of nodes in the radial direction, and "*tr*" is the total number of time steps. MATLAB-based code is written to solve discretized algebraic equations (Eq. (7), Eq. (8), and Eq. (9)).

$$M_{(i+1,n)} = M_{(i,n)} \tag{7}$$

$$M_{(i,n)} = \left(\frac{1}{1+S}\right) \times M_{(i-1,n)} + \left(\frac{S}{1+S}\right) \times M_{air} \tag{8}$$

$$M_{(i,n+1)} = F \times \left[\left(\frac{dr}{2 \times r(i)}\right) + 1\right] \times M_{(i+1,n)} + \left(1 - 2 \times F\right) \times M_{(i,n)} + F \times \left[1 - \left(\frac{dr}{2 \times r(i)}\right)\right] \times M_{(i-1,n)} \tag{9}$$

Where, $S = \frac{dr \times h_m}{D}$ and $F = \frac{D \times dt}{dr^2}$, since explicit based FDM is used to discretize equations, thus, above equations are stable when $F \leq 0.5$. Further, effective moisture diffusivity (*D*) is calculated using temperature dependent Arrhenius equation (Eq. (10)) (S. Kumar et al., 2024).

$$D = 7.45 \times 10^{-4} \times e^{\left(\frac{-6264.3}{T}\right)} \tag{10}$$

where "*T*" is the drying temperature in K and unit of "*D*" is m^2/s.

Table 1. Values of parameters considered for modeling

Parameters	Values/expressions
Radius of tea particle (*R*)	0.001 m
Moisture content of air (M_{air})	0.032 kg/kg of dry air (*RH* of 5% at 100°C)
Water vapor diffusivity in air (D_a) (Temple & Van Boxtel, 1999)	2.74×10^{-5} m^2/s
Initial moisture content (MC_o) (d.b.) (Sanyal, 2016)	1.85 g/g or 65% (w.b.)
Temperature of the surface of the particle	30°C

The input parameters required to model the drying particle are given in Table 1. The mean radius of the cylindrical particle is taken as 0.001 mm. The water vapor diffusivity in air is taken as 2.74×10^{-5} m^2/s. All the thermophysical properties of the air are considered at the mean film temperature (i.e. average of temperatures of air and surface of particle). The mass transfer coefficient (in natural convection) is calculated using Eq. (5) and Eq. (6) by considering the dimension of the particle and thermophysical properties of air. The surface temperature of the particle is considered as 30°C. The initial moisture content present in the tea is considered as 1.85 g/g (d.b.) or 65% (w.b.). The desired moisture content in tea should be less than 5% (w.b.). Further, MATLAB-based codes are developed to solve Eq. (7), Eq. (8), and Eq. (9). Grid and time independent tests are performed to ensure accurate and reliable results through modeling. Grid and time independent tests result in the optimum number of nodes and optimum time step size. For the grid independence test, the whole dimension of radius of particle is divided into 10, 20, 30, and 40 number of nodes. The values of moisture content inside particle corresponding to the number of nodes of 10, 20, 30, and 40 are plotted (Figure (3)). The results of moisture distribution corresponding to nodes of 30 and 40 are overlapping to each other for whole radius of particle. Thus, the number of nodes 30 is found to be best value for further consideration to compute the drying outputs. Further, for the time independence test, whole drying duration is divided into time steps of sizes 0.05 s, 0.1 s, 0.2 s, 0.3 s, and 0.4 s; and moisture distribution plots corresponding to all the time step sizes are plotted (Figure (4)). It is found that moisture distribution plots for all sizes of time steps are overlapping, thus, a suitable value of time step i.e., 0.3 s is considered among all the time steps for further study. The reason for choosing a time step size of 0.3 s is to avoid any instability in the computation due to higher values of time step sizes. Thus, the number of nodes of 30 and time step size of 0.3 s are selected for further studying the drying behavior of particle. The accuracy of the present model for drying of cylindrical object is tested with experimental work of Simal et al. (1998) (Simal et al., 1998). The modeling results for broccoli stem considering current model are validated with the experimental results of drying broccoli stem at 90°C (as shown in Figure (5)) and found out good agreement (R^2= 0.9983) between the two.

Figure 3. Grid independence test.

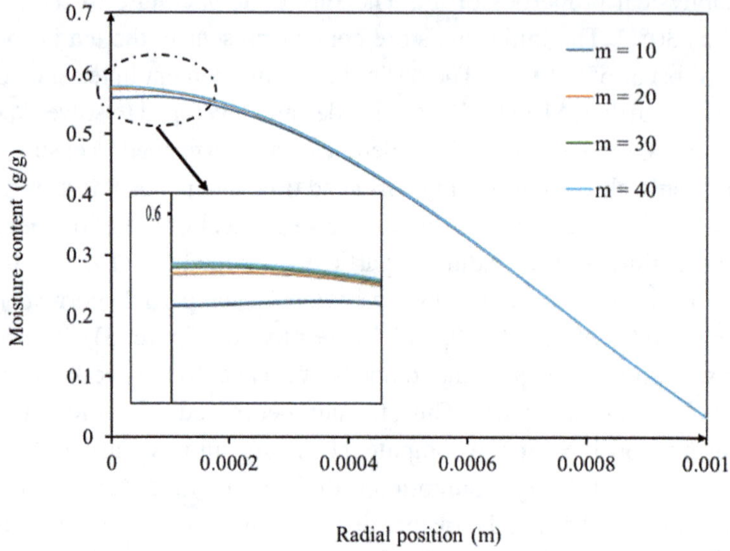

Figure 4. Time independence test.

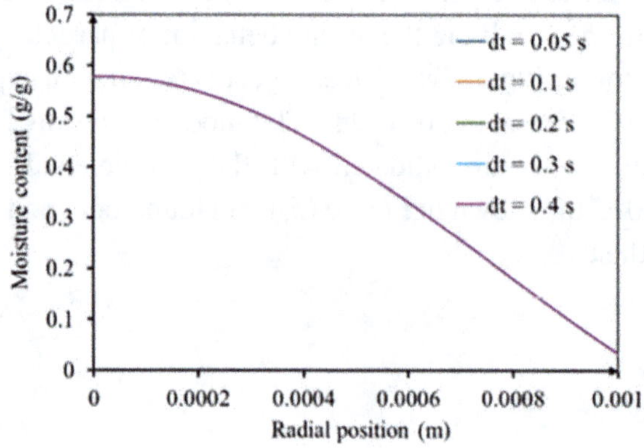

Figure 5. Validation of the present numerical model.

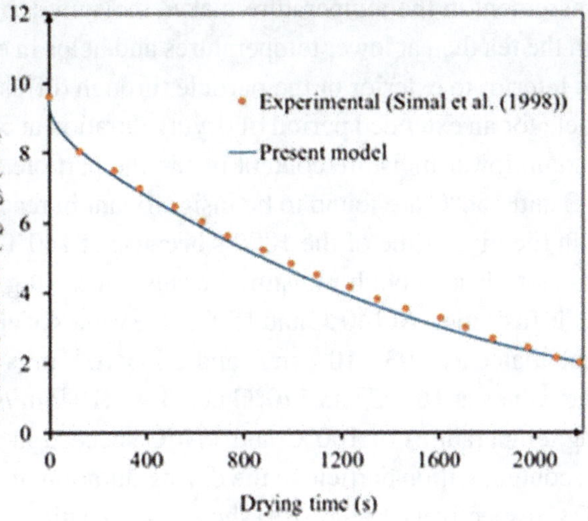

3. RESULTS AND DISCUSSION

The drying behavior of orthodox tea particle under convective drying in a single tray dryer is studied. Drying air temperatures of 140°C, 150°C, 160°C, and 170°C are chosen to perform drying study. The drying curves such as moisture distribution, mean moisture content, and moisture ratio are plotted in Figure (6) and Figure (7) to understand the moisture transport in the particle with time at different temperatures. Moisture distribution inside the particle at 140°C, 150°C, 160°C, and 170°C for drying times of 600 s, 900 s, 1200 s, 1500 s, and 1800 s are plotted in Figure (6). The moisture distribution in the particle corresponding to 140°C and 150°C is not uniform for all the drying durations. However, moisture distribution at 160°C and 170°C is found uniform corresponding to the drying durations of 1800 s, and between 1500 s – 1800 s, respectively. The variation in the moisture content from center to periphery of the particle are found to be between 0.082 – 0.032 g/g (d.b.) corresponding to 160°C, however, at 170°C these variations are 0.059 – 0.032 g/g (d.b.) and 0.042 – 0.032 g/g (d.b.) for drying durations of 1500 s and 1800 s, respectively. The reason behind finding large variation in the distribution of moisture content in the particle during initial drying durations is difficulty in extracting the bound moisture (chemically and physically bound) from the interior of the particle. Higher temperatures achieve uniformity in moisture content distribution faster than lower temperature (as clearly vivid in Figure (6)). Further, common findings from the modeling results show that on increasing air temperature drying becomes fast. This is because increment in the temperature makes moisture transport (moisture diffusion) faster in the tea than at lower temperatures and helps in extracting bound moisture from the interior to exterior of the particle through diffusion. In addition, exposing tea particle for an extended period of drying duration at certain drying air temperature can attain lower moisture content inside the particle. Drying air temperatures of 140°C and 150°C are found to be insignificant in reaching the desired moisture content in the given time of the 1800 s because at 140°C and 150°C, the tea particle struggles to lose enough moisture because of having lesser values of effective moisture diffusivities. At 140°C and 150°C, the values of effective moisture diffusivities are calculated as 1.93×10^{-10} m^2/s and 2.77×10^{-10} m^2/s which are lower as compared to the values at 160°C and 170°C i.e., 3.90×10^{-10} m^2/s and 5.41×10^{-10} m^2/s. Thus, drying temperatures of 160°C and 170°C succeed in reaching the acceptable moisture content within particle in the drying duration of 1800 s (approx.) and 1500 s – 1800 s, respectively. Figure 7(a) shows the variation of mean moisture content with the drying time for all the considered drying temperatures. Drying at 160°C and 170°C gives mean moisture contents of 0.063 g/g (d.b.) and 0.038 g/g (d.b.), respectively in 1800 s. However, results for moisture content corresponding to 140°C and 150°C are found to be 0.27 g/g (d.b.) and 0.132 g/g (d.b.) in 1800 s, which are far away from meeting the acceptable moisture content. Further, variations of moisture ratio are plotted at temperatures of 140°C, 150°C, 160°C, and 170°C as shown in Figure 7(b). Moisture ratios are calculated by dividing moisture content (d.b.) at any drying time with initial moisture content (d.b.) of particle. Drying at higher temperatures shows faster moisture ratio reduction as compared to lower drying temperatures. The reason for the faster reduction in moisture at elevated temperatures is to have faster moisture diffusion in the particle. Moisture ratios corresponding to 160°C and 170°C are found to be 0.017 and 0.003, respectively for 1800 s. However, moisture ratios corresponding at 140°C and 150°C are found to be 0.130 and 0.055. Thus, from the above results drying temperatures of 160°C and 170°C are found to be acceptable drying temperatures for drying of orthodox tea in a single tray dryer for drying durations of 1800 s (approx.), and between 1500 s - 1800 s, respectively.

Figure 6. Distribution of moisture content in the particle at temperatures of (a) 140°C (b) 150°C (c) 160°C (d) 170°C.

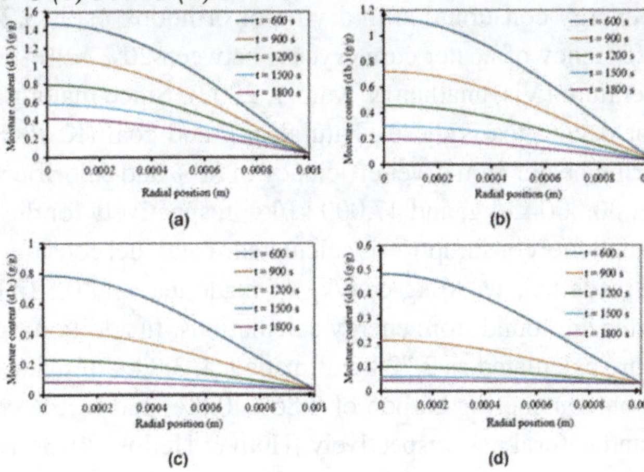

Figure 7. Variation of (a) mean moisture content, (b) moisture ratio with time for 140°C, 150°C, 160°C, and 170°C.

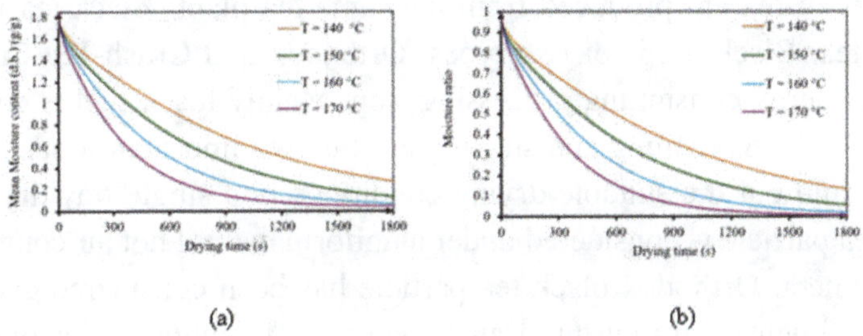

4. RECOMMENDATIONS AND ENERGY CONSUMPTION

The current modeling results can be utilized for designing single tray tea dryer, or solar energy based indirect single tray tea dryer in natural mode of operation for drying of orthodox tea. For solar energy-based tea dryer, air heated by using solar collectors flows to dryer in a natural mode of flow. The current modeling results can also be used to design slow speed conveyor tea dryer for drying of orthodox tea because slow speed conveyor can be treated as moist tea material with fixed tray.

The typical energy consumption in drying of orthodox tea is 3.24 kWh/kg of made tea and efficiency of heater cum dryer is between 20% – 40% as reported in the previous literature (Visvanathan & Kumar, 2002). Since majorly types of fuel consumption for drying tea is diesel, natural gas, and coal (K. R. Kumar et al., 2023). Considering heater cum dryer efficiency of 30% and calorific value of fuels as 45,000 kJ/kg, 30,000 kJ/kg, and 47,000 kJ/kg, respectively for diesel, coal, and natural gas, the total fuel consumption is calculated. Total fuel consumption of 0.864 kg diesel/kg of made tea, 1.296 kg coal/kg of made tea, and 0.827 kg of natural gas/kg of made tea are found from energy calculations. In addition, corresponding CO_2 emissions are calculated as 2.72 kg of made tea, 3.8 kg of made tea, and 2.21 kg of made tea on considering carbon of 0.86%, 0.8%, and 0.73% corresponding to diesel, coal, and natural gas, respectively (Huth & Heilos, 2013). As fossil fuels usage is one of the primary reasons for greenhouse emissions in tea estates, this necessitates the adoption of renewable energy for tea drying.

5. CONCLUSIONS

All types of tea are produced from the same tea plant. Black tea is the most consumed tea. Black tea is of two types: Orthodox and Crush-Tea-Curl. Drying is the most energy-consuming processing step. Mainly fossil fuel is consumed in tea factories. Casehardening and stewing are the two important issues during tea drying. To find out the suitable drying conditions in a single tray dryer, a moist orthodox tea particle is considered under a uniform natural hot air convective drying environment. Orthodox black tea particle has been considered cylindrical in shape. One dimensional moisture transfer equation in cylindrical coordinate with convective boundary condition is solved for drying of orthodox tea particle. Drying air temperatures of 140°C, 150°C, 160°C, and 170°C are chosen to perform drying study. Drying at 140°C and 150°C has shown an insignificant effect to reach the desired moisture content in 1800 s. However, orthodox particle when subjected to 160°C and 170°C reaches acceptable moisture content in the drying durations of 1800 s (approx.) and between 1500 s – 1800 s, respectively. This demonstrates that higher drying temperatures are both effective and efficient because of higher value of effective diffusion coefficients. Drying at 160°C and 170°C gives final mean moisture contents of 0.063 g/g (d.b.) and 0.038 g/g (d.b.), respectively in 1800 s. Moisture ratios corresponding to 160°C and 170°C are found to be 0.017 and 0.003, respectively in 1800 s. Additionally, it was recommended that the current modeling results can be utilized for designing single tray tea dryer, or solar energy based indirect single tray tea dryer in natural mode of operation for drying of orthodox tea. Further, results for total fuel consumption are found to be 0.864 kg diesel/kg of made tea, 1.296 kg coal/kg of made tea, and 0.827 kg of natural gas/kg of made tea from energy calculations. In addition to that CO_2 emissions are calculated as 2.72 kg of made tea, 3.8 kg of made tea, and 2.21 kg of made tea corresponding to diesel, coal, and natural gas, respectively.

REFERENCES

Arunsandeep, G., Lingayat, A., Chandramohan, V. P., Raju, V. R. K., & Reddy, K. S. (2018). A numerical model for drying of spherical object in an indirect type solar dryer and estimating the drying time at different moisture level and air temperature. *International Journal of Green Energy*, *15*(3), 189–200. DOI: 10.1080/15435075.2018.1433181

Çengel, Y. A., & Ghajar, A. J. (2015). Heat and Mass Transfer: Fundamentals & Applications. In *McGraw-Hill Education* (5th ed.). McGraw-Hill Education.

Huth, M., & Heilos, A. (2013). Fuel flexibility in gas turbine systems: Impact on burner design and performance. In *Modern Gas Turbine Systems: High Efficiency, Low Emission, Fuel Flexible Power Generation*. Woodhead Publishing Limited., DOI: 10.1533/9780857096067.3.635

Kaya, A., Aydin, O., & Dincer, I. (2007). Numerical modeling of forced-convection drying of cylindrical moist objects. *Numerical Heat Transfer Part A*, *51*(9), 843–854. DOI: 10.1080/10407780601112753

Kumar, K. R., Dashora, K., Krishnan, N., Sanyal, S., Chandra, H., Dharmaraja, S., & Kumari, V. (2021). Feasibility assessment of renewable energy resources for tea plantation and industry in India - A review. *Renewable & Sustainable Energy Reviews*, *145*. Advance online publication. DOI: 10.1016/j.rser.2021.111083

Kumar, K. R., Dashora, K., Kumar, S., Dharmaraja, S., Sanyal, S., Aditya, K., & Kumar, R. (2023). A review of drying technology in tea sector of industrial, non-conventional and renewable energy based drying systems. *Applied Thermal Engineering, 224*(December 2022), 120118. https://doi.org/DOI: 10.1016/j.applthermaleng.2023.120118

Kumar, S., Kumar, K. R., & Dashora, K. (2024). Impact study of operating parameters on drying evolution of spherical tea particle under convective influence. *Journal of Food Process Engineering*, *47*(8), 1–11. DOI: 10.1111/jfpe.14702

Kumari, V., Chandra, H., Aditya, K., Kumar, K. R., Dashora, K., Krishnan, N., Sanyal, S., Dharmaraja, S., & Kumar, S. (2021). Feasibility study on renewable energy system in tea (Camellia sinensis) estates of North-East India. *Indian Journal of Agricultural Sciences*, *91*(11), 1631–1635. DOI: 10.56093/ijas.v91i11.118574

Liu, F., Zhu, H., Li, C., Wang, Y., Zhang, J., Tang, X., Zhang, T., & Liu, Y. (2025). A combined drying process involving hot air and roasting for improving summer congou black tea quality. *Food Research International, 201*(August 2024), 115584. https://doi.org/DOI: 10.1016/j.foodres.2024.115584

Lu, M., Jiang, Y., Zhao, W., Zhang, J., Chen, Z., & Ning, J. (2025). Effects of different drying methods on color difference, taste and chemical components of Yunnan Congou black tea. *Lebensmittel-Wissenschaft + Technologie, 223*(March), 117696. DOI: 10.1016/j.lwt.2025.117696

Sanyal, S. (2016). *Tea Manufacturing Manual* (2nd ed.). Tea Research Association.

Shomali, A., & Abbasi Souraki, B. (2019). Experimental investigation and mathematical modeling of drying of green tea leaves in a multi-tray cabinet dryer. *Heat and Mass Transfer, 55*(12), 3645–3659. DOI: 10.1007/s00231-019-02662-6

Simal, S., Rossello, C., Berna, A., & Mulet, A. (1998). Drying of shrinking cylinder-shaped bodies. *Journal of Food Engineering, 37*(4), 423–435. DOI: 10.1016/S0260-8774(98)00095-8

Tea Board India. (2022). *Global Tea.* https://www.teaboard.gov.in/pdf/Global_Tea_Website_1005_pdf7206.pdf

Temple, S. J., & Van Boxtel, A. J. B. (1999). Thin layer drying of black tea. *Journal of Agricultural Engineering Research, 74*(2), 167–176. DOI: 10.1006/jaer.1999.0448

Visvanathan, C., & Kumar, S. (2002). *Small medium scale Industries in Asia (Energy and Environment): Tea Sector.* Regional Energy Resources Information Center (RERIC), Asian Institute of Technology, Pathumthani, Thailand. https://doi.org/DOI: 10.13140/2.1.3285.0720

Zeng, Z., Han, C., Li, H., Chen, T., Zeng, J., Fang, Z., Li, B., & Wu, W. (2024). Modeling and evaluation in transient analysis of industrialized counter-flow tea drying system. *Case Studies in Thermal Engineering, 53*(November 2023), 103781. https://doi.org/DOI: 10.1016/j.csite.2023.103781

Zeng, Z., Han, C., Wang, Q., Yuan, H., Zhang, X., & Li, B. (2024). Analysis of drying characteristic, effective moisture diffusivity and energy, exergy and environment performance indicators during thin layer drying of tea in a convective-hot air dryer. *Frontiers in Sustainable Food Systems, 8*(April), 1–13. DOI: 10.3389/fsufs.2024.1371696

Chapter 7
Evaluation of Low GWP Hydrocarbons-CO2 Blends for Refrigeration Applications

Umamaheshz Boddepalli
https://orcid.org/0009-0006-0196-5949
Indian Institute of Technology, Kharagpur, India

Rohit Roy
https://orcid.org/0000-0002-6567-1185
Indian Institute of Technology, Kharagpur, India

Sourav Mitra
Indian Institute of Technology, Kharagpur, India

Ramgopal Maddali
Indian Institute of Technology, Kharagpur, India

ABSTRACT

The present theoretical study reports the characteristics of refrigerant blends of hydrocarbons and CO2.The thermodynamic, environmental and flammability characteristics of these blends are obtained through suitable mixture models. The safety class of blends of propane, propylene, isobutane and dimethyl ether are categorised as per the ASHRAE conditions. This study also presents the thermodynamic performance of these blends in a simple VCRC using standard AHRI cooling conditions with a constant cooling capacity of 3 Ton. The results indicated that the (COP) of the zeotropic blends is low compared to the pure hydrocarbons with a constant tempera-

DOI: 10.4018/979-8-3373-3621-3.ch007

ture heat exchange fluid. This paper also analysed the environmental performance (GWP, TEWI) of the selected pure natural refrigerants and their blends. The effect of emission factor on indirect warming impact is also discussed. The results describe that among the pure refrigerants R-600a performance is best and CO2 performance is least. However, none of the blend performance is better than the R-32. This study also included the analysis of thermodynamic properties of the mixtures.

1. INTRODUCTION

Refrigerant is the pure substance or mixture which is used in refrigeration systems to extract heat from a low temperature source and reject the heat to a high temperature heat sink with the help of external work. The desirable properties of an ideal refrigerant are: high latent heat of vaporisation, low specific heat, low boiling point, low global warming potential (GWP), low corrosive resistance, low viscosity and zero ozone depleting potential (ODP). But existence of an ideal refrigerant with all the properties is not possible (McLinden et al., 2020). With the present achievable climate targets, searching for zero ODP and low GWP substances that can be used as refrigerants is the primary objective of research in the field of refrigeration. Montreal protocol 1987, which is the binding agreement between the countries ratified to phase out the high ODP substances ("Montreal Protocol on Substances That Deplete the Ozone Layer Final Act 1987," 1989). Kigali agreement, an amendment to the Montreal protocol sets the deadline to reduce the usage of the high GWP hydrofluorocarbons (HFCs) (Environemental Protection Agency, 2016). The global cooling pledge at the conference of parties-28 (COP-28) in UAE targeted to reduce cooling related emissions by 68% of 2022 levels by 2050 (Efficiency, n.d.). European union F-gas regulation targeted to reduce the use of fluorinated gases by 72-73% by 2030 and 78% by 2050 (Gschrey & Zeiger, 2015). The Paris accord which is a legally binding agreement targeted to limit the global average temperature well below 2°C, above pre-industrial levels and pursue efforts to limit the temperature increase to 1.5°C above pre-industrial levels (Delbeke et al., 2019). Therefore, to achieve sustainable and climate friendly cooling, it is necessary to shift from the artificial, high GWP refrigerants which pose severe damage to environment to climate friendly low GWP refrigerants.

CO_2, ammonia, and hydrocarbons are some examples of natural refrigerants. Ammonia is used widely in cold storage applications, but has high toxicity(Pearson, 2008). Synthetic refrigerants like Hydrofluoroolefins (HFOs) with zero ODP, low GWP and low flammability can be viable refrigerants to replace the HFCs. But when these refrigerants are released into the atmosphere, they have the potential to form the trifluoracetic acid (TFA), which shows some adverse effects on the aquatic environment. Also, combustion of these gasses' releases highly poisonous gases like hydrogen fluoride and carbonyl fluoride (Nations & Programme, 2016). CO_2 with easy availability, low GWP, zero ODP and excellent heat transfer properties can be an excellent candidate in almost all refrigeration applications. However, despite its merits, CO_2 is currently limited to fewer applications only due to its low critical temperature (31.1°C), high working pressure (7.38 MPa), and low coefficient of performance (COP) at higher ambient temperature(Bansal, 2012).

Hydrocarbons as natural refrigerants with zero ODP, and low GWP, offer viable solution to replace the high GWP HFCs. However, flammability restricts their widespread usage. The American Society of Heating, Refrigerating and Air-conditioning Engineers (ASHRAE) placed these hydrocarbons under the highly flammable(A3) safety class (ASHRAE, 2023). On the basis of one negative aspect, we cannot avoid these refrigerants usage in the field of refrigeration and air-conditioning. Hence one possibility is to blend these flammable refrigerants with the non-flammable CO_2 so that the flammability can be reduced and the critical temperature can be increased. In order to assess the performance of these blends, it is essential to calculate the properties of the binary mixtures using suitable models. Literature on blends of hydrocarbons with CO_2 for refrigeration and air conditioning applications is very scarce.

1.1 Flammability Analysis

ASHRAE classified the flammability of refrigerants based on the three parameters (Kennoy, 2018), namely lower flammability limit (LFL), heating value and laminar burning velocity (LBV). The classification based on these properties is shown in Table 1.

Table1. Parameters for the safety classification of refrigerants.

Class	Flame propagation at 60°C and 50% RH	Lower flammability limit at 23°C or 60°C & 50% RH	Net calorific value	Laminar Burning Velocity at 25 °C in dry air
A1	None	N/A	N/A	N/A
A2	Yes	>0.1kg/m³	<19 kJ/g	>10 cm/s
A2L	Yes	>0.1kg/m³	<19 kJ/g	<10 cm/s
A3	Yes	< 0.1kg/m³	>19 kJ/g	>10 cm/s

1.2 Lower Flammability Limit (LFL)

Lower flammability limit is defined as the minimum concentration of the fuel below which it is too lean to burn. Previous experimental studies on the variation of flammability of flammable gases with CO_2 addition concluded that with addition of CO_2 the LFL of the mixture increases and they proposed the mathematical correlation equation 2 to calculate the LFL of hydrocarbons on mixing with inert gas (Gant et al., 2011). LFL of the mixture can be calculated with the Le Chatelier's rule (Le Chatelier, 2004).

$$LFL_{mixture} = \frac{100}{\Sigma \frac{C_{fuel}}{LFL_{fuel}}} \quad (1)$$

LFL_{fuel} can be calculated by the given approximation, which is useful to calculate the LFL of the fuel with the addition of the CO_2.

$$LFL_{fuel} = \frac{C_1}{\left(\frac{C_1}{L_1} - 0.01094 * C_{in}\right)} \quad (2)$$

Where,

C_1 = Proportion by % volume of fuel,
L_1 = Lower flammability limit of fuel,
C_{in} = Proportion by % volume of Inert gas (CO_2).

1.3 Laminar Burning Velocity (LBV)

LBV indicates how fast the flame propagates in the unburnt fuel. Experimental studies are carried out on the effect of CO_2 on the laminar burning velocity of the propane-air mixture and flame stability of isobutane. These studies concluded that with increasing CO_2 concentration the LBV decreases, with addition of 0 to 10% of CO_2 the LBV reduced by 2%, but with 70 to 80% the reduction is around 33%. The study based on the 1-dimensional, pre-mixed CANTERA code on the effect of CO_2 on the LBV of flammable gases reported that at 82.5% concentration of CO_2 the flame ceases to exist. The study also proposed a mathematical correlation to estimate the LBV of the propane and isobutane with CO_2 addition, which are given by equation (3) and equation (4) and Table 3 (Yelishala et al., 2019),(Wang et al., 2019).

Experimental studies reveal the dilution effect of CO_2 on the LBV of dimethyl-ether (DME)-air mixture and reported that with addition of CO_2 the LBV of the DME reduces. It is also stated that CO_2 has more dilution effect than nitrogen (Mohammed et al., 2017). Literature on experimental studies also indicate the effect of CO_2 on the LBV of propane, and conclude that with increasing CO_2 concentration the LBV of the propane decreases (Zhang et al., 2024).

Correlation to estimate laminar burning velocity (S_u) of Propane and Carbon dioxide mixture,

$$S_u = S_{u0}(1 + a(\varnothing - 1) + b(\varnothing - 1)^2)\left(\frac{T_u}{T_{u0}}\right)^{\alpha_0+\alpha_1\varnothing}\left(\frac{P}{P_0}\right)^{\beta_0+\beta_1\varnothing}(1 - cD)^{d+e\varnothing} \tag{3}$$

Table 2. Power law fit coefficients of equation (3)

S_{u0} (cm/s)	a	b	c	d	e	α_0	α_1	β_0	β_1
38.853	0.690	-3.137	1.206	-0.011	0.268	2.035	0.275	-0.347	0.074

Correlation to estimate laminar burning velocity of isobutane and carbon dioxide mixture,

$$S_u = S_{u0}(1 + a(\varnothing - 1) + b(\varnothing - 1)^2)\left(\frac{T_u}{T_{u0}}\right)^{\alpha_0+\alpha_1\varnothing}\left(\frac{P}{P_0}\right)^{\beta_0+\beta_1\varnothing}(1 - cD)^{d+e\varnothing} \tag{4}$$

Table 3. Power law fit coefficients of equation (4)

S_{u0} (cm/s)	a	b	c	d	e	α_0	α_1	β_0	β_1
32.655	0.549	-3.083	1.18	-0.022	0.292	1.875	-0.063	00.371	0.106

T_{uo} is reference temperature, equal to 298 K, P_0 is reference pressure, equal to 1 atm., D is CO_2 mole fraction (0% < D < 80%), P is pressure (0.5 atm. < P < 4.2 atm.), T_u is temperature (298 K < T_u < 580 K) and ∅is equivalence ratio (0.8 < ∅ < 1.2).

The thermodynamic, safety and environmental aspects of the selected refrigerants are listed in Table 4.

Table 4. Thermodynamic, safety and environmental aspects of selected Hydrocarbons, CO_2 and R-32

Refrigerant	Critical temperature Tc (°C)	Critical pressure Pc (bar)	Laminar burning velocity (cm/s)	Lower flammability limit(kg/m³)	ASHRAE number	GWP-100
Propane	97	43	38.6	0.037	R290	3
Propylene	91	46	39.3 (Yelishala et al., 2020)	0.037	R1270	2
Isobutane	134	34	32.655	0.043	R600a	3
Dimethyl Ether	127	36	42.1	0.062	RE170	1
Carbon dioxide	31	74	-	-	R744	1
R-32	78.26	57.82	6.7 (Hegetschweiler et al., 2023)	0.38	R-32	677

1.4. Environmental Aspects

The environmental aspects are indicated in terms of GWP and total environmental warming impact (TEWI). All hydrocarbons have GWP well below 4. GWP of the mixtures can be easily calculated with the mass weighted average of all refrigerants (Agarwal & Clark, 2016).

GWP of mixture = proportion by % mass of component $A \times GWP$ of A + proportion by % mass of component $B \times GWP$ of B (5)

Total Environmental Warming Impact (TEWI): TEWI reflects the total global warming potential of the refrigeration equipment during its lifetime. This includes direct warming impact and indirect warming impact. Direct warming impact relates to the warming potential of the refrigerant due to leakage and disposal at the end of the life. On the other hand, indirect warming relates to the emissions released during the generation of the power, which is required to run the system. The method for calculating TEWI is given below (Jonathan Fryer, Kevin Lee, 2012).

$$TEWI = GWP_{direct} + GWP_{indirect} \tag{6}$$

$$GWP_{indirect} = E_{annual} \times \beta \times \mathrm{n} \tag{7}$$

$$GWP_{direct} = (\mathrm{GWPxmx}\, L_{annual} {}^{*} n) + \mathrm{GWPxmx}\left(1 - \alpha_{recovery}\right) \tag{8}$$

Where, E_{annual} is annual energy consuption, B is emissivity factor and m is the mass of refrigerant charge. The initial refrigerant charge amount is assumed as 0.25 kg per kW (Islam et al., 2021). In the above equation, n is system operating life and L_{annual} is annual leakage of refrigerant. For chiller annual leakage is reported as 7%. $\alpha_{recovery}$ is recovery rate after the life of the system and is 70% if mass of refrigerant charge <100 kg, 90% if mass of refrigerant charge >100 kg (Jonathan Fryer, Kevin Lee, 2012).

Emission intensity factor is defined as the mass of CO_2 emitted by the power generator to produce one kWh of electrical power supplied to the refrigeration system. Depending upon the power generation system, the emissivity factor is different for different countries. Among the developed countries France has very low emmision factor due to its high dependency on the nuclear energy. The emission intensity factor of different countries is listed below (CEA, 2011)(International Energy Institute (IEI), 2023).

Figure 1. Emission intensity factor of different countries per kWh power generation.

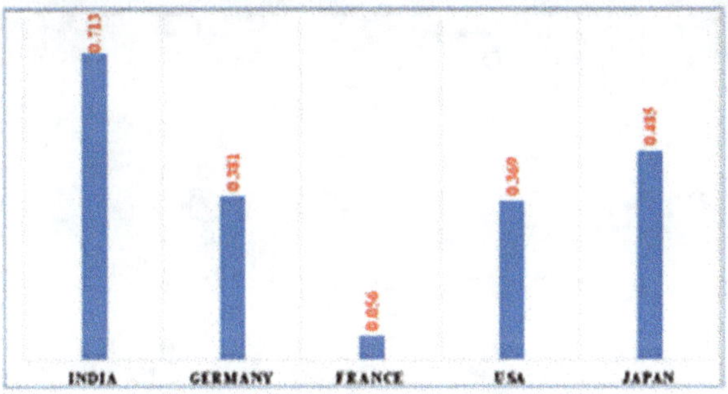

Compared to the developed nations, India has high emission factor due to the relatively low efficiency of the coal-based power plants.

1.5. Analysis of Refrigerant Mixtures

None of the pure refrigerants are ideal refrigerants. One refrigerant may have some merits at the same time some drawbacks too. Hence mixtures will provide an opportunity to get the optimum properties of its constituents. Mixtures can be broadly classified as zeotropic, azeotropic and near-azeotropic mixtures based on the temperature over which the isobaric phase change occurs. In zeotropic mixtures the phase change occurs at varying temperatures, whereas in azeotropic mixtures, the phase change occurs at same temperature. Zeotropic mixture shows temperature glide (TG), which is defined as the difference between the dew point and bubble point temperatures during the isobaric phase change.

1.6. Analysis of Vapour Compression Refrigeration System

The Vapour Compression Refrigeration System (VCRS) is the most commonly used refrigeration system due to many practical reasons. The vapour compression refrigeration cycle mainly consists of compressor, condenser, expansion valve and evaporator. VCRS can be operated as either subcritical or transcritical. Subcritical cycle indicates phase change process happens during heat extraction as well as heat rejection processes. But in trascritical cycle the phase change process occurs during heat extraction in evaporator only. During heat rejection, only sensible cooling of the high-pressure gas will take place. The simple VCRS along with the corresponding cycle diagram on pressure-enthalpy chart (P-h) are shown in Figure 2 and Figure 3.

Figure 2. Schematic diagram of water-cooled VCRS.

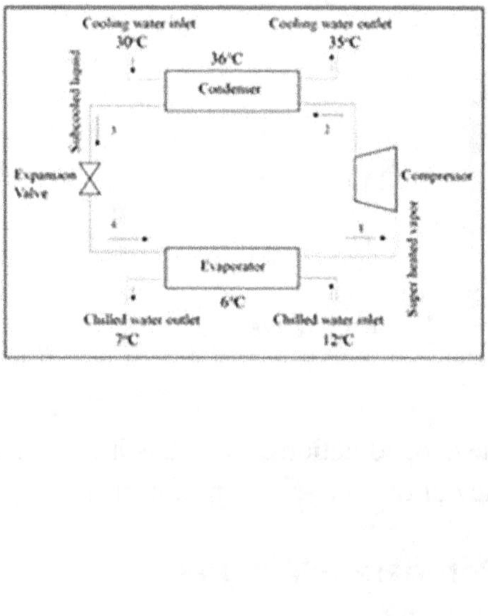

Figure 3. Temperature-entropy diagram of single stage transcritical CO_2 system.

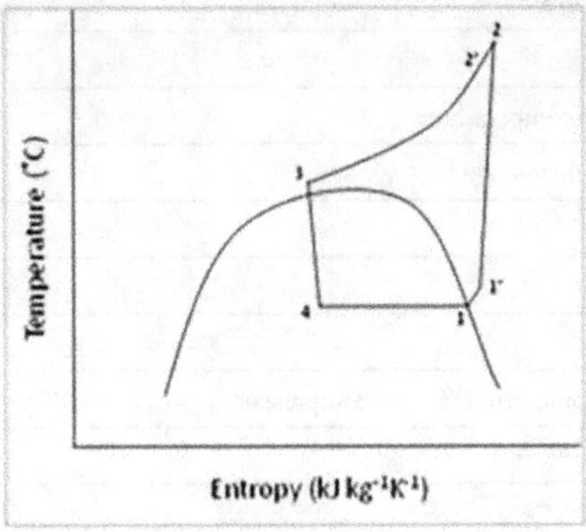

Figure 3: 4-1: evaporation; 1-1[1] suction gas super heating; 1[1]-2: adiabatic, non-reversible compression to gas cooler pressure with 69% efficiency; 2-3: isobaric heat rejection; 3-4 expansion from super critical pressure to subcritical pressure.

2. METHOD AND MATERIALS

Based on the availability of binary mixture models in REFPROP, propane, propylene, dimethyl ether and isobutane have been considered for the theoretical analysis. All the gases in the group 18 of the periodic table will serve as the flame suppressants. But among these gasses, helium, neon, argon and nitrogen have very low critical temperatures. On the other hand, CO_2 is freely available and has better critical temperature compared to the inert, noble gases. So, CO_2 is considered as the flame suppressing refrigerant for the calculation of flammability of the above selected hydrocarbons. The operating parameters to analyse the vapour compression refrigeration cycle have been taken as per the Air-conditioning, Heating and Refrigeration Institute (AHRI) conditions (Islam et al., 2021). The adiabatic non-reversible isentropic efficiency of compressor is assumed as 69% (Poggi et al., 2008).Constant temperature external energy fluid is taken as the energy exchanger in both the heat exchangers. The operating parameters as per the assumed conditions are listed in table 5. The properties of these zeotropic blends are calculated using REFPROP-MATLAB interface.

Table 5. Operating parameters of VCRS.

Condenser water inlet temperature	$t_{c,wi}$	30 °C
Chilled water inlet temperature	$t_{e,wo}$	12 °C
Evaporator/condenser approach temperature	$t_{e,ap}, t_{c,ap}$	1 °C
Water outlet, inlet temperature difference	$\Delta t_e, \Delta t_c$	5 °C
Evaporator temperature	t_e	6 °C
Condenser temperature	t_c	36 °C
Evaporator cooling capacity	Q_e	3 Ton
Adiabatic non-reversible isentropic efficiency of compressor	η_{isen}	69%
Initial refrigerant charge	m	0.25 kg/kW

3. MATHEMATICAL FORMULATION

Coefficient of performance reflects the performance of the VCRS. It is defined as the ratio between refrigeration effect to the work input to the compressor.

$$W_c = m(h_{1^1} - h_2) \text{ - Work input to the compressor} \tag{9}$$

$$Q_e = m(h_{1^1} - h_4) \text{ - Refrigeration effect} \tag{10}$$

$$Q_c = m(h_{2^1} - h_3) \text{ - Heat rejected by condenser} \tag{11}$$

$$COP = \frac{Refrigerationeffect(RE)}{energyrequiredtorunthesystem} = \frac{(h_{1^1} - h_4)}{(h_{1^1} - h_2)} \tag{12}$$

Where, m is mass flow rate, $h_1^{\,1}$ is enthalpy at compressor inlet, h_2 enthalpy at compressor exit, h_3 enthalpy at condenser exit, h_4 is enthalpy at evaporator inlet.

Volumetric refrigeration capacity (VRC): Volumetric refrigeration capacity indicates the compactness of the compressor. High volumetric capacity means smaller size of the compressor. Refrigerants with low volumetric refrigeration capacity require bigger size compressor which results higher initial cost, higher weight and reduced COP due to higher mechanical losses.

$$\text{VRC} = \frac{Refrigerationeffect(RE)}{Specificvolumeofcompressorinlet} = \frac{(h_{1^1} - h_4)}{v_{1^1}} \tag{13}$$

4. RESULTS AND DISCUSSION

The flame characteristics, namely lower flammability limit, net calorific value, laminar burning velocity of propane are calculated. As expected, with increasing CO_2 concentration, the flame characteristic calorific value and laminar burning velocity decreased, and the lower flammability limit increased. At 62.5% CO_2, the propane mixture safety class shifts from A3 to A2, and at 82.5% CO_2 the laminar burning velocity is almost negligible, and it can be classified as A2L. At 63.5% CO_2, isobutane shifts its safety class from A3 to A2, and at 83.7% CO_2 the laminar burning velocity is almost negligible, and it can be classified as A2L. With increasing CO_2, the GWP value also decreased from 3 to 1.33. Compared to other hydrocarbons at 40.5% CO_2 only dimethyl-ether shifts its safety class from A3 to A2. At 61.5% CO_2, propylene shifts its safety class from A3 to A2. With increasing CO_2, the GWP value is reduced from 2 to 1.2. Literature data is not available to calculate the laminar burning velocity of the propylene and dimethyl ether. Table 6 shows the theoretically calculated safety class of the hydrocarbon blends with CO_2 per the ASHRAE data.

Table 6. Safety class of selected hydrocarbons with CO_2 addition.

CO_2 (%vol)	Safety class			
	Propane	Isobutane	Dimethyl ether	propylene
0	A3	A3	A3	A3
20	A3	A3	A3	A3
40	A3	A3	A3	A3
42.5	A3	A3	A2	A3
61.5	A3	A3	-	A2
62.5	A2	A3	-	A2
63.5	A2	A2	-	-
70	A2	A2	-	-
82.5	A2L	A2	-	-
83.7	A2L	A2L	-	-

-literature data not available to calculate laminar burning velocity. So further safety class is not given.

Figure 4 shows the change of lower flammability limit of selected hydrocarbons with CO_2 addition. The dilution effect of CO_2 on dimethyl-ether is more when compared to the other hydrocarbons, followed by propylene, propane and isobutane.

Figure 4. Change of lower flammability limit of selected hydrocarbons with CO_2.

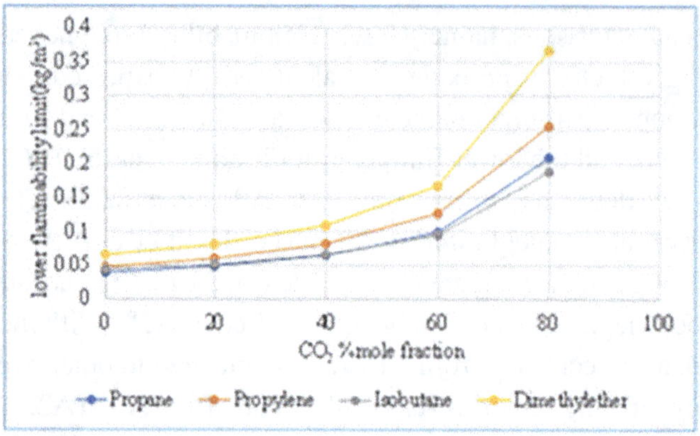

Figure 5. Change of laminar burning velocity of propane and Isobutane with CO_2.

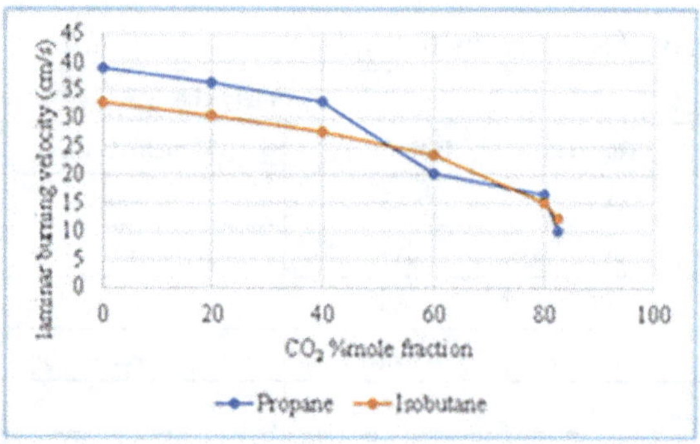

Figure 5 shows the theoretically calculated laminar burning velocity of the propane and isobutane with the addition of CO_2. As shown, initially pure propane burning velocity is higher than isobutane, with addition of CO_2 the laminar burning velocity significantly reduced. At 82.5% of CO_2 mole fraction both show almost similar values. The variation of net calorific value of hydrocarbons with CO_2 addition has been studied and are shown in Figure 6.

Figure 6. Variation of net calorific value of proposed hydrocarbons with CO_2 addition.

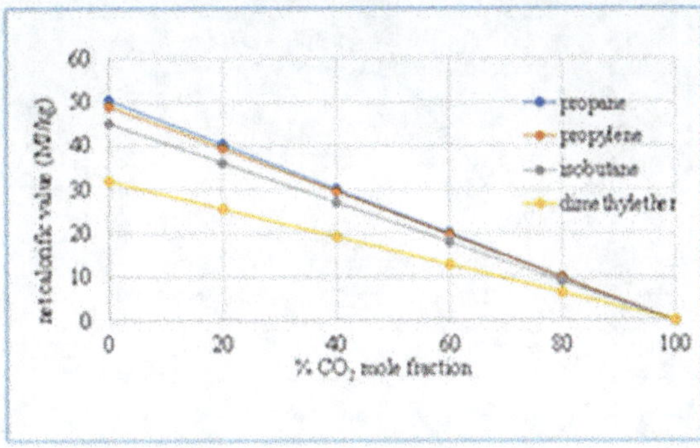

From the theoretical analysis, it is clear that the flammability of the hydrocarbons can be reduced thereby making them more acceptable in the refrigeration system with some precautionary measures. However, it is essential to find their thermodynamic performance as it affects the environmental properties. Hence a simple vapour compression refrigeration cycle analysis was performed as per the AHRI conditions. The performance of the transcritical CO_2 system significantly depends on the compressor discharge pressure and it has to be optimized for maximum COP. For the considered operating parameters the discharge pressure is optimised to obtain the best COP. This value is compared with some existing optimum pressure correlations, which are shown in Table 7.

Table 7. Comparison of optimum COP with literature data

Author	Correlation for optimum pressure (MPa)	Optimum COP	% Deviation of COP
(Qi et al., 2013)	$P_{opt} = 4.9 + 2.25t_c - 0.17t_{ev} + 0.002t_c^2$ $-10 \le t_{ev} \le 10, 30 \le t_c \le 50$	4.07	0
(Sarkar et al., 2004)	$P_{opt} = 132.3 - 8.4t_c + 0.3t_c^2 - 27.7$ $\times 10^{-4}t_c^3 25 \le t_c \le 45$	4.05	0.5%

Figure 7. Variation of COP of transcritical CO_2 system with discharge pressure.

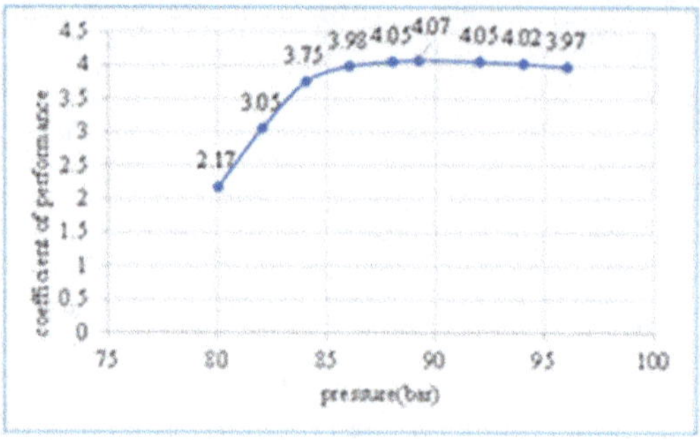

For the assumed evaporator and condenser temperatures and using the MATLAB-REFPROP interface, the variation of COP with the gas cooler (discharge) pressure has been studied and shown in Figure 7. The maximum COP obtained is 4.07 at 87.69 bar. Among the hydrocarbon refrigerants, isobutane (R-600a) shows marginally better results compared to the other hydrocarbon refrigerants. COP of all the selected hydrocarbons is better than R-32, which is a HFC and is currently used in commercial systems.

Figure 8. Coefficient of Performance of different refrigerants.

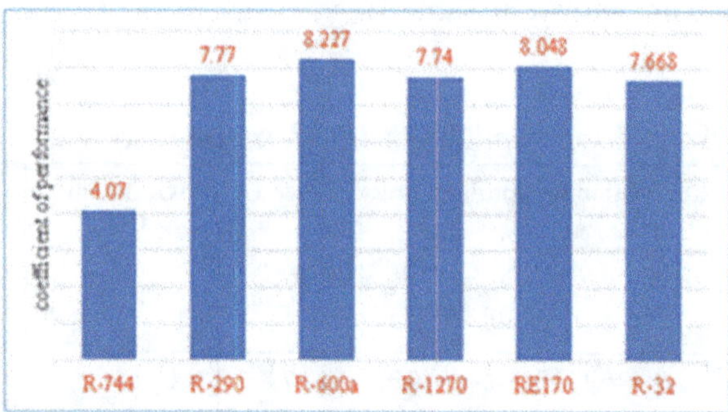

The total environmental warming impact (TEWI) of the refrigeration system with the selected refrigerants is summarised in Figure 9. The indirect warming impact depends on the refrigerant performance. The coefficient of performance of the R-744 is very less, hence its indirect warming impact is more compared to the other refrigerants. On the other hand, direct warming impact mostly depends on the annual leakage rate and recovery rate of the refrigerant at the end of the system life. For simplification the annual leakage rate is considered as 7% and recovery rate as

70%. The annual leakage rate and refrigerant recovery rate in turn depend on the initial charge. The required refrigerant amount depends on the type of refrigerant, operating parameters and system components internal volume. In this study for simplification a fixed amount of the refrigerant charge is considered for all refrigerants. However, the allowable charge for the flammable refrigerants depends on the lower flammability limit and for hydrocarbons the maximum allowable charge is 300g(Colbourne et al., 2020).The direct warming impact of the natural refrigerants is very less due to their low GWP values. The selected hydrocarbon refrigerants are flammable but the performance is good. On the other hand, CO_2 performance is poor but is non-flammable. So, as a trade-off, it may be preferable to use CO_2 and hydrocarbon mixtures as refrigerants, provided the low COP issue is taken through advanced cycles.

Figure 9. Comparison of TEWI of pure hydrocarbons with R-32.

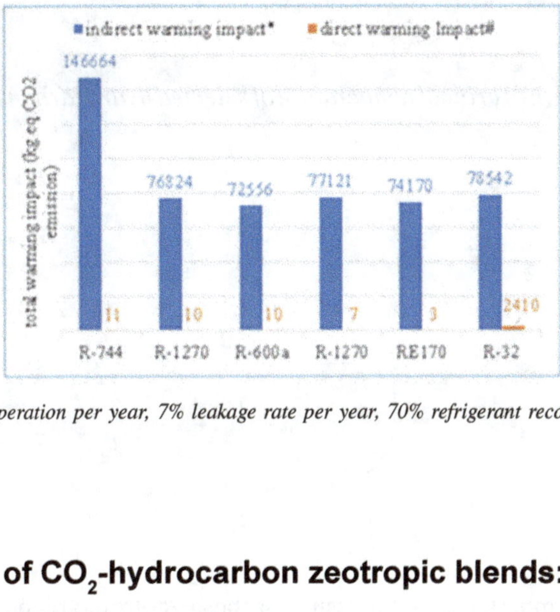

(3650 hours of operation per year, 7% leakage rate per year, 70% refrigerant recovery rate and 15-year system lifetime)

4.1 Analysis of CO_2-hydrocarbon zeotropic blends:

The temperature glide is calculated using MATLAB-REFPROP interface at the evaporator pressure of CO_2 for the taken conditions (4 MPa) and the results are shown in Figure 10. All the proposed blends exhibit the temperature glide above 2K, so the blends can be categorised as the zeotropic blends. Among these blends, isobutane-CO_2 shows highest temperature glide. The temperature glide can be beneficial if the external heat source/heat sink temperature also varies during heat transfer. Temperature glide is detrimental if the external heat source/sink temperature variation is small.

Figure 10. Variation of temperature glide at evaporator pressure(4MPa) with CO_2 mole fraction.

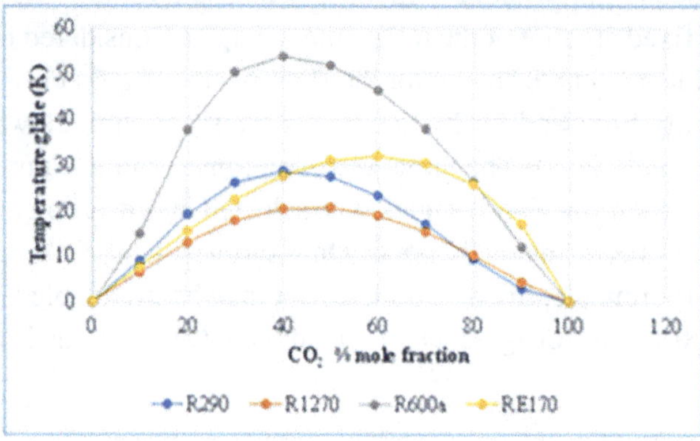

Figure 11. Variation of critical temperature of selected hydrocarbon blends with CO_2.

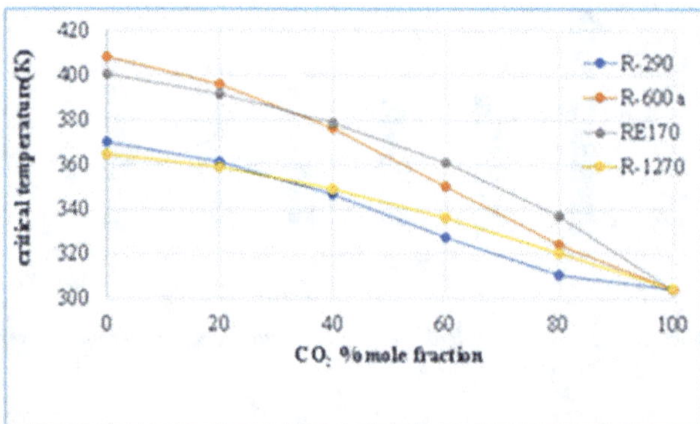

The critical temperature and pressure of these zeotropic blends have been calculated using REFPROP and are shown in Figure 11 and Figure 12, respectively. It can be observed that for isobutane and propane after 40% addition of CO2 there is significant reduction of critical temperature. For dimethyl-ether and propylene the reduction is nearly linear.

Figure 12. Variation of critical pressure of selected hydrocarbons with CO_2.

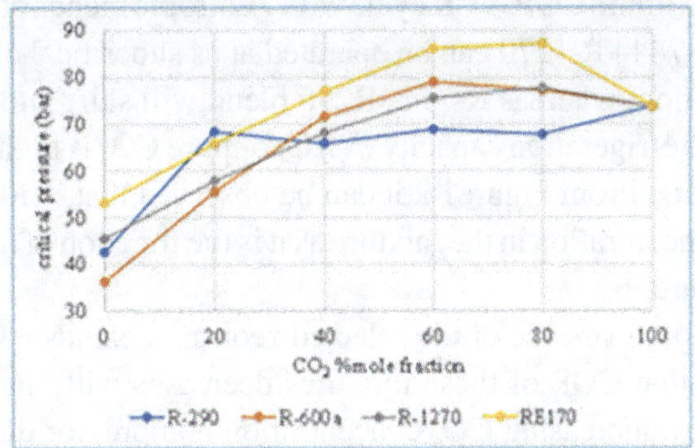

Figure 13. Comparison between critical and optimum discharge pressures of mixtures.

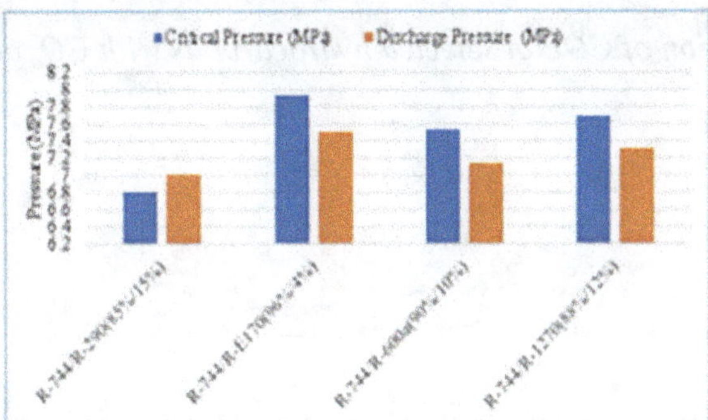

Figure 14. Variation of volumetric refrigeration capacity of selected blends with CO_2.

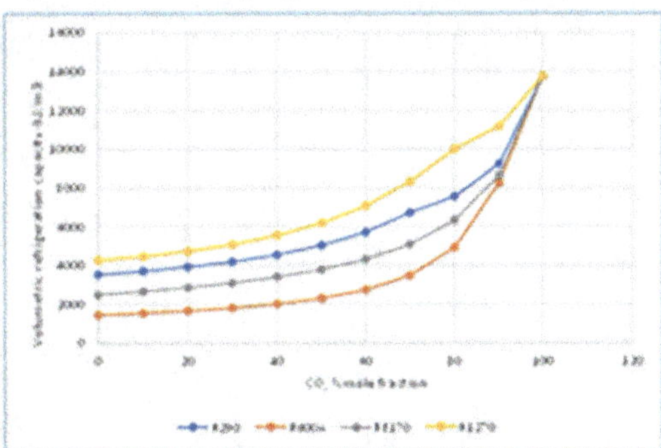

Figure 13 shows the comparison between critical pressure and discharge pressure of mixtures for maximum COP. VCR cycle with Zeotropic blends of R-744+R-600a, R-744+RE170, R-744+R-1270 can be operated at as subcritical cycles for the taken operated condition, whereas R-744+R290 blend will shift into the transcritical cycle. Volumetric refrigeration capacity (VRC) of pure CO_2 is higher than the other selected refrigerants. From Figure14, it can be observed that, VRC increases with increasing CO_2 concentration in the mixture as it is the function of number of carbon atoms in the mixture.

Coefficient of performance of the selected zeotropic blends is shown in Figure 15. Results show that COP of these mixtures decreases with increasing the CO_2 concentration. With increasing CO_2 concentration compressor discharge pressure increases and thereby COP reduces. Initially RE170 performance is better but at the end R1270 shows better performance compared to the other blends. R-600a+R-744 performance is lowest. Figure 16 shows the COP variation with discharge pressure of compressor.

Figure 15. Variation of COP of selected hydrocarbons with CO_2 addition

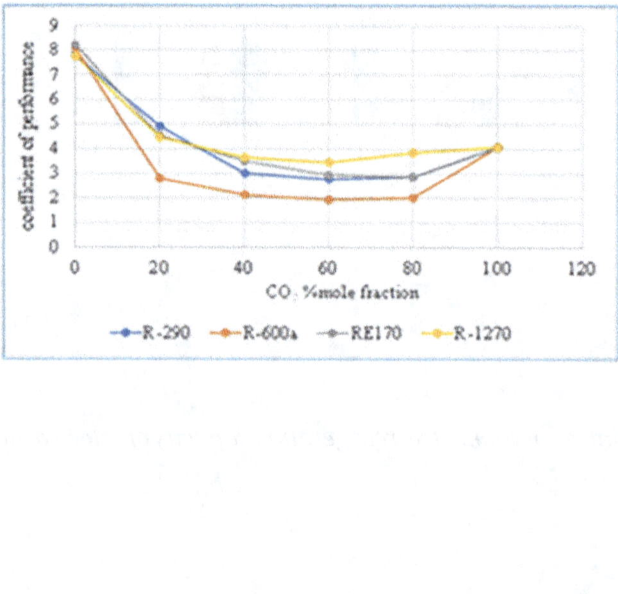

Figure 16. Variation of COP with discharge pressure for dimethyl ether and CO_2 mixture.

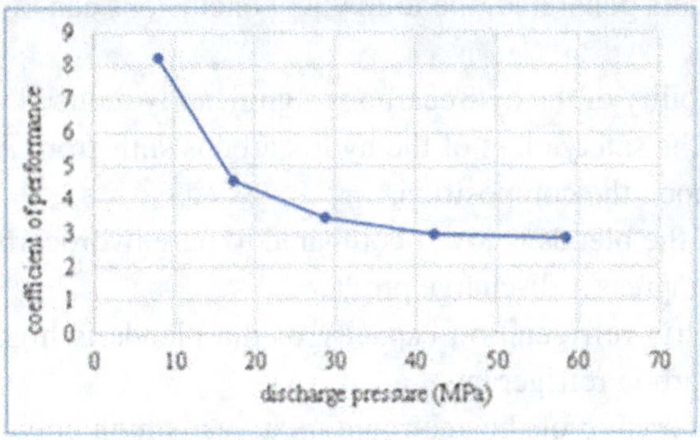

Figure 17. Comparison of TEWI of selected zeotropic mixtures with R-32.

(3650 hours of operation per year, 7% leakage rate per year, 70% refrigerant recovery rate and 15-year system lifetime)

The total warming impact of the selected zeotropic blends with best performance has been shown in Figure 17. Due to the low performance of these blends the indirect warming impact due to power utilisation of these blends is more compared to R-32. However, the direct warming impact of these blends is very less, even lesser than the pure refrigerants. However, with reducing emissivity factor the indirect warming impact can be reduced, simultaneously the TEWI of these mixtures can also be reduced.

5. CONCLUSIONS

Based on the data generated, the following conclusions can be drawn:

- The flammability of the hydrocarbons significantly reduced with the addition of CO_2 and the safety class of the hydrocarbons shift from A3 to A2 or A2L depending upon the composition.
- The COP of the blends is lower compared to pure hydrocarbons, mainly due to higher compressor discharge pressure.
- The volumetric refrigeration capacity of the blends is higher compared to pure hydrocarbon refrigerants
- Critical temperature of the zeotropic mixtures is high compared to the pure R-744, hence the cycle can be operated in subcritical mode.
- The TEWI of the zeotropic blends is more than the pure hydrocarbons due to low COP
- Thus, in order to make the systems with hydrocarbon-CO_2 blends energy efficient, it is essential to explore advanced cycles so that they can compete with pure hydrocarbons in terms of energy consumption while surpassing them in terms of personal and environmental safety.

ACKNOWLEDGEMENTS

This theoretical investigation is part of the project "Development of low GWP chemicals and their blends as alternatives to HCFCs and HFCs" sponsored by the Ministry of Environment (Forest and Climate Change, Govt of India, Ozone Cell, 1st Floor, 9 Institutional Area, Lodhi Road, New Delhi - 100 003). The authors would like to acknowledge the financial support from the Ministry of Environment, New Delhi, India.

REFERENCES

Agarwal, R. S., & Clark, E. (2016). Refrigerant Blends : Calculating Global Warming Potentials. *Ozon Action FACT SHEET, 290*, 2. http://www.unep.fr/ozonaction/information/mmcfiles/7786-e-Calculating_GWPofBlends_post_Kigali.pdf

ASHRAE. (2023). Update on new refrigerants designations and safety classifications. *ANSI/ASHRAE Standard 34-2019, August*, 1–5. https://www.ashrae.org/technical-resources/standards-and-guidelines/ashrae-refrigerant-designations

Bansal, P. (2012). A review - Status of CO 2 as a low temperature refrigerant: Fundamentals and R&D opportunities. *Applied Thermal Engineering*, *41*, 18–29. DOI: 10.1016/j.applthermaleng.2011.12.006

CEA. (2011). CO 2 Baseline Database for the Indian Power Sector User Guide. *Report by Ministry of Power Govt of India, January*, 1–34. http://cea.nic.in/reports/others/thermal/tpece/cdm_co2/user_guide_ver9.pdf

Colbourne, D., Suen, K. O., Li, T. X., Vince, I., & Vonsild, A. (2020). General framework for revising class A3 refrigerant charge limits – a discussion. *International Journal of Refrigeration, 117*, 209–217. DOI: 10.1016/j.ijrefrig.2020.04.024

Kennoy, D.H. (2018). ANSI/ASHRAE Addendum g to ANSI/ASHRAE Standard 34-2016. *Ashrae, 8400*, 1–6. www.ashrae.org

Delbeke, J., Runge-Metzger, A., Slingenberg, Y., & Werksman, J. (2019). The paris agreement. *Towards a Climate-Neutral Europe: Curbing the Trend*, 24–45. https://doi.org/DOI: 10.4324/9789276082569-2

Efficiency, E. (n.d.). *Sustainable Cooling Towards a Cool COP28 " Cool " COP28 : Objectives.*

Environemental Protection Agency. (2016). The Montreal Protocol on Substances that Deplete the Ozone Layer: The Kigali Amendment. *The Kigali Amendment (2016): The Amendment to the Montreal Protocol Agreed by the Twenty-Eighth Meeting of the Parties (Kigali, 10-15 October 2016), September 1987.*

Fryer, J., & Lee, K. T. Q. (2012). Methods of calculating Total Equivalent Warming Impact (TEWI). *The Australian Institute of Refrigeration Air Conditioning and Heating*, 21.

Gant, S. E., Pursell, M. R., Lea, C. J., Fletcher, J., Rattigan, W., Thyer, A. M., & Connolly, S. (2011). Flammability of hydrocarbon and carbon dioxide mixtures. *Process Safety and Environmental Protection, 89*(6), 472–481. DOI: 10.1016/j.psep.2011.06.017

Gschrey, B., & Zeiger, B. (2015). F-Gas Regulation: Technical Advice to Member States on implementing Article 7(2). *Öko-Recherche, 7*(517). https://ec.europa.eu/clima/policies/f-gas/docs/151023_hfc23_byproduction_en.pdf

Hegetschweiler, M. J., Berger, L., Hesse, R., Beeckmann, J., Bariki, C., Pitsch, H., & Linteris, G. T. (2023). Data reduction considerations for the burning velocity of spherical constant volume flames of R32 (CH2F2) with air. *Combustion and Flame, 254*, 112807. DOI: 10.1016/j.combustflame.2023.112807

International Energy Institute (IEI). (2023). Statistical Review of World Energy 2023. *BP Energy Outlook, 2023*(70), 8–20.

Islam, M. A., Mitra, S., Thu, K., & Saha, B. B. (2021). Study on thermodynamic and environmental effects of vapor compression refrigeration system employing first to next-generation popular refrigerants. *International Journal of Refrigeration, 131*(August), 568–580. DOI: 10.1016/j.ijrefrig.2021.08.014

Le Chatelier, M. H. (2004). Note on mine gas measurement by flammability limits. *Process Safety Progress, 23*(3), 172–174. DOI: 10.1002/prs.10031

McLinden, M. O., Seeton, C. J., & Pearson, A. (2020). New refrigerants and system configurations for vapor-compression refrigeration. *Science, 370*(6518), 791–796. DOI: 10.1126/science.abe3692 PMID: 33184206

Mohammed, A. N., Juhany, K. A., Kumar, S., Kishore, V. R., & Mohammad, A. (2017). Effects of CO2/N2 dilution on laminar burning velocity of stoichiometric DME-air mixture at elevated temperatures. *Journal of Hazardous Materials, 333*, 215–221. DOI: 10.1016/j.jhazmat.2017.03.042 PMID: 28359037

Montreal protocol on substances that deplete the Ozone layer final act 1987. (1989). *Journal of Environmental Law, 1*(1), 128–136. https://doi.org/DOI: 10.1093/jel/1.1.128

Nations, U., & Programme, E. (2016). *Sources, fates, toxicity, and risks of trifluoroacetic acid and its salts : relevance to substances regulated under the Montreal and Kyoto protocols A report prepared by the UNEP Environmental Effects*. 1–26.

Pearson, A. (2008). Refrigeration with ammonia. *International Journal of Refrigeration, 31*(4), 545–551. DOI: 10.1016/j.ijrefrig.2007.11.011

Poggi, F., Macchi-Tejeda, H., Leducq, D., & Bontemps, A. (2008). Refrigerant charge in refrigerating systems and strategies of charge reduction. *International Journal of Refrigeration, 31*(3), 353–370. DOI: 10.1016/j.ijrefrig.2007.05.014

Qi, P. C., He, Y. L., Wang, X. L., & Meng, X. Z. (2013). Experimental investigation of the optimal heat rejection pressure for a transcritical CO2 heat pump water heater. *Applied Thermal Engineering, 56*(1–2), 120–125. DOI: 10.1016/j.applthermaleng.2013.03.045

Sarkar, J., Bhattacharyya, S., & Gopal, M. R. (2004). Optimization of a transcritical CO2 heat pump cycle for simultaneous cooling and heating applications. *International Journal of Refrigeration, 27*(8), 830–838. DOI: 10.1016/j.ijrefrig.2004.03.006

Wang, Z., Yelishala, S. C., Yu, G., Metghalchi, H., & Levendis, Y. A. (2019). Effects of Carbon Dioxide on Laminar Burning Speed and Flame Instability of Methane/Air and Propane/Air Mixtures: A Literature Review. *Energy & Fuels, 33*(10), 9403–9418. DOI: 10.1021/acs.energyfuels.9b02346

Yelishala, S. C., Kannaiyan, K., Wang, Z., Metghalchi, H., Levendis, Y. A., & Sadr, R. (2020). Thermodynamic study on blends of hydrocarbons and carbon dioxide as zeotropic refrigerants. *Journal of Energy Resources Technology, Transactions of the ASME, 142*(8), 1–10. https://doi.org/DOI: 10.1115/1.4045930

Yelishala, S. C., Wang, Z., Metghalchi, H., Levendis, Y. A., Kannaiyan, K., & Sadr, R. (2019). Effect of Carbon Dioxide on the Laminar Burning Speed of Propane-Air Mixtures. *Journal of Energy Resources Technology, Transactions of the ASME, 141*(8). https://doi.org/DOI: 10.1115/1.4042411

Zhang, Z., Ito, M., & Hihara, E. (2024). *Laminar Burning Velocity in R290 / R32, R290 / R1234yf, and R290 / HFO-1123 Binary Mixtures : Comparing Experimental Results to Mixing Rule Calculations AGC Inc. Chemicals Company Research & Development General Division*. The University of Toky.

Chapter 8
Experimental Evaluation of Performance and Emission Characteristics of Phytol-Diesel Blends in a Single-Cylinder Compression Ignition Engine

S. K. Mohammad Shareef
https://orcid.org/0000-0002-2634-9273
CVR College of Engineering, Hyderabad, India

P. Srinivas Reddy
https://orcid.org/0000-0003-4632-1291
CVR College of Engineering, Hyderabad, India

B. Abhiram
CVR College of Engineering, Hyderabad, India

ABSTRACT

Conventional fossil fuels are rapidly depleting due to population growth and rising energy consumption, necessitating the search for renewable energy sources. Among these, biodiesel has gained attention for its potential as a sustainable substitute when blended with diesel in optimal ratios. Recent advancements in nanotechnology have introduced nano-additives as catalysts to enhance engine efficiency and

DOI: 10.4018/979-8-3373-3621-3.ch008

reduce emissions. This study focuses on improving diesel engine performance and reducing emissions by blending alcohol with diesel in varying proportions and incorporating nanoparticles to enhance thermal conductivity. A diesel engine is used to investigate combustion characteristics, engine performance, and emissions. Key challenges, such as fuel burning rate, ignition lag, and emissions like NOx, HC, and CO, are addressed. The study highlights the potential of nanoparticles to improve thermal efficiency and explores the comparative benefits of alcohol-diesel blends, contributing to the development of sustainable energy solutions.

1. INTRODUCTION

Diesel engines are extensively utilized in agriculture, transportation, and industry due to their exceptional reliability, combustion efficiency, and power output. However, the growing concerns over pollution and emission challenges have underscored the need for innovative technologies that prioritize low emissions and superior fuel efficiency in internal combustion engine research (Pradelle et al., 2019). Consequently, numerous alternative fuels have become the focus of extensive studies, including biodiesel, alcohols, ethers, 2,5-dimethylfuran, natural gas (Papagiannakis et al., 2017), Fischer–Tropsch diesel (Geng, 2020), and others. Among these alternatives, biodiesel remains the most widely adopted.

Alcohol, particularly, has gained significant traction in gasoline engines due to its favourable cetane and octane numbers (No, 2016). More recently, researchers have explored its application in compression ignition engines by blending it with biodiesel to address biodiesel's high viscosity (Atmanli, 2016). Alcohols are categorized into lower alcohols (spanning from 1-carbon methanol to 2-carbon ethanol) and higher alcohols (ranging from 3-carbon propanol to 20-carbon phytol). The properties of these alcohols—such as higher cetane numbers, calorific value, density, and lower latent heat of evaporation—are influenced by the number of carbon atoms. Additionally, higher-carbon alcohols exhibit reduced corrosion risk and increased oxygen content. N-pentanol, a five-carbon straight-chain alcohol, enhances the atomization quality of biodiesel blends due to its lower kinematic viscosity compared to biodiesel (Liang, 2021). Moreover, n-pentanol improves the low-temperature flow properties of biodiesel blends (Jin, 2019), and its compatibility with diesel and biodiesel increases with a greater number of carbon atoms (Atmanli, 2016). Studies have shown that adding 1-pentanol to diesel results in good stability without phase separation (Atmanli & Yilmaz, 2018). However, Babu and Anand (Babu & Anand, 2017) noted that while CO and NOx emissions increased, THC decreased with n-pentanol addition to biodiesel blends, attributed to the additional carbon and oxygen atoms provided by n-pentanol. Despite these benefits, blends with n-pentanol exhibit higher brake-specific fuel consumption (BSFC) and lower indicated thermal efficiency (ITE) due to its lower heating value, indicating a trade-off between emission reduction and engine performance (Appavu, 2019; Ashok, 2019; Atmanli & Yilmaz, 2020; Pan, 2019). Additionally, higher-carbon alcohols, such as n-butanol, have been linked to increased production of polycyclic aromatic hydrocarbons (PAH) (Yilmaz & Davis, 2016).

Second-generation biofuels represent a pivotal component of global strategies to develop clean, renewable energy sources, not only enhancing energy security but also improving sustainability in transport applications. Recent studies highlight the role of alcohol- and biodiesel-based blends in reducing greenhouse emissions and improving combustion characteristics under real engine conditions (Chivu et al., 2024; Sethin et al., 2024; Suhara et al., 2024). Such efforts underscore the need for integrated approaches that simultaneously advance biofuel development and optimize engine operation.

A significant challenge lies in ensuring their compatibility with existing fuel infrastructure, primarily designed for compression-ignition (CI) engines. Incorporating second-generation biofuels as blending agents for petro-diesel offers a promising pathway to extend the utility of conventional diesel. A coordinated effort to simultaneously advance second-generation biofuel development and optimize engine operation concepts is essential. This integrated approach could accelerate the adoption of second-generation biofuels as effective blending agents for CI engine applications.

Biodiesels from different feedstocks, such as soy, rapeseed, algae, animal fats, and waste oils, have been extensively investigated as blending agents for CI operation (Altin et al., 2001; Barnwal & Sharma, 2005; Dwivedi & Sharma, 2014; El-Araby et al., 2018; Geo et al., 2012; Makareviciene et al., 2023). Cuphea methyl ester has also been recently investigated by the U.S. Department of Agriculture as a viable blending agent since its physical properties are like those of diesel (Da Silva & De Matos, 2009; Gedam & Sampathkumaran, 1986). Synthetic fuels, such as hydro-treated vegetable oil (HVO), have received attention in European nations because of their compatibility with the diesel infrastructure and their favorable CI engine characteristics (Dantas Neto et al., 2014; Dwivedi, 2020). Bio-derived alcohols, such as ethanol and butanol, have been investigated in the past decade as blending agents for diesel fuel. Hansen et al. (Hansen et al., 2005) performed research on ethanol–diesel blends and concluded that further studies are needed to ensure long-term engine durability when these fuel blends are used. Rizopoulos et al. (Rakopoulos et al., 2010; Rakopoulos et al., 2008) blended butanol and diesel at 8% and 16% by volume and analyzed the combustion behavior of a six-cylinder direct-injection engine. They found an increase in ignition delay and reductions in the maximum cylinder pressure and temperature. Iso-pentanol was also recently found to be suitable as a blending agent with gasoline for homogenous charge compression ignition (HCCI) engines (Hoang & Pham, 2025). The above literature search reveals that there has been a significant amount of research on incorporating first-generation biofuels in the CI engine infrastructure.

The main objective of this research is to assess second-generation biofuels as potential blending agents for compression-ignition (CI) engines. Among long-chain alcohols, which offer higher energy content compared to ethanol, butanol, and iso-pentanol, phytol ($C_{20}H_{40}O$)—a heavy alcohol—emerges as a promising candidate. Researchers at the Argonne National Laboratory (ANL) have engineered strains of photosynthetic bacteria capable of producing phytol. However, this process remains confined to laboratory scales, producing insufficient quantities for large-scale engine testing. Consequently, the phytol samples used in this study were sourced externally. As an oxygenated fuel, phytol is anticipated to emit fewer particulates (Buchner et al., 2017; Liger et al., 2002; Warrick et al., 2011) than conventional petro-diesel, making it a cleaner alternative. Its physical and chemical properties, such as cetane number, heat of combustion, and density, closely resemble those of diesel, indicating its potential as a blending agent (De Laurier, 1994; Gaikwad et al., 2024; Hoang & Pham, 2025; Mueller & De Laurier, 2001). However, its high viscosity and low vapor pressure necessitate the exploration of diesel–phytol blends rather than using pure phytol as a standalone fuel.

The global pursuit of sustainable alternatives to fossil fuels (Liu et al., 2022; Makareviciene, 2013; Rachman & Prasetyo, 2022; Salvi & Subramanian, 2015; Teixeira, 2021) has intensified, with efforts focusing on fuels that can integrate seamlessly into existing energy infrastructures. At ANL, researchers have developed bioengineered strains that produce phytol, a hydrogenated diterpene alcohol, through photosynthetic bacterial processes. Phytol's properties, including its cetane number, density, and heat of vaporization, align closely with diesel (Balakrishnan, 2007; Johari et al., 2022; Rachman & Prasetyo, 2014; Sazali, 2020; Shatnawi et al., 2018), reinforcing its viability as a diesel blending agent for CI engines. This study aims to evaluate the feasibility of using phytol–diesel blends, specifically P2, P4, P6, and P8, representing 2%, 4%, 6%, and 8% phytol by volume, respectively. Performance and emissions testing were conducted using a single-cylinder engine under standard operating conditions to compare the blends' characteristics.

Phytol, also known as florasol, is an acyclic hydrogenated diterpene alcohol with diverse applications, including serving as a precursor for synthetic Vitamin E and Vitamin K production. In ruminant animals, the fermentation of plant material in the gut releases phytol, a component of chlorophyll, which is subsequently converted into phytanic acid and stored in fat tissues. Additionally, shark liver oil (Al-Harrasi, 2019; Islam, 2018; Khan, 2014) contains phytol, which can be converted into pristane, highlighting its versatile biological and industrial relevance.

Phytol ($C_{20}H_{40}O$) tetramethyl-2-hexadecen-1-ol

Figure 1. Chemical structure of phytol

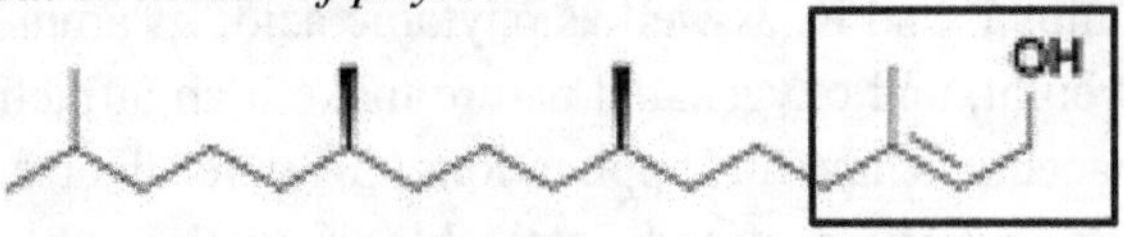

Properties of Phytol:
FORMULA:$C_{20}H_{40}O$
DENSITY: 850kg/m^3
Molar mass: 296.53 g/mol
ChemSpider: 203 to 207^{0}c (397 to 399^{0}F; 476 to 477 K) at 10 mmHg
3D Model (JSmol): Interactive image
PubChem: 5280435

Since being a component of chlorophyll, phytol is the most prevalent acyclic diterpene (3,7,11,15-tetramethylhexadec-2-en-1-ol) found through environment and it can be produced by practically organism such as plants, algae, and bacteria (cyanobacteria) .

Commercial Applications

Phytol is widely utilized in the fragrance industry and serves as an ingredient in cosmetics, shampoos, toilet soaps, household cleaners, and detergents. Additionally, it is used as a diluent in some Cannabis distillate vapes. The global consumption of phytol is estimated to range between 0.1 and 1.0 metric tons annually.

This study adopts a systematic approach to evaluate the feasibility of using phytol as a blending agent with diesel fuel in compression ignition (CI) engines. Four blends of phytol and diesel were prepared, containing 2%, 4%, 6%, and 8% phytol by volume, denoted as P2, P4, P6, and P8, respectively. The initial phase of the study focused on analysing the physical and chemical properties of these blends. The heat of combustion, density, and cetane number of all blends were found to be comparable, indicating minimal variation. However, significant differences were observed in vapor pressure and viscosity, with the vapor pressures of the blends being calculated rather than measured, using simple mixing rules.

1.1 Methods of Phytol Production

Phytol ($C_{20}H_{40}O$) is a branched-chain acyclic diterpene alcohol that serves as the side chain of chlorophyll molecules and plays a significant role as a biochemical precursor to vitamins E and K, as well as phytanic acid. Its amphiphilic structure, high molecular weight, and oxygenated nature make it an attractive candidate for industrial, pharmaceutical, and fuel applications. Given its diverse applications, the development of efficient, low-cost, and sustainable production methods for phytol has become an area of growing research interest. Broadly, phytol can be obtained from natural extraction, chemical synthesis, by-product recovery, and biotechnological approaches, each with its own challenges and opportunities.

1.1.1. Historical Background of Phytol Discovery

Phytol was first identified in the late 19th century by K. L. Reinke during studies of chlorophyll degradation products. Richard Willstätter, who won the Nobel Prize in 1915 for his work on chlorophyll, also confirmed that the long hydrophobic side chain attached to the porphyrin head group was phytol. Early work focused on isolating it in pure form for chemical characterization rather than large-scale use. By the 20th century, phytol became a known precursor for vitamins E and K, which created demand in pharmaceuticals. However, its limited availability from plant extraction restricted commercial expansion.

1.1.2. Natural Extraction from Chlorophyll-Containing Biomass

The most traditional and direct method of obtaining phytol is through the degradation of chlorophyll in plant or algal biomass. Chlorophyll molecules consist of a porphyrin head group with a hydrophobic phytol tail esterified to it. Hydrolysis of the chlorophyll ester bond releases free phytol.

a. **Plant sources:**

Higher plants, grasses, and leafy vegetables contain substantial amounts of chlorophyll, which in turn can serve as a reservoir for phytol. Extraction is typically performed using solvent systems such as acetone, ethanol, or methanol, followed by alkaline hydrolysis to release phytol from chlorophyll. Early studies demonstrated that wheatgrass, spinach, and nettle leaves yield significant phytol upon chlorophyll breakdown.

b. **Algal sources:**

Algae and cyanobacteria are particularly promising due to their high chlorophyll content and rapid biomass productivity. Marine microalgae such as Chlorella, Scenedesmus, and Nannochloropsis have been reported to yield phytol concentrations sufficient for industrial extraction. Large-scale algal cultivation systems offer an advantage because they can be coupled with wastewater treatment and carbon sequestration, enhancing sustainability.

c. **Limitations:**

Despite its simplicity, natural extraction faces economic and scalability challenges. Large biomass requirements, seasonal availability, and solvent-intensive processing make the approach costly. Moreover, crude extracts often contain chlorophyll degradation products such as pheophytin, requiring purification steps that reduce yield and increase cost.

1.1.3. By-Product Recovery from Artemisinin Extraction

One of the most promising industrially relevant approaches is the recovery of phytol as a by-product from Artemisia annua processing. Artemisinin, a sesquiterpene lactone used in antimalarial drugs, is extracted from A. annua leaves. During this process, chlorophyll-rich residues accumulate in large quantities. Instead of being discarded or incinerated, these residues can be hydrolyzed to release phytol.

Process Description:

- Chlorophyll by-products are adsorbed on silica gel columns during artemisinin separation.
- Strong polar solvents such as ethanol or methanol are used to elute chlorophyll.
- Alkaline hydrolysis (NaOH or KOH solutions) cleaves the ester bond, yielding phytol and chlorophyllin.
- Solvents such as petroleum ether or ethyl acetate are used to extract phytol, while chlorophyllin remains in the aqueous phase.

Advantages:

- Utilizes waste material from an existing large-scale industry.
- Reduces environmental burden from burning chlorophyll residues.
- Provides a low-cost route since raw materials are abundantly available.

Industrial Potential:

China alone produces thousands of tons of artemisinin annually, generating several hundred tons of chlorophyll by-products. Converting this waste into phytol could significantly lower production costs and support circular economy models.

1.1.4. Chemical Synthesis Routes

Synthetic chemistry offers another pathway to phytol production, though it is rarely pursued on a commercial scale due to cost and complexity.

a) **Synthesis from Isoprenoid Intermediates:**

Phytol belongs to the isoprenoid family, derived from the head-to-tail polymerization of isoprene units. Laboratory synthesis has been demonstrated using geranylgeraniol as a precursor. Hydrogenation and selective modifications yield phytol or its derivatives.

b) **Catalytic Hydrogenation of Phytol Derivatives:**

Partial hydrogenation of phytadiene or phytone intermediates also yields phytol. These routes typically employ metal catalysts such as palladium or platinum under controlled conditions.

c) **Limitations:**

Chemical synthesis is energy-intensive, requires expensive catalysts, and involves multi-step reactions with chirality control issues. It is therefore not economically viable for large-scale production compared to natural or by-product methods.

1.1.5. Biotechnological and Metabolic Engineering Approaches

Recent advances in metabolic engineering have opened new possibilities for phytol production through genetically engineered microorganisms.

a) **Engineered Photosynthetic Bacteria:**

Researchers at Argonne National Laboratory (ANL) have demonstrated the use of genetically modified cyanobacteria that can produce phytol directly from CO_2 through the methylerythritol phosphate (MEP) pathway. By optimizing enzyme expression, strains have been engineered to accumulate phytol precursors such as geranylgeranyl pyrophosphate, which can be converted to phytol through reductive steps.

b) **Yeast-Based Production:**

Yeasts such as Saccharomyces cerevisiae have been engineered to host the isoprenoid biosynthetic pathway, producing phytol from simple sugars. While yields remain low, ongoing research into pathway optimization and fermentation strategies shows promise.

c) **Advantages and Challenges:**

Advantages: Renewable, sustainable, and scalable; avoids reliance on agricultural biomass.

Challenges: Low titters, product toxicity to host cells, need for downstream purification.

Biotechnological production is not yet commercially feasible but represents a long-term sustainable solution that could align with future bioeconomy goals.

1.1.6. Sustainability and Commercial Prospects

- **Environmental Impact:** Recovery from waste streams reduces environmental load. Biotech methods further contribute by sequestering CO_2.
- **Cost Considerations:** Extraction from artemisinin by-products is currently the most cost-effective. Synthetic methods remain prohibitively expensive.
- **Applications:** Phytol has applications in pharmaceuticals (precursor to vitamins), cosmetics (fragrance and detergent additive), and as explored here, fuel blending for CI engines due to its diesel-like cetane number and density.
- **Commercial Demand:** Global phytol consumption is estimated at 0.1–1.0 metric tons annually, mostly for specialty applications. Scaling for fuel applications would require breakthroughs in cost and yield.

1.1.7. Future Directions in Phytol Production

- **Integration with Biofuel Platforms:** Coupling phytol extraction with biodiesel and bioethanol plants could improve economics.
- **Green Solvents and Catalysts:** Replacement of petroleum solvents with ionic liquids or deep eutectic solvents for phytol extraction.
- **Hybrid Approaches:** Combining algae cultivation with biorefinery models to co-produce phytol, lipids, and proteins.
- **Techno-Economic Analysis:** Comparative lifecycle studies of different production methods are needed to assess feasibility for large-scale applications.

1.1.8. Global Market and Cost Considerations

- **Market Size**: Current global phytol demand is estimated at ~500–1,000 kg/year, dominated by pharmaceuticals and cosmetics. For biofuel applications, demand would rise by several orders of magnitude.
- **Production Cost**: Depending on the method, costs range from USD \$300–600/kg (extraction-based) to >\$1000/kg (synthetic). For fuels, costs must drop below \$2/kg to be competitive with diesel.
- **Major Producers:** Limited industrial suppliers exist, with much of the supply concentrated in Europe and Asia. Companies such as BASF, Tokyo Chemical Industry (TCI), and Sigma-Aldrich sell phytol in laboratory to pilot-scale quantities

1.1.9. Challenges in Scaling Phytol for Biofuel

- **Low Yield per Biomass:** Even chlorophyll-rich plants yield limited phytol per kg of biomass (~0.2–0.5%).
- **Energy Balance:** Solvent extraction and distillation consume significant energy, raising net CO_2 emissions.
- **Competition with High-Value Markets:** Pharma and cosmetics markets can pay higher margins, diverting phytol from energy applications.
- **Fuel Property Limitations:** Despite good cetane and density, phytol's high viscosity (~63 cSt at 25 °C) and low vapor pressure restrict its direct use, requiring blending.

1.1.10. Emerging Research Directions

- Genetic Engineering of Crops: Designing high-chlorophyll crops or engineered plants with higher phytol accumulation.
- Synthetic Biology: Development of modular pathways for phytol biosynthesis in E. coli and yeast to increase titers beyond 1 g/L.
- Co-Production Models: Integrating phytol recovery with algal biofuel refineries, where lipids are converted to biodiesel and chlorophyll residues yield phytol.
- Life Cycle Assessment (LCA): Early LCA studies show that phytol from waste chlorophyll could have ~40–60% lower GHG footprint compared to petroleum diesel, provided solvent recovery efficiencies are >90%.

1.1.11. Industrial and Commercial Relevance of Phytol

- Pharmaceuticals: Precursor to Vitamin E (tocopherols) and Vitamin K1 (phylloquinone). Both are essential for human health, making phytol strategically important in nutraceutical industries.
- Cosmetics & Fragrances: Used as an aromatic ingredient in perfumes, soaps, shampoos, and detergents due to its grassy, floral odor.
- Food Industry: Approved as a food additive and flavoring agent in small quantities.
- Biomedicine: Recent studies highlight anti-inflammatory, anti-cancer, and antimicrobial properties of phytol derivatives.
- Biofuel Blending: Due to a high cetane number (~46) and energy content close to diesel, phytol is investigated as an advanced biofuel component.

1.1.12. Advances in Green Extraction Technologies

Conventional solvent extraction (ethanol, petroleum ether) is effective but unsustainable. Researchers are exploring:

- Supercritical CO_2 extraction: Provides solvent-free phytol recovery with tunable selectivity; however, high pressure (200–400 bar) makes it energy-intensive.
- Ionic Liquids (ILs) and Deep Eutectic Solvents (DES): Emerging "green" solvents capable of dissolving chlorophyll and releasing phytol under mild conditions. These reduce volatility and improve recyclability.
- Microwave- and Ultrasound-Assisted Extraction (MAE & UAE): Improve yields by enhancing cell wall disruption and solvent penetration.

Case Example: Islam et al. (2018) reported that microwave-assisted hydrolysis of chlorophyll-rich biomass shortened reaction times by 50% compared to conventional methods, while increasing phytol yield.

Phytol can be produced through multiple pathways, each with distinct advantages and challenges. Traditional natural extraction is straightforward but limited in scale. By-product recovery from artemisinin extraction offers immediate industrial viability and aligns with circular economy principles. Chemical synthesis remains largely academic due to cost constraints, while biotechnological production represents the most sustainable future option, albeit requiring further optimization. For the current study, phytol was externally sourced, but future research should emphasize scalable, eco-friendly production methods to support its application as a renewable blending agent for diesel engines.

2. LITERATURE REVIEW

The worldwide demand for energy in recent years has shown major fluctuations due to increasing fuel prices and the depletion of conventional fossil resources. These issues have placed a heavy economic burden on importing nations and intensified the global search for sustainable alternatives to petroleum-based fuels. The primary objective of such research is to identify renewable and efficient substitutes capable of ensuring long-term energy security, reducing dependence on crude oil, and improving air quality.

The choice of alternative fuels is guided by several criteria—cost-effectiveness, availability, ignition quality, calorific value, and emission behavior. An ideal substitute fuel should ignite easily, deliver comparable performance, generate fewer exhaust pollutants, and require minimal modification to existing engines. As petroleum reserves decline, vegetable oils, alcohols, and other bio-derived fuels have gained prominence as promising alternatives to diesel.

Combustion of fossil fuels remains one of the main causes of increased greenhouse gas emissions and global warming. Consequently, reducing engine exhaust emissions has become a key challenge for energy and environmental engineers. Incomplete combustion of the air–fuel mixture leads to higher concentrations of carbon monoxide (CO), hydrocarbons (HC), and nitrogen oxides (NOx), which depend strongly on the design of the combustion chamber and the air–fuel ratio.

2.1 Vegetable Oils

Vegetable oils—such as rapeseed, olive, sunflower, and pongamia have been extensively studied as potential substitutes for diesel fuel. These oils are renewable, biodegradable, and can be derived from a wide variety of crops. Early studies demonstrated that vegetable oils possess heating values close to diesel, but their high viscosity results in poor atomization and incomplete combustion. To overcome this limitation, transesterification and blending with diesel have been explored as effective methods to reduce viscosity and improve spray characteristics.

Research on tree-borne oils such as Pongamia pinnata has confirmed their viability for large-scale biofuel production. These oils offer acceptable ignition quality and energy content, though issues related to deposit formation and injector fouling remain. Straight vegetable oil (SVO) has also been investigated as a low-energy-input biofuel option, but its long-term use tends to cause carbon buildup and lubricant contamination inside diesel engines. Despite these drawbacks, short-term tests have shown that vegetable oils can deliver satisfactory performance with reduced emissions when operated under optimized conditions.

2.2 Phytol as an Emerging Biofuel

Among the various bio-derived alcohols, phytol—a long-chain oxygenated alcohol—has recently emerged as a promising second-generation biofuel. It offers combustion properties similar to diesel, with a comparable cetane number and energy content but lower sulphur and aromatic content. Studies have indicated that blending small proportions of phytol with diesel can improve combustion efficiency and reduce particulate matter without requiring engine modifications.

Phytol's oxygen content enhances the oxidation of carbonaceous species during combustion, leading to lower CO and HC emissions. Additionally, its high boiling point and low vapour pressure promote stable combustion at elevated temperatures. However, at higher blending ratios, its high viscosity and lower heating value can adversely affect fuel atomization and increase specific fuel consumption.

Recent advances in biofuel technology have reinforced the importance of developing sustainable long-chain alcohols like phytol. These fuels not only improve combustion stability but also align with global decarbonization goals by reducing greenhouse gas emissions. Several contemporary investigations have highlighted phytol's potential as a renewable blending agent that combines the advantages of oxygenated fuels with the performance characteristics of diesel.

2.3 Inferences from the Literature Review

From the literature, several conclusions can be drawn:

1. Phytol is a viable alternative to diesel due to its physicochemical similarity and oxygen-rich structure.
2. Raw vegetable oils are unsuitable for direct use in diesel engines because of their high viscosity, which limits atomization.
3. Transesterification effectively reduces viscosity in bio-oils, while methyl esters are generally preferred over ethyl esters for stability.
4. Phytol–diesel blends can achieve lower CO and HC emissions with negligible engine modifications, though NOx mitigation requires further optimization.

3. EXPERIMENTAL INVESTIGATION

An experimental setup is made with necessary instrumentation to evaluate the performance and emissions of the diesel engine at different operating conditions. In this chapter, the apparatus used to conduct the experiments and the procedures followed in this study are described.

3.1. Materials and Methods

In this study the four stroke single cylinder diesel engine will run with different proportions of Diesel-Phytol fuel to evaluate the performance with emissions.

Figure 2: Four stroke single cylinder diesel engine.

3.2. Engine Specifications

Table 1. Kirloskar Four stoke single cylinder diesel engine specifications.

Cylinder	Single Cylinder
Stroke	4 Stroke
Brand	Kirloskar
Compression Ratio	16.51
Fuel Tank Capacity	6.5 L
Cubic Capacity	0.553 L
Bore X Stroke	80 mm X 110 mm
Speed	1500 rpm
Torque at Full Load	2.387 Kg-m
Crank Shaft Center Height	203 mm

continued on following page

Table 1. Continued

Cylinder	Single Cylinder
Type of Fuel Injection	Direct Injection
Overloading Capacity of Engine	10% of rated output
Lub Oil Sump Capacity	3.7 L
Size of bare engine	617X504X843 mm
Dry engine weight	130 Kg
Power Take-off	Flywheel End

3.3. Blending Process

Selecting the optimal combination of components to produce the final finished product is a critical economic consideration for both traders and refiners. Blending is a complex process that goes beyond simple mixing, as modern automotive fuel may consist of up to 15 different hydrocarbon streams, each influencing the final specification and overall cost.

Blending can be carried out through various processes, including in-line blending via a manifold system, batch blending in tanks, and onboard blending into marine vessels. Each method has its unique advantages, challenges, and requires specific expertise for efficient execution:

- In-line blending: Gasoline, distillates, jet fuel, and kerosene are blended by injecting proportionate amounts of each component into the main stream, where turbulence ensures thorough mixing.
- Additives: Components such as octane enhancers, metal deactivators, antioxidants, anti-knock agents, gum and rust inhibitors, and detergents are added during or after blending to impart specific properties not inherent in hydrocarbons.
- Onboard blending: This method allows suppliers and trading organizations to prepare cargo to required specifications without the need for onshore facilities, with specialists using laboratory hand blends and blending-specific software to achieve optimal results.

In this study, phytol and diesel were blended using a magnetic stirrer, and five different samples were prepared for experimentation.

Figure 3. Magnetic stirrer.

3.4. Fuel Preparation

Phytol blended with diesel at different proportions.

SAMPLE 1: Pure diesel.
SAMPLE 2: 2 wt.% of phytol added to diesel.
SAMPLE 3: 4 wt.% of phytol added to diesel.
SAMPLE 4: 6 wt.% of phytol added to diesel.
SAMPLE 5: 8 wt.% of phytol added to diesel.

Figure 4. Phytol alcohol

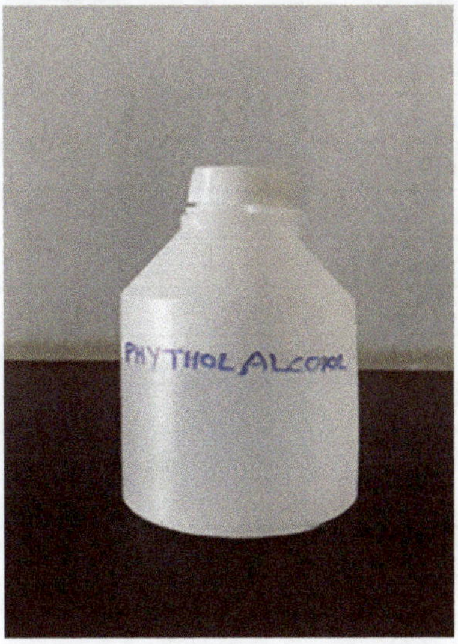

Figure 5. Blended samples of Phytol and diesel.

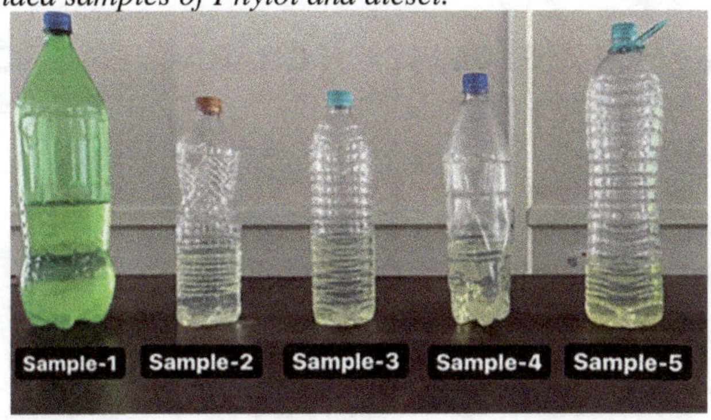

From the figure, sample 1 is 500 ml of diesel; sample 2 is 500 ml of diesel + 10 ml of phytol; sample 3 is 500 ml of diesel + 20 ml of phytol; sample 4 is 500 ml of diesel + 30 ml of phytol; sample 5 is 500 ml of diesel + 40 ml of phytol.

Table 2: Comparison of physical and chemical properties of phytol and diesel fuels

S.NO	Fuel property	Diesel	Phytol
1	Carbon content[wt%]	86.64	80.62
2	Hydrogen content[wt%]	13.01	13.5
3	Oxygen content[wt%]	0	6.05
4	Molecular weight[g/mole]	~170	296.54
5	Sulphur content[ppm]	11.2	<10
6	Heat of combustion[kJ/kg]	45,500	43,600
7	Heat of vaporization[kJ/kg]	361	130
8	Cetane number	47.7	45.9
9	Density @ 25^0 C[Pa]	849.2	850.9
10	Vapour pressure @ 25^0C[cSt]	1000	<1
11	Viscosity @ 25^0C [cSt]	3.775	63.54

Load testing was limited to 0, 3, 6, and 9 kg due to engine test-rig capacity constraints. This limitation is acknowledged and will be addressed in future studies with extended full-load testing

4.PERFORMANCE AND EMISSION CHARACTERISTICS

4.1 Sample 1

4.1.1 Performance Test

Table 3. Performance test readings for sample 1.

LOADING CONDITIONS	B.P(kW)	B.S.F.C(g/kW.hr)	B.The.E	Mech.E
NO LOAD	15.12	402.7	12	28.7
3 kg	15.79	398.7	14.3	29.2
6 kg	15.92	392.4	16.7	29.9
9 kg	15.1	387.7	19.2	31.4

4.1.2 Graph Represents Loading Conditions vs Brake Power

Figure 6. Variation of Brake Power with Load for Sample 1 (Diesel).

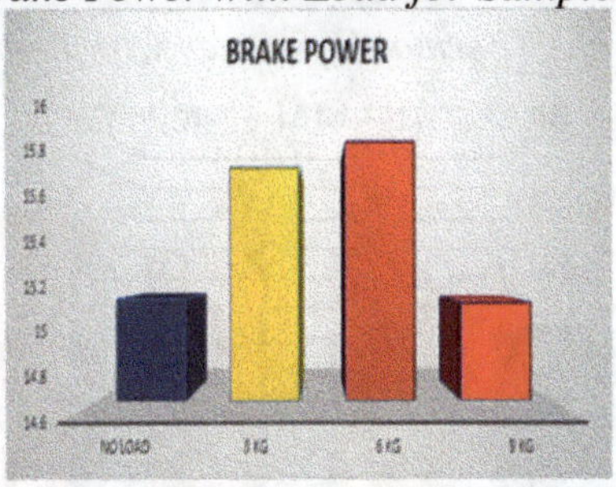

4.1.3 Graph Represents Loading Conditions vs B.S.F.C

Figure 7. Variation of Brake Specific Fuel Consumption (BSFC) with Load for Sample 1 (Diesel).

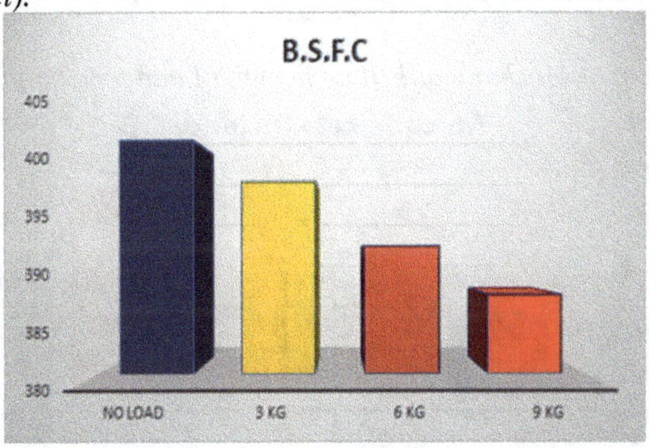

4.1.4 Graph Represents Loading Conditions vs Brake Thermal Efficiency.

Figure 8. Variation of Brake Thermal Efficiency with Load for Sample 1 (Diesel).

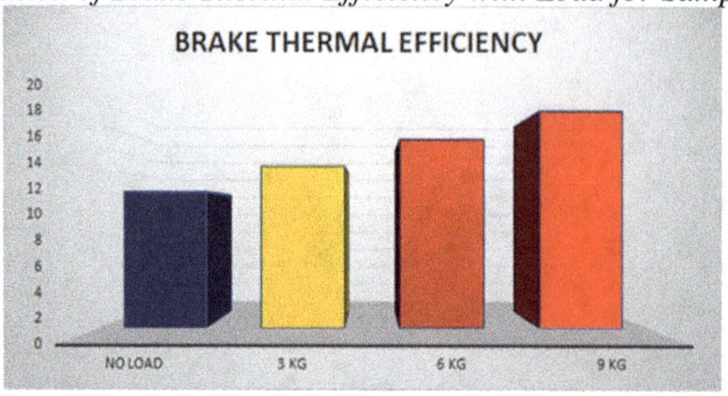

4.1.5 Graph Represents Loading Conditions vs Mechanical Efficiency

Figure 9. Variation of Mechanical Efficiency with Load for Sample 1 (Diesel).

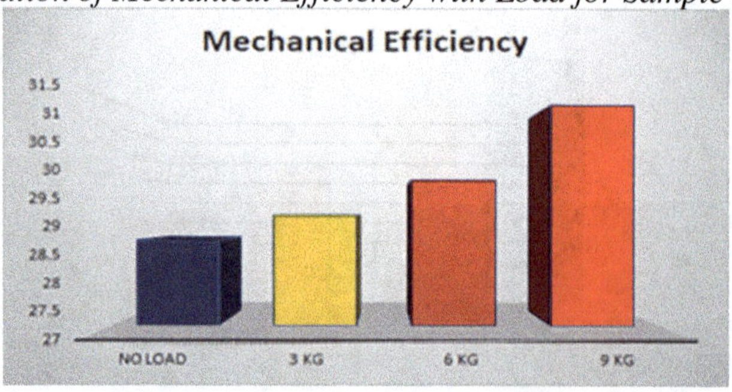

4.2. Emissions

Table 4. Emission characteristics of sample 1.

LOADING CONDITIONS	CO %	H.C(ppm)	NOX
NO LOAD	0.015	18	152
3 kg	0.017	18	177
6 kg	0.018	17	182
9 kg	0.02	15	194

4.2.1. Graph Represents Loading Conditions vs Carbon Monoxide

Figure 10. Variation of Carbon Monoxide (CO) Emissions with Load for Sample 1 (Diesel).

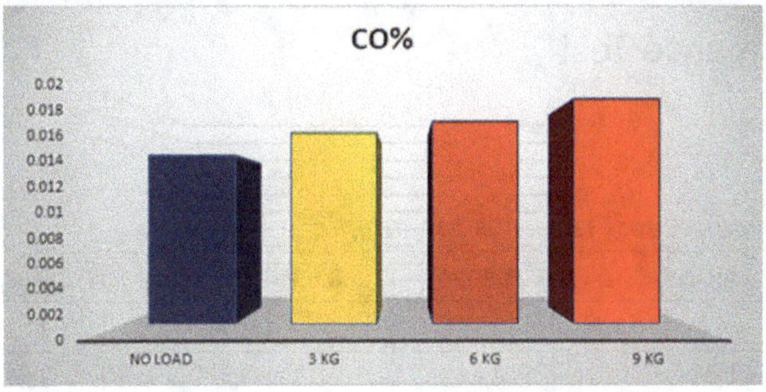

4.2.2. Graph Represents Loading Conditions vs Hydrocarbons (HC)

Figure 11. Variation of Hydrocarbon (HC) Emissions with Load for Sample 1 (Diesel)

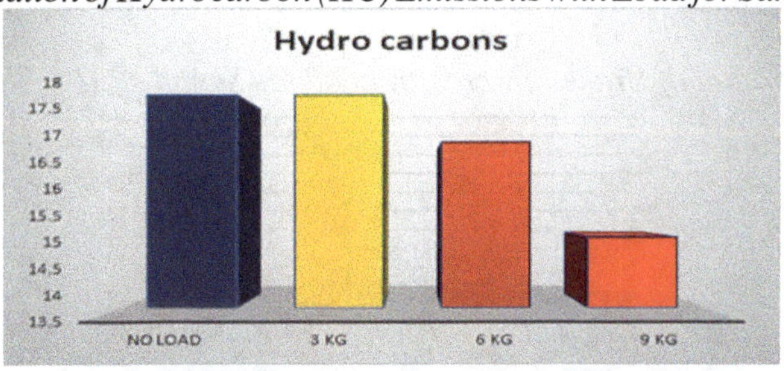

4.2.3. Graph Represents Loading Conditions vs Nitric Oxide (Nox)

Figure 12. Variation of Nitric Oxide (NOx) Emissions with Load for Sample 1 (Diesel)

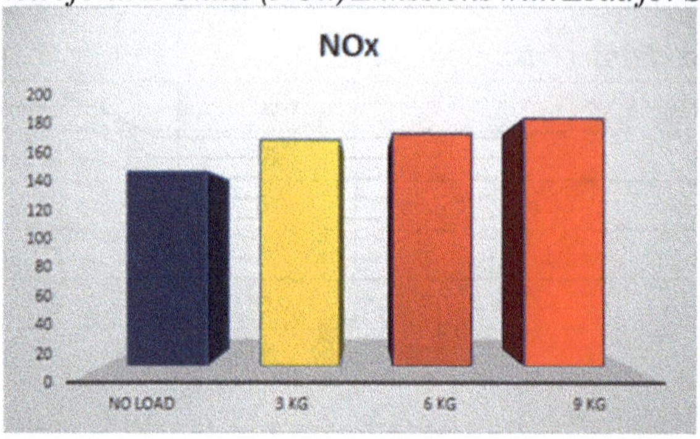

4.3. Sample 2

4.3.1. Performance Test

Table 5. Performance test readings for sample 2.

LOADING CONDITIONS	B.P(kW)	B.S.F.C(g/kW.hr)	B.The.E	Mech.E
NO LOAD	15.37	411.7	11.8	28.2
3 kg	15.42	397.7	15.7	28.8
6 kg	15.87	395.4	15.9	29.1
9 kg	15.9	397.7	16	32.1

4.3.2. Graph Represents Loading Conditions vs Brake Power

Figure 13. Variation of Brake Power with Load for Sample 2 (P2 Blend).

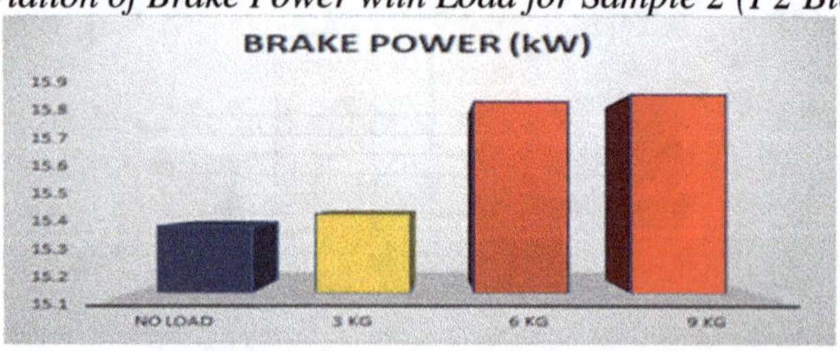

4.3.3. Graph Represents Loading Conditions vs B.S.F.C

Figure 14. Variation of Brake Specific Fuel Consumption (BSFC) with Load for Sample 2 (P2 Blend).

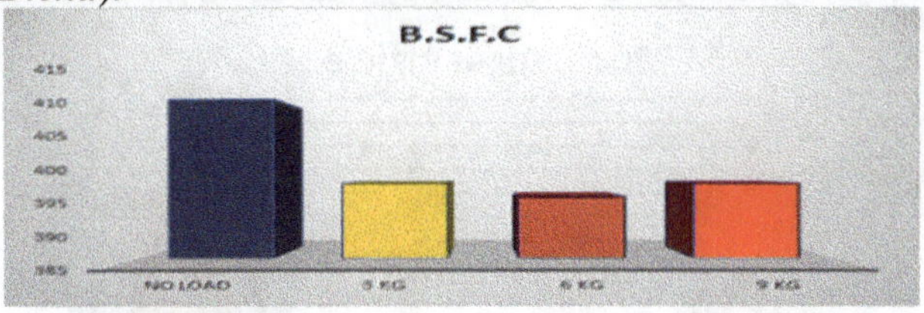

4.3.4. Graph Represents Loading Conditions vs Brake Thermal Efficiency.

Figure 15. Variation of Brake Thermal Efficiency with Load for Sample 2 (P2 Blend).

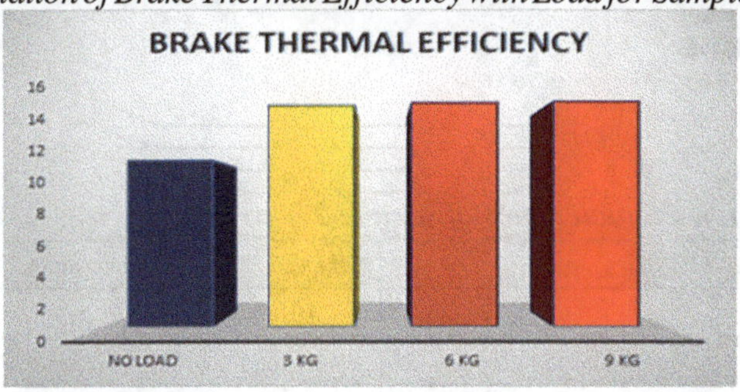

4.3.5. Graph Represents Loading Conditions vs Mechanical Efficiency

Figure 16. Variation of Mechanical Efficiency with Load for Sample 2 (P2 Blend).

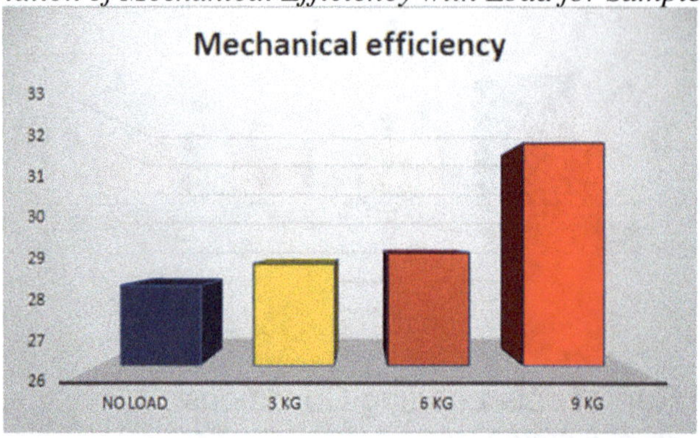

4.4. Emissions

Table 6. Emission characteristics of sample 2.

LOADING CONDITIONS	CO %	H.C (ppm)	NOX
NO LOAD	0.018	16	166
3 kg	0.016	15	173
6 kg	0.014	13	192
9 kg	0.015	11	193

4.4.1. Graph Represents Loading Conditions vs Carbon Monoxide

Figure 17: Variation of Carbon Monoxide (CO) Emissions with Load for Sample 2 (P2 Blend).

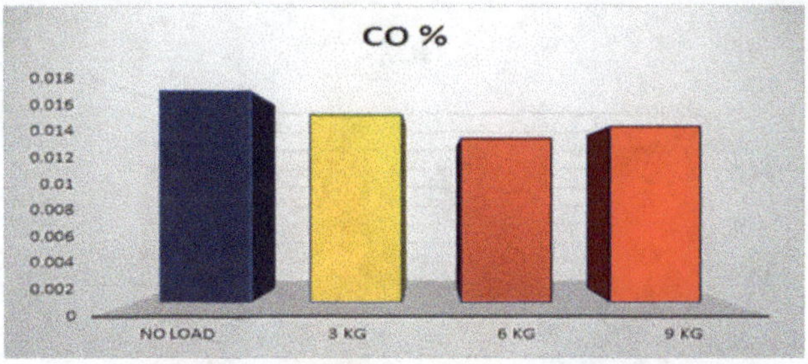

4.4.2. Graph Represents Loading Conditions vs Hydrocarbons (HC)

Figure 18: Variation of Hydrocarbon (HC) Emissions with Load for Sample 2 (P2 Blend).

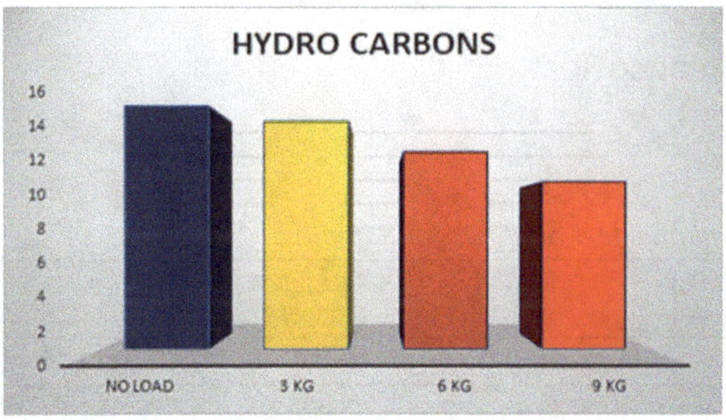

4.4.3. Graph Represents Loading Conditions vs Nitric Oxide (Nox)

Figure 19: Variation of Nitric Oxide (NOx) Emissions with Load for Sample 2 (P2 Blend).

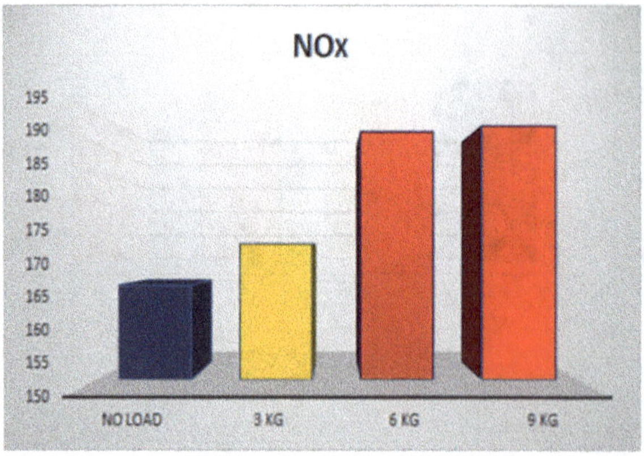

4.5. Sample 3

4.5.1. Performance Test

Table 7. Performance test readings for sample 3.

LOADING CONDITIONS	B.P(kW)	B.S.F.C(g/kW.hr)	B.The.E	Mech.E
NO LOAD	14.78	412.8	11.2	27.9
3 kg	15.21	388.8	15.2	28.2
6 kg	15.57	376.2	16	29.9
9 kg	16.2	375.3	16.1	31

4.5.2. Graph Represents Loading Conditions vs Brake Power

Figure 20. Variation of Brake Power with Load for Sample 3 (P4 Blend)

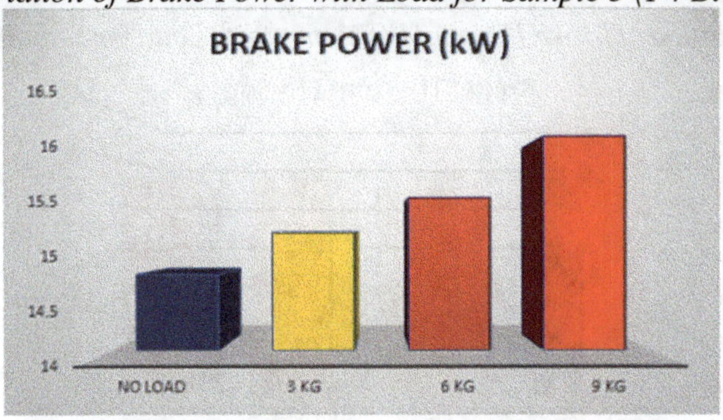

4.5.3. Graph Represents Loading Conditions vs B.S.F.C

Figure 21. Variation of Brake Specific Fuel Consumption (BSFC) with Load for Sample 3 (P4 Blend).

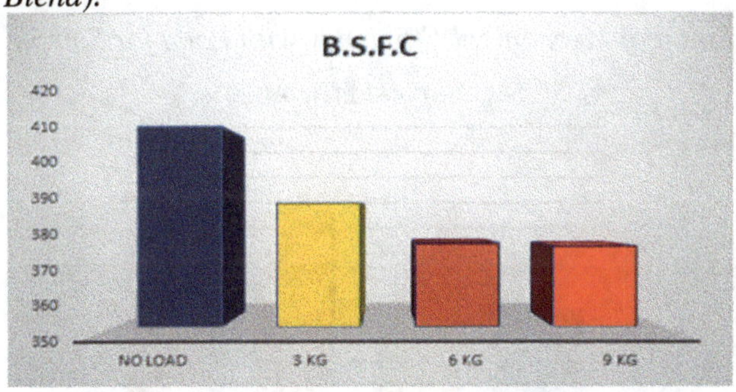

4.5.4. Graph Represents Loading Conditions vs Brake Thermal Efficiency

Figure 22. Variation of Brake Thermal Efficiency with Load for Sample 3 (P4 Blend).

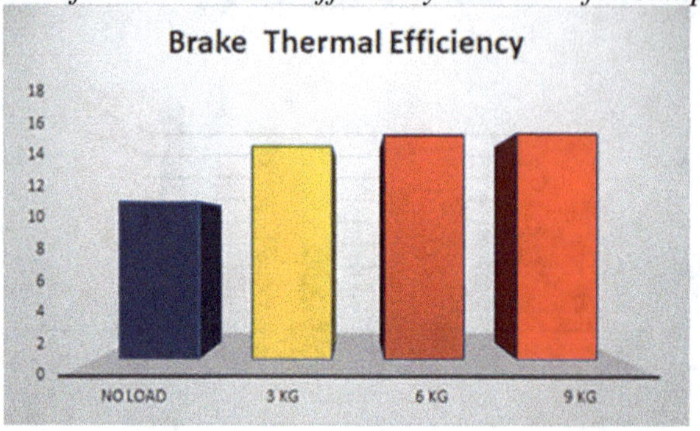

4.5.5. Graph Represents Loading Conditions vs Mechanical Efficiency

Figure 23. Variation of Mechanical Efficiency with Load for Sample 3 (P4 Blend).

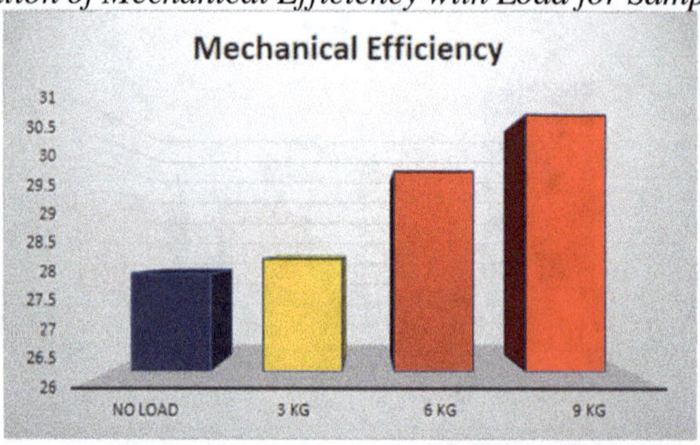

4.6. Emissions

Table 8. Emission characteristics of sample 3.

LOADING CONDITIONS	CO %	H.C(ppm)	NOX
NO LOAD	0.019	14	171
3 kg	0.02	13	182
6 kg	0.021	13	199
9 kg	0.02	10	210

4.6.1. Graph Represents Loading Conditions vs Carbon Monoxide

Figure 24. Variation of Carbon Monoxide (CO) Emissions with Load for Sample 3 (P4 Blend).

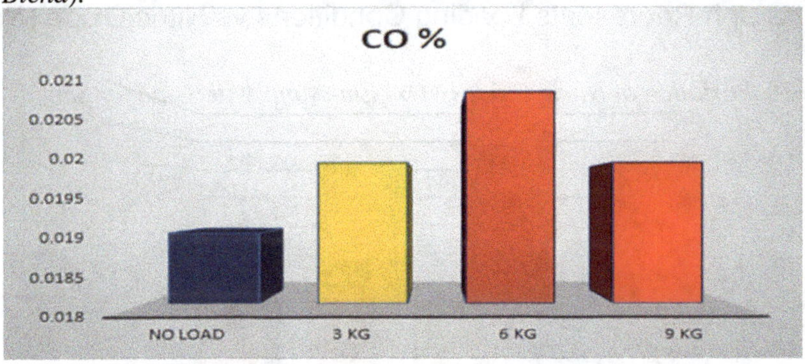

4.6.2. Graph Represents Loading Conditions vs Hydrocarbons (HC)

Figure 25. Variation of Hydrocarbon (HC) Emissions with Load for Sample 3 (P4 Blend).

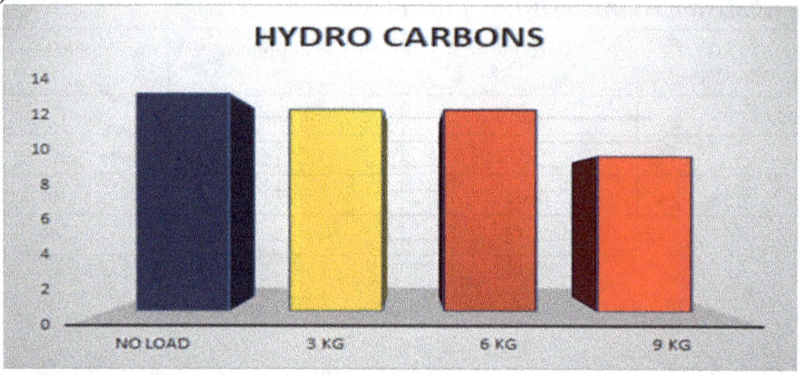

4.6.3. Graph Represents Loading Conditions vs Nitric Oxide (Nox)

Figure 26. Variation of Nitric Oxide (NOx) Emissions with Load for Sample 3 (P4 Blend).

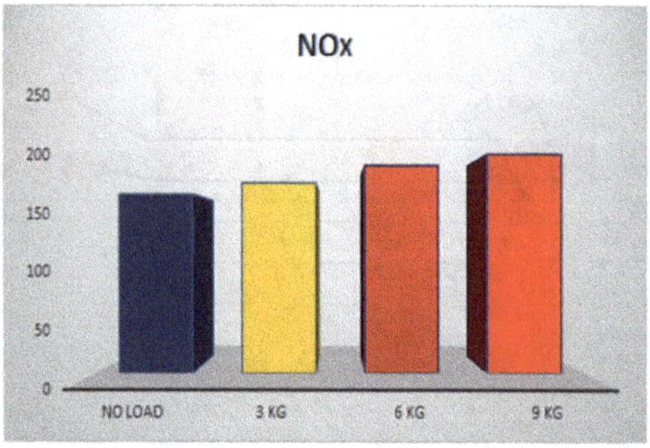

4.7sample 4

4.7.1. Performance Test:

Table 9: Performance test readings for sample 4.

LOADING CONDITIONS	B.P(kW)	B.S.F.C(g/kW.hr)	B.The.E	Mech.E
NO LOAD	14.66	400.1	10.9	26.7
3 kg	15.11	398.7	12.2	27.2
6 kg	15.67	381.2	14.9	28.4
9 kg	16.01	377.4	15.7	34.5

4.7.2. Graph Represents Loading Conditions vs Brake Power

Figure 27. Variation of Brake Power with Load for Sample 4 (P6 Blend).

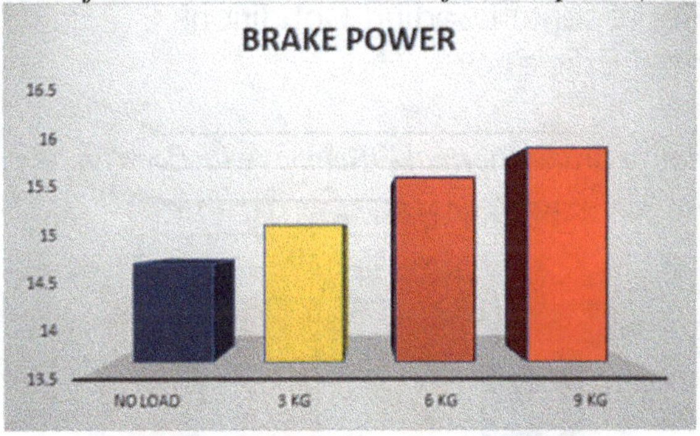

4.7.3. Graph Represents Loading Conditions vs B.S.F.C

Figure 28. Variation of Brake Specific Fuel Consumption (BSFC) with Load for Sample 4 (P6 Blend).

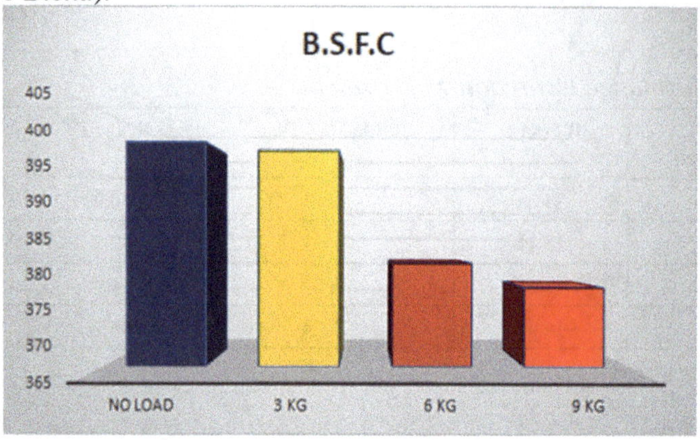

4.7.4. Graph Represents Loading Conditions vs Brake Thermal Efficiency

Figure 29. Variation of Brake Thermal Efficiency with Load for Sample 4 (P6 Blend).

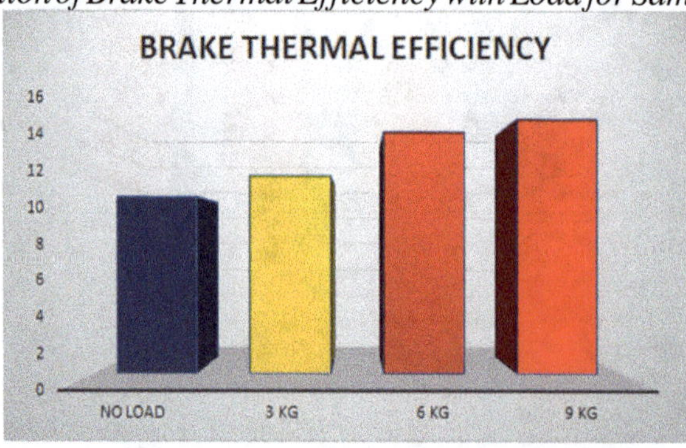

4.7.5. Graph Represents Loading Conditions vs Mechanical Efficiency

Figure 30. Variation of Mechanical Efficiency with Load for Sample 4 (P6 Blend).

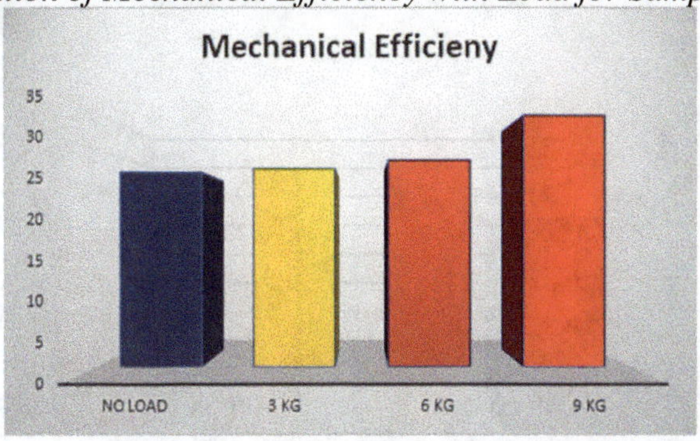

4.8 Emissions

Table 10: Emission characteristics of sample 4.

LOADING CONDITIONS	CO %	H.C(ppm)	NOX
NO LOAD	0.021	11	210
3 kg	0.022	11	250
6 kg	0.026	9	350
9 kg	0.02	8	520

4.8.1. Graph Represents Loading Conditions vs Carbon Monoxide

Figure 31. Variation of Carbon Monoxide (CO) Emissions with Load for Sample 4 (P6 Blend).

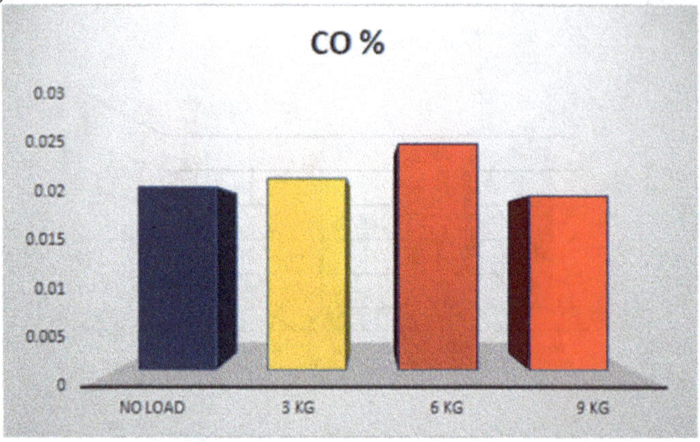

4.8.2. Graph Represents Loading Conditions vs Hydrocarbons

Figure 32. Variation of Hydrocarbon (HC) Emissions with Load for Sample 4 (P6 Blend).

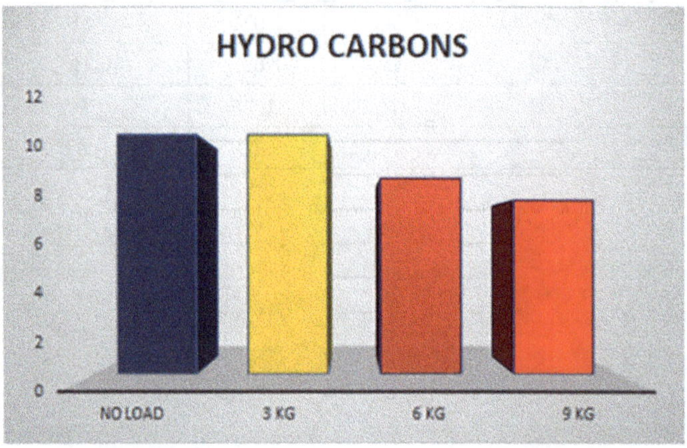

4.8.3. Graph Represents Loading Conditions vs Nitric Oxide (Nox)

Figure 33. Variation of Nitric Oxide (NOx) Emissions with Load for Sample 4 (P6 Blend).

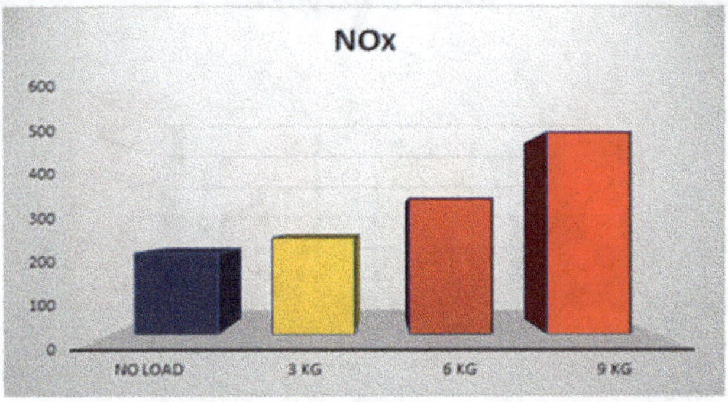

4.9 Sample 5

4.9.1 Performance Test

Table 11. Performance test readings for sample 5.

LOADING CONDITIONS	B.P(kW)	B.S.F.C(g/kW.hr)	B.The.E	Mech.E
NO LOAD	14.81	410.2	11.5	25.8
3 kg	15.72	407.8	12.7	26.9
6 kg	15.99	402.2	13.8	27.7
9 kg	16.61	389.7	16.1	29.1

4.9.2. Graph Represents Loading Conditions vs Brake Power

Figure 34. Variation of Brake Power with Load for Sample 5 (P8 Blend).

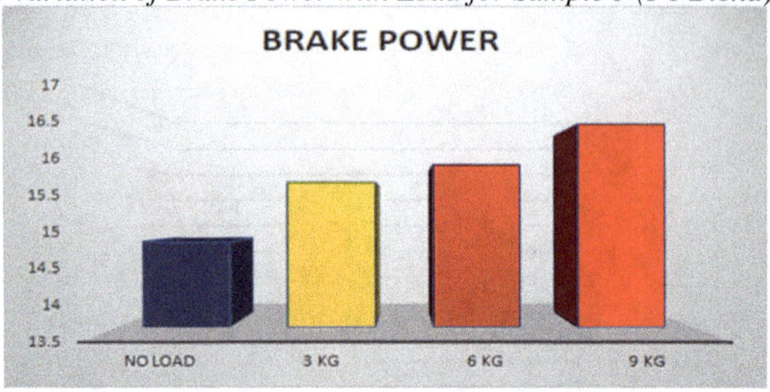

4.9.3. Graph Represents Loading Conditions vs B.S.F.C

Figure 35. Variation of Brake Specific Fuel Consumption (BSFC) with Load for Sample 5 (P8 Blend).

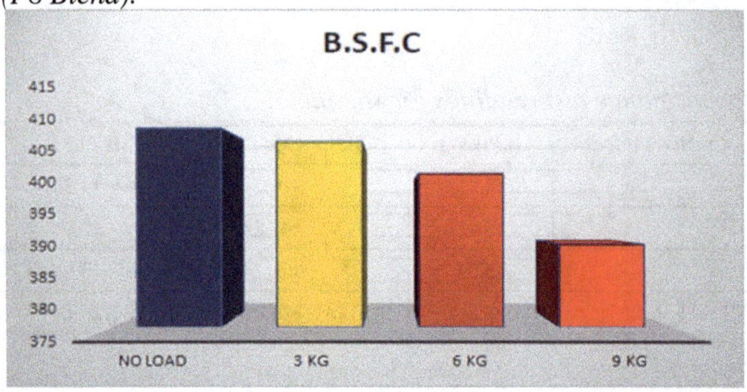

4.9.4. Graph Represents Loading Conditions vs Brake Thermal Efficiency

Figure 36. Variation of Brake Thermal Efficiency with Load for Sample 5 (P8 Blend).

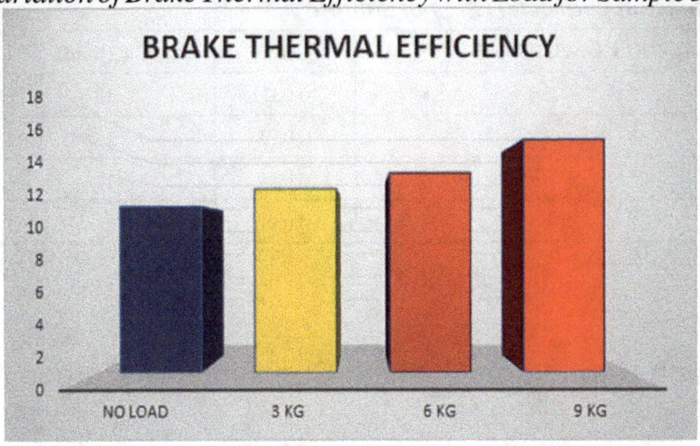

4.9.5. Graph Represents Loading Conditions vs Mechanical Efficiency

Figure 37. Variation of Mechanical Efficiency with Load for Sample 5 (P8 Blend).

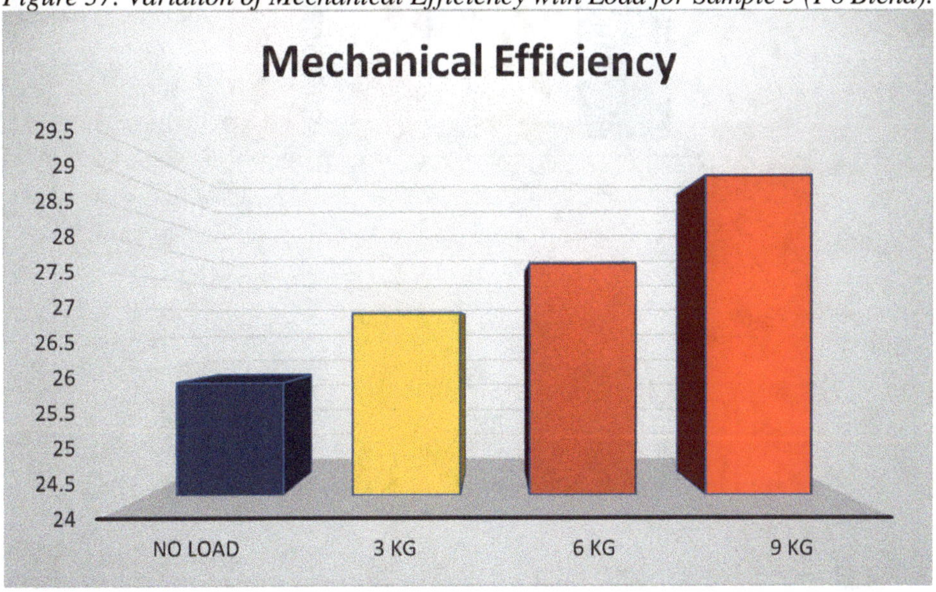

4.10. Emissions

Table 12: Emission characteristics of sample 5.

LOADING CONDITIONS	CO %	H.C(ppm)	NOX
NO LOAD	0.031	13	240
3 kg	0.039	12	300
6 kg	0.044	11	380
9 kg	0.048	8	610

4.10.1. Graph Represents Loading Conditions vs Carbon Monoxide

Figure 38. Variation of Carbon Monoxide (CO) Emissions with Load for Sample 5 (P8 Blend).

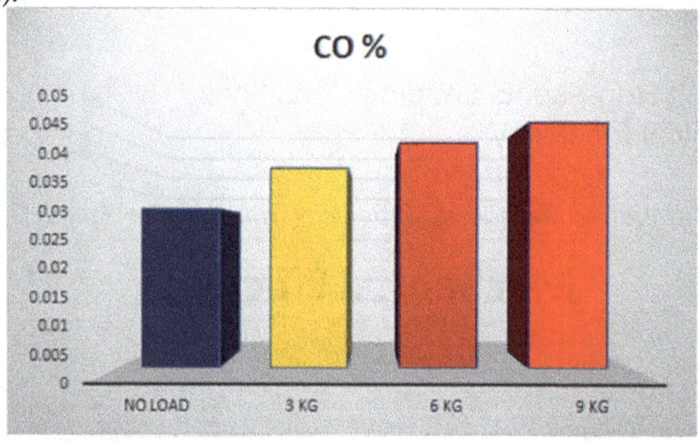

4.10.2. Graph Represents Loading Conditions vs Hydrocarbons

Figure 39. Variation of Hydrocarbon (HC) Emissions with Load for Sample 5 (P8 Blend).

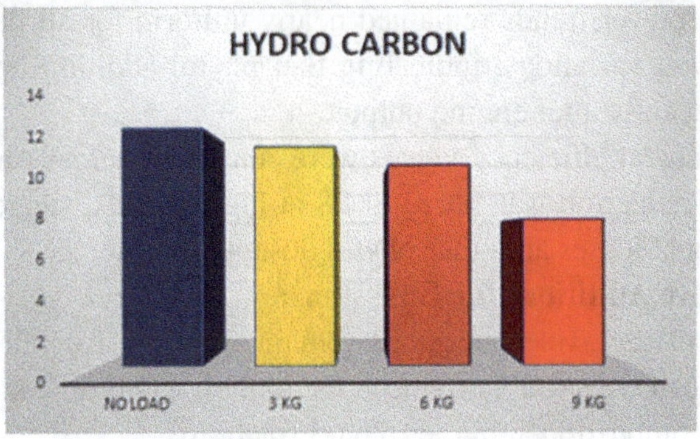

4.10.3. Graph Represents Loading Conditions vs Nitric Oxide (Nox)

Figure 40. Variation of Nitric Oxide (NOx) Emissions with Load for Sample 5 (P8 Blend).

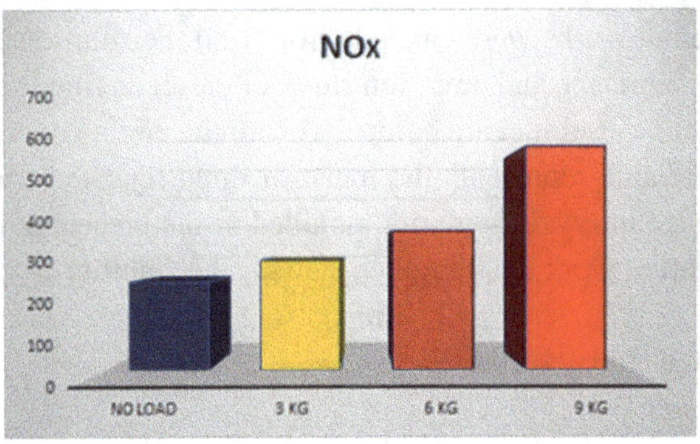

5. RESULTS AND DISCUSSION

Although individual graphs are presented for each blend, the following discussion draws direct comparisons across samples to highlight overall performance and emission trends,

Comparative Additions for Performance:

- When comparing all blends, P2 exhibited the closest brake thermal efficiency to diesel across all loads, whereas P6 and P8 showed reduced efficiency, particularly beyond 6 kg load.
- Brake power trends remained nearly uniform for all blends, with only marginal variations, indicating that phytol addition up to 8% did not significantly alter engine output.
- Mechanical efficiency values were consistent across samples, though a slight reduction was noted in P8 at higher loads, suggesting increased frictional losses at higher phytol dosages.

Comparative Additions for Emissions:

- Carbon monoxide emissions remained lowest for P2 at medium and high loads, while P8 produced the highest CO levels, indicating incomplete combustion at higher phytol fractions.
- Hydrocarbon emissions decreased progressively with increasing phytol content, with P6 and P8 showing the lowest HC values, reflecting the oxygenated nature of phytol.
- NOx emissions showed a rising trend with phytol concentration: P2 and P4 behaved similarly to diesel, whereas P6 and especially P8 recorded substantial increases at higher loads.

The experimental works were carried out to find the solution for objectives of present work. Performance and emission study of diesel engine fuelled with different blends of phytol-diesel fuel are compared with the engine operated pure diesel.

Each reported value represents the mean of three repeated trials. While error bars and confidence intervals were not included in the present study, they will be incorporated in future work to strengthen statistical reliability.

Performance characteristics:

Brake power across blends remained nearly constant, indicating that phytol addition up to 8% does not significantly alter power output due to comparable calorific value and density to diesel.

From the experimental results of performance and emission studies, suitable blend ratio of phytol-diesel fuel optimized. A detailed discussion of the performance, emission and studies are as follows:

Performance Results:

- **For Sample 1:** When 500 ml of pure diesel is used we got an avg brake power of 15.5kW, brake thermal efficiency of 15.55%, and mechanical efficiency of 29.8%.
- **For Sample 2:** When 500 ml of pure diesel and 10 ml of phytol is used we got an avg brake power of 15.64kW, brake thermal efficiency of 14.85%, and mechanical efficiency of 29.5%.
- **For Sample 3:** When 500 ml of pure diesel and 20 ml of phytol is used we got an avg brake power of 15.44kW, brake thermal efficiency of 14.62%, and mechanical efficiency of 29.25%.
- **For Sample 4:** When 500 ml of pure diesel and 30 ml of phytol is used we got an avg brake power of 15.36kW, brake thermal efficiency of 13.42%, and mechanical efficiency of 29.2%.
- **For Sample 5:** When 500 ml of pure diesel and 40 ml of phytol is used we got an avg brake power of 15.78kW, brake thermal efficiency of 13.52%, and mechanical efficiency of 27.37%.

The variation in brake thermal efficiency across blends can be explained by the dual influence of phytol's oxygenated structure and its physical properties. At lower blending ratios (P2), the additional oxygen content promotes more complete combustion, leading to efficiency values comparable to or slightly higher than neat diesel. However, as the phytol fraction increases (P6 and P8), its relatively lower heating value reduces the energy released per unit of fuel, while higher viscosity limits atomization and mixing with air. These factors collectively result in reduced combustion efficiency at higher blends. This trend reflects a typical trade-off observed in oxygenated fuel blends, where the benefits of enhanced combustion are counterbalanced by energy density and flow property limitations.

Emissions Results:

- **For Sample 1:** When 500 ml of pure diesel is used an avg of 0.017% of Carbon Monoxide(CO), 17 ppm of Hydrocarbons (HC), and 176 ppm of Nitric Oxide (NOx) emissions were released.
- **For Sample 2:** When 500 ml of pure diesel and 10 ml of phytol is used an avg of 0.015% of Carbon Monoxide(CO), 13.75 ppm of Hydrocarbons (HC), and 181 ppm of Nitric Oxide (NOx) emissions were released.
- **For Sample 3:** When 500 ml of pure diesel and 20 ml of phytol is used an avg of 0.02% of Carbon Monoxide(CO), 12.5 ppm of Hydrocarbons (HC), and 190.5 ppm of Nitric Oxide (NOx) emissions were released.
- **For Sample 4:** When 500 ml of pure diesel and 30 ml of phytol is used an avg of 0.02% of Carbon Monoxide(CO), 9.75 ppm of Hydrocarbons (HC), and 332.5 ppm of Nitric Oxide (NOx) emissions were released.

- **For Sample 5:** When 500 ml of pure diesel and 40 ml of phytol is used an avg of 0.04% of Carbon Monoxide(CO), 11 ppm of Hydrocarbons (HC), and 382.5 ppm of Nitric Oxide (NOx) emissions were released.

6. CONCLUSION

This work performs a study of a biologically derived heavy alcohol called phytol ($C_{20}H_{40}O$) and its potential for displacement of Petro diesel. First a property analysis of pure diesel, and 2%, 4%, 6% and 8% phytol/diesel blends were performed. The assessment of phytol's physical and chemical properties revealed that it may be a suitable agent for blending with diesel fuel for CI engine applications.

The performance test of all the samples resulted to the values of brake power, brake thermal efficiency, and mechanical efficiency. The difference between these parameters for all samples are negligible.

NOx and CO emissions resulting from the phytol/diesel blends for sample 2 and 3 are similar to those exhibited by the pure diesel (sample 1).NOx emissions of the blends of sample 4 and 5 make a lot of difference when compared with others. When comparing all the samples in the context of emissions, the emissions released by sample 2 are nearly identical with slightly lower values when compared with all other samples. A consistent trend in CO emissions is not apparent.

According to experimental work, differences noticed in performance and emissions between diesel and phytol/diesel blends are small – indicating that phytol may be suitable as a blending agent with diesel.

Further study should be done for proper blending of phytol with diesel for improvement in performance of diesel engine.

Some of the blends which can be used in place of phytol are ethanol blends, oxygenated blends, bio diesel blends etc.

7. FUTURE SCOPE

Fuel additives are chemical compounds designed to enhance the performance of fuels and address deficiencies in motor vehicles. They can be broadly categorized into three types: performance additives, distribution additives, and fuel quality additives.

Performance additives -improve engine performance by

1. Removing and preventing deposits in fuel injectors, valves, and combustion chambers.
2. Enhancing lubricity.

3. Improving combustion efficiency.
4. Enhancing low-temperature properties.
5. Reducing undesirable emissions.

Distribution additives are address issues such as corrosion, foaming, filter plugging, and microbial contamination. Fuel quality additives are maintaining fuel quality by inhibiting oxidation, reducing degradation, and improving long-term storage stability. Selecting the right combination of additives is crucial for ensuring engine durability and reducing tailpipe emissions.

The use of fuel additives has surged due to stringent environmental regulations aimed at limiting greenhouse gas (GHG) emissions. Regulatory bodies like the U.S. Environmental Protection Agency (EPA), the California Air Resources Board (CARB), and the European Union (EU) have driven the push for cleaner, more efficient fuels. Additional factors contributing to the growth of fuel additive usage include the adoption of gasoline direct injection (GDI) engines, the demand for ultra-low sulfur diesel fuel (ULSD), and the overall expansion of the automotive industry. This growth is expected to continue into the next decade.

Current Applications of Fuel Additives

According to the U.S. EPA, the typical passenger vehicle emits about 4.6 metric tons of carbon dioxide per year.2 With passenger vehicles composing 93% of the estimated 1.2 billion vehicles globally, the amount of carbon emissions from passenger vehicles alone climbs to a concerning total.3 Carbon emissions have risen rapidly since the Industrial Revolution, with global GHG emissions from the transportation sector at around 4.62 metric gigatons in 2016.4 Consequently, fuel additive manufacturers have focused on improving their additive formulations to meet fuel economy and emissions standards of commercial fuel blends.

The future of fuel additives will be shaped by sustainability targets, regulatory pressures, and techno-economic feasibility studies beyond 2025. For phytol–diesel blends specifically, future investigations should explore extended engine speeds, full-load testing, long-duration endurance studies, and comparative economic feasibility.

In addition to the above, further experimental investigations should explore the influence of varying engine speeds and operating temperatures on the performance and emission behavior of phytol–diesel blends. Since real-world diesel engines operate under a wide range of speeds and thermal conditions, assessing these parameters would provide deeper insights into ignition delay, brake thermal efficiency, and pollutant formation trends. Moreover, endurance tests under variable operating regimes will help to establish the long-term stability and compatibility of phytol as a blending agent. These future studies will extend the current baseline findings and support the practical application of phytol–diesel blends in commercial compression ignition engines.

In addition, future investigations should incorporate a comparative economic analysis between phytol production costs and conventional petroleum diesel prices. Such an evaluation, combined with life cycle and techno-economic assessments, will be essential for determining the commercial feasibility of phytol–diesel blends.

REFERENCES

Al-Harrasi, A.. (2019). Extraction and characterization of phytol from algal biomass. *Journal of Applied Phycology*, *31*, 1397–1407. DOI: 10.1007/s10811-018-1674-5

Altin, R., Çetinkaya, S., & Yücesu, H. S. (2001). Potential of using vegetable oil fuels as fuel for diesel engines. *Energy Conversion and Management*, *42*(5–6), 529–538. DOI: 10.1016/S0196-8904(00)00076-1

Appavu, V.. (2019). Performance evaluation of biodiesel–alcohol fuel blends in diesel engines. *Applied Energy*, *239*, 1–10. DOI: 10.1016/j.apenergy.2019.01.039

Ashok, B.. (2019). Assessment of long-chain alcohol biodiesel blends. *Renewable Energy*, *139*, 1354–1362. DOI: 10.1016/j.renene.2019.03.026

Atmanli, A. (2016). Comparative analysis of n-pentanol and diesel–biodiesel–alcohol blends. *Fuel*, *177*, 219–225. DOI: 10.1016/j.fuel.2016.03.029

Atmanli, A. (2016). Influence of higher alcohols on diesel/biodiesel performance. *Fuel*, *182*, 237–244. DOI: 10.1016/j.fuel.2016.04.050

Atmanli, A., & Yilmaz, N. (2018). Phase stability and combustion evaluation of 1-pentanol–diesel blends. *Energy Conversion and Management*, *166*, 332–339. DOI: 10.1016/j.enconman.2018.04.066

Atmanli, A., & Yilmaz, N. (2020). Energy and emission analysis of alcohol–biodiesel fuels. *Energy Reports*, *6*, 110–119. DOI: 10.1016/j.egyr.2019.11.010

Babu, D., & Anand, K. (2017). Experimental study of n-pentanol addition on emission characteristics of CI engines. *Fuel*, *196*, 486–495. DOI: 10.1016/j.fuel.2017.01.065

Balakrishnan, J. (2007). "Fuel cell technology." *Proceedings of ICIAFS*.

Barnwal, B. K., & Sharma, M. P. (2005). Prospects of biodiesel production from vegetable oils in India. *Renewable & Sustainable Energy Reviews*, *9*(4), 363–378. DOI: 10.1016/j.rser.2004.04.013

Buchner, A. J., Honnery, D., & Soria, J. (2017). Stability and three-dimensional evolution of a transitional dynamic stall vortex. *Journal of Fluid Mechanics*, *823*, 1–15. DOI: 10.1017/jfm.2017.379

Chivu, R. M., Roman, C., & Porumb, R. (2024). Assessment of engine performance and emissions with eucalyptus oil blends. *Energies*, *17*. Advance online publication. DOI: 10.3390/en17010013

Da Silva, F. J. A., & De Matos, J. E. X. (2009). A note on the potential of CNSL in fuel blends for engines in Brazil. *Revista Tecnologia, 14*(2), 53–58.

Dantas Neto, A. A., Silva, J. P., & Santos, G. A. (2014). Effect of biodiesel/diesel-based microemulsions on the exhaust emissions of a diesel engine. *Brazilian Journal of Petroleum and Gas, 8*(2), 1–11.

De Laurier, J. D. (1994). An ornithopter wing design. *Journal of Aircraft, 31*(6), 1318–1324. DOI: 10.2514/3.46704

Dwivedi, G., & Sharma, M. P. (2014). Potential and limitation of straight vegetable oils as engine fuel – An Indian perspective. *Renewable & Sustainable Energy Reviews, 38*, 1188–1208. DOI: 10.1016/j.rser.2014.05.015

Dwivedi, G. C. S. (2020). Hydrotreated vegetable oil and synthetic diesel for CI engines. *Energy Reports, 6*, 185–196. DOI: 10.1016/j.egyr.2020.02.032

El-Araby, R., Amin, A., El Morsi, A. K., El-Ibiari, N. N., & El-Diwani, G. I. (2018). Study on the characteristics of palm oil–biodiesel–diesel fuel blend. *Egyptian Journal of Petroleum, 27*(4), 519–528. DOI: 10.1016/j.ejpe.2017.07.001

Gaikwad, A., Yoon, S., & Park, S. (2024). Combustion enhancement and emission reduction in RCCI engine using green-synthesized CuO nanoparticles with Cymbopogon martinii methyl ester and phytol blends. *Industrial Crops and Products, 198*. Advance online publication. DOI: 10.1016/j.indcrop.2023.116627

Gedam, P. H., & Sampathkumaran, P. S. (1986). Cashew nut shell liquid: Extraction, chemistry and applications. *Progress in Organic Coatings, 18*(3–4), 179–205. DOI: 10.1016/0033-0655(86)80019-6

Geng, P.. (2020). Performance analysis of Fischer–Tropsch diesel blends in diesel engines. *Renewable Energy, 147*, 1721–1730. DOI: 10.1016/j.renene.2019.10.066

Geo, E., Mrad, N., Tazerout, M., & Aloui, F. (2012). Experimental analysis of biofuel as an alternative fuel for diesel engines. *Applied Energy, 96*, 282–289. DOI: 10.1016/j.apenergy.2012.02.027

Hansen, A. C., Zhang, Q., & Lyne, P. W. L. (2005). Ethanol–diesel fuel blends—A review. *Progress in Energy and Combustion Science, 31*(5), 419–445. DOI: 10.1016/j.pecs.2005.03.001 PMID: 15474927

Hoang, A. T., & Pham, V. V. (2025). Iso-pentanol–gasoline blends for HCCI engines. *Fuel Communications, 8*, 100083. DOI: 10.1016/j.fuelcom.2025.100083

Hoang, A. T., & Pham, V. V. (2025). *Optimizing combustion efficiency and emission reduction in alcohol-based fuel blends*. Fuel Communications., DOI: 10.1016/j.fuelcom.2025.100094

Islam, M.. (2018). Microwave-assisted hydrolysis of chlorophyll-rich biomass for phytol production. *Bioresource Technology*, *258*, 203–212. DOI: 10.1016/j.biortech.2018.03.051

Jin, C.. (2019). Low-temperature properties of biodiesel–pentanol blends. *Energy Procedia*, *159*, 224–229. DOI: 10.1016/j.egypro.2018.12.021

Johari, A., Singh, S., & Vidya, S. (2022). Engine performance analysis for diesel engine using hydrogen as an alternative fuel. *Materials Today: Proceedings*, *50*, 4478–4484. DOI: 10.1016/j.matpr.2021.12.462

Khan, R. A.. (2014). Industrial and biological relevance of phytol and its derivatives. *Industrial Crops and Products*, *53*, 29–37. DOI: 10.1016/j.indcrop.2013.12.024

Liang, X.. (2021). Effect of n-pentanol addition on atomization and combustion in biodiesel engines. *Applied Thermal Engineering*, *192*. Advance online publication. DOI: 10.1016/j.applthermaleng.2021.116930

Liger, M., Pornsin-Sirirak, N., Tai, Y.-C., Ho, S., & Ho, C.-M. (2002). "Large-area electrostatic-valved skins for adaptive flow control on ornithopter wings." *AIAA Conference Proceedings*.

Liu, W., Sun, L., Li, Z., Fujii, M., & Dong, L. (2022). Trends and future challenges in hydrogen production and storage research. *Environmental Science and Pollution Research International*, *29*, 33520–33541. DOI: 10.1007/s11356-022-19691-1 PMID: 32529621

Makareviciene, V.. (2013). Biodiesel from algae: Promising sustainable alternative fuel. *Environmental Science and Biotechnology Reviews*, *12*, 55–78.

Makareviciene, V., Skorupskaite, V., & Andruleviciute, V. (2023). Biodiesel fuel from microalgae—Promising alternative for the future. *Reviews in Environmental Science and Biotechnology*, *22*, 215–232. DOI: 10.1007/s11157-022-09634-3

Mueller, T. J., & De Laurier, J. D. (2001). "An overview of micro air vehicle aerodynamics." In *AIAA Fixed and Flapping Wing Aerodynamics*, pp. 15–52.

No, S. Y. (2016). Application of alcohols as alternative fuels for internal combustion engines. *Renewable & Sustainable Energy Reviews*, *54*, 1331–1345. DOI: 10.1016/j.rser.2015.10.044

Pan, M.. (2019). Combustion analysis of diesel–alcohol blends. *Energy, 172*, 154–163. DOI: 10.1016/j.energy.2019.01.055

Papagiannakis, R. G., Dinos, N., & Rakopoulos, C. D. (2017). Combustion characteristics of natural gas and alternative fuels in CI engines. *Fuel, 206*, 18–28. DOI: 10.1016/j.fuel.2017.05.094

Pradelle, F., Quiles, C., Martinez-Hernandez, E., & Feijoo, G. (2019). Recent developments in biodiesel combustion and emission control. *Energy Conversion and Management, 183*, 1–14. DOI: 10.1016/j.enconman.2019.01.012

Rachman, M. A., & Prasetyo, J. (2014). "Electricity from bio-waste-based gas hydrogen as a source of renewable energy biomass." *Proceedings of ICIEV.*

Rachman, M. A., & Prasetyo, J. (2022). Investigating the effects of biodiesel produced from *Crambe abyssinica* on combustion, performance and emissions. *Renewable Energy Research Journal, 10*(2), 123–134.

Rakopoulos, C. D., Rakopoulos, D. C., & Giakoumis, E. G. (2010). Comparative study of alcohol–diesel fuel blends. *Energy Conversion and Management, 51*(12), 2850–2858. DOI: 10.1016/j.enconman.2010.07.014

Rakopoulos, C. D., Rakopoulos, D. C., Hountalas, D. T., & Giakoumis, E. G. (2008). Combustion and performance characteristics of butanol–diesel blends in CI engines. *Energy Conversion and Management, 49*(5), 1090–1099. DOI: 10.1016/j.enconman.2007.10.007

Salvi, B. L., & Subramanian, K. A. (2015). Sustainable development of road transportation sector using hydrogen energy system. *Renewable & Sustainable Energy Reviews, 51*, 1281–1296. DOI: 10.1016/j.rser.2015.07.009

Sazali, N. (2020). Emerging technologies by hydrogen: A review. *International Journal of Hydrogen Energy, 45*(23), 12034–12045. DOI: 10.1016/j.ijhydene.2020.02.035

Sethin, A., Thirumalini, S., & Prakash, S. (2024). Effects of blended diesel–biodiesel fuel on emissions of a CRDI engine. *ACS Omega, 9*(3), 1005–1015. DOI: 10.1021/acsomega.4c00012

Shatnawi, N. A. Q., Aljaberi, M., & Aljaberi, N. (2018). "Hydrogen-based energy storage systems: a review." *Proceedings of ICRERA.*

Suhara, A., Kumar, P., & Reddy, A. "Biofuel sustainability: Review of progress and performance." *Green Energy and Sustainability*, 2024. URL: https://www.sciencedirect.com

Teixeira, A. C. R.. (2021). Alternative fuel technologies emissions for heavy-duty trucks: A review. *Environmental Science and Pollution Research International*, *28*, 34297–34334. DOI: 10.1007/s11356-021-14013-y PMID: 33686605

Warrick, D., Tobalske, B., Powers, D., & Dickinson, M. "The aerodynamics of hummingbird flight." *AIAA Conference Proceedings*, 2011.

Yilmaz, N., & Davis, S. M. (2016). PAH formation tendencies in higher-alcohol biodiesel combustion. *Fuel Processing Technology*, *150*, 123–131. DOI: 10.1016/j.fuproc.2016.02.011

Chapter 9
Geometry-Driven Optimization of Orifices for Multidisciplinary Engineering Applications:
Optimization of Orifices for Viscous Fluids

Santosh Kumar Panda
https://orcid.org/0000-0003-0606-6548
GIET University, India

Kali Charan Rath
GIET University, India

Balaji Kumar Choudhury
https://orcid.org/0000-0001-7587-3998
Parala Maharaja Engineering College, Berhampur, India

ABSTRACT

Flow measurement is a critical component across numerous industrial sectors. The orifice meter has become a preferred alternative to traditional wellhead chokes, offering enhanced control and precision in measuring mass flow rate. It is suitable for fluid flow measurements, utilizing in pressure drop and the discharge coefficient. This chapter explores orifice meters in detail, focusing on the impact of various shapes, sizes, classifications, and operational parameters on their performance for high viscous fluid flow applications.This chapter considered different orifices such as sector, crescent, segmental to determine the flow charteristics on the effect of

DOI: 10.4018/979-8-3373-3621-3.ch009

input parameters. It examines how changes in shape factor, width factor, and non-dimensional flow coefficient affect velocity, pressure drop and discharge coefficient of upside and downside of the flow. This study presents vital approaches into the more efficient design for high viscous flow application of orifice meters.

INTRODUCTION

Various types of flow measurement devices are utilized across different industries, including turbine meters, vortex shedding meters, Coriolis meters, and magnetic resonance meters, each suited to specific flow applications. Among these, the orifice meter is a common device used alongside instruments like venturimeters, venturi tubes, flow nozzles, and density meters to determine flow rates by analyzing fluid behavior. The orifice meter operates based on Bernoulli's principle, which explains the correlation between pressure and velocity in terms of energy within a flowing fluid. Bernouli's principle says the total energy (summation of pressure energy, kinetic and potential) in a location is constant. When fluid passes through an orifice, a pressure drop occurs, allowing for the calculation of mass or volumetric flow rate. This measurement technique is widely employed in sectors such as mining, chemical processing, oil and gas, nuclear energy, refrigeration, electricity generation, and the food industry. It is suitable for both single and multi-phase flows, making it a versatile and critical tool in these fields. In the petroleum industry, wellhead chokes are another flow regulation device inured to manage production rates from wells. However, these chokes often face challenges such as pressure shocks and downside turbulence, which can lead to inaccurate flow rate readings and potential damage to both the surface equipment and the reservoir, as noted by Gassan (1991). To mitigate these issues, orifice meters are implemented as an effective alternative for more accurate flow measurement and reduced operational disturbances.

Flow Measuring Device

Flow measurement devices play an important role for the determination of flow profile for a fluid. Types of flow measuring devices are described as follows

a. *Orifice Plates*

    The most common and simple flow measuring device is orifice which will be discussed in detail during this book chapter.

b. *Weir*

A weir is an obstruction used as an orifice for open fluid flow like reservoirs, dams, lakes, ponds and canals. Normally water (incompressible fluid) is used for weir flow for the irrigation and drainage application.

c. *Nozzle*

A nozzle is a flow passage that has varying cross section area used to increase the flow velocity with the expense of pressure drop. The types of nozzles are convergent, divergent and convergent – divergent used as per the required applications. The other application is to regulate the mass flow rate, speed, velocity, change of flow direction, etc. of a flowing stream.

d. *Venturimeter and Venturi Tubes*

Venturimeter is an obstruction type flow meter similar to a nozzle having a convergence and divergence part used to measure the flow with the help of Δp. The Δp is less in the venturimeter compared to the orifice meter but has a more complicated design structure.

e. *Pitot Tubes*

A pitot tube is a flow (velocity) measuring device used to measure the diversity along with static and dynamic pressures at stagnation point of flowing fluid. Bernoulli's theorem is applied to predict the velocity of the flow for small and larger ducts. After calculating the velocity with the multiplication area, mass and volume flow rate is determined.

f. *Rotating Meters*

Rotating meters are classified into two types: Turbine meter and paddle wheel meter. Both the meter uses the linear velocity of a rotating element to measure the flow with the impact of fluid flowing velocity or actual flow rate of flow. This blade or impeller is mounted on a shaft with the support of a bearing at the center and locked with a key. When the fluid affects the tip of the blade, the impeller and shaft etc., rotates with the fluid force. The rotating blades create angular velocity and are converting to flow quantity with the calculation. This device is more accurate for gas, liquid and gas-liquid flow.

g. *Vortex Shedding Flow Meters*

Bluff body with vortex, vortices frequency and wake, concept is used for measurement of flow velocity. The vortices are directly proportional to velocity of flow and mass flow rate.

h. *Magneto Flow Meters*

Magnetic resonance meters drive on the principle of induction. A conductive liquid surrounded with a magnetic field along the flow to measurable signal which is relative to flow velocity.

i. *Coriolis effect Meters*

This type of flow meters is affected with the use of the Coriolis effects (means motion mechanic of the fluid) which is directly related to measure the mass and volume flow rate.

j. *Hot Wire/Film probes*

The low magnitude flow velocity of air, liquid and gas flow rate is quantified with the help of hot wire or hot film anemometer. The principle of this type of flow device is that a resistance wire is placed along the flow direction with a constant current flow and the velocity will be measured with the resistance of the wire. Based on the property of the material the velocity of the flow will be measured. 2 – Wire and 3 – wire anemometer is a more popular measuring device for the velocity measurement.

Introduction to Orifice Plate

The operational parameters of fluid flow are classified into two types such as positive displacement and Inferential. Positive displacement fluid flow meters which consist of parts, such as the primary component, which has direct contact with the fluid interface and the secondary part through which fluid passed gives an output signal as volume flow rate. Positive displacement flow meters are very accurate instruments used for low flow rates. These devices are unaffected by the fluid properties i.e. density, viscosity and floating nature. Reciprocating piston, rotating piston, rotating disk, sliding and rotating vanes, and gear and lobed impeller are the categorists of positive displacement type appropriate for smaller flow rates. The advantages of such a flow meter are that it is used for high viscous and corrosive types of fluids. Orifice Plates, Flow Nozzles and Venturi Tubes, Pitot Tubes, Rotating meters, Vortex Shedding Meters, Magnetic Meters, Coriolis Meters are known as Inferential types. The flow meter works with the principle of flow phenomenon. Nozzle and venurimeter is the primary interfacial device with 60% higher flow rate than the orifice meter for a particular flow condition. For a particular pressure drop, a smaller diameter nozzle or venturimeter gives a better flow rate but the devices are complicated and more expensive than orifice meters. In the comparison of the percentage of p of different interfacial type flow meters along the orifice flow for a particular diameter ratio, and found orifice plate has utmost pressure drop and guides better results for flow accuracy Emerson, (2017) over other flow measuring devices. The discharge through an orifice is presented in equation (1) which describes Bernoulli's equation with no energy loss during the flow.

Figure 1. Measuring system used with orifice.

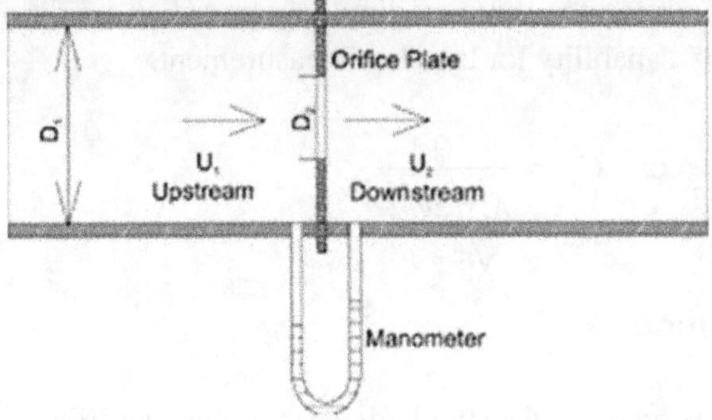

The Figure 1 illustrates the geometric configuration of an orifice mounted within a pipeline. As fluid flows through the system, it first approaches the orifice from the up side. Once it reaches a certain level, the fluid continues past the orifice and moves downstream. During this transition, there is a sharp drop in pressure accompanied by an increase in velocity. After some distance downside of the orifice, the pressure begins to recover in what is referred to as the expansion zone. A differential manometer is typically used to measure the Δp along the orifice by observing the height difference of the manometric fluid. Pressure tapping points are positioned at a distance of $2D_1$ upside and $0.5D_1$ downside from the orifice plate, where D_1 represents the pipe diameter. The narrowest flow area within the orifice meter is called the vena contracta. One of the key advantages of the orifice meter compared to other flow measurement devices is its ability to precisely determine flow characteristics, making it useful for calibration purposes. Orifice meters are simple in design, easy to operate, cost effective, and contain no moving components, which enhance reliability. Thanks to this straightforward construction, they do not require frequent calibration and can operate effectively under a wide range of temperature and pressure conditions. However, there are certain limitations. The output from an orifice meter is inherently nonlinear due to the turbulence created downside. Additionally, over time, the sharp edges of the orifice may wear down, altering the flow profile. Other influencing factors, such as the pipe configuration, flow characteristics, and type of fluid, can affect the flow coefficient making this an ongoing area of interest in research and development.

Orifice devices have more reliability, range ability and repeatability for wider applications. The flow coefficients of the orifice meter are having uncertainty of ± 0. 5 wide range of orifice diameter ratio. This uncertainty further enhances if the diameter ratio is below 0.2 and higher than 0.7. Similarly, the uncertainty gives minimum within a Re range of 4000 to 100000, which is the wide range and reliability for the flow and geometrical application. The operational life time of orifice is 5000 to 17000 m^3/hour with maintaining ± 1% error. Due to simple construction, orifice plates are giving the same reading for same flow conditions, which gives more repeatability capability for the flow measurement.

$$C_d = \frac{\dot{Q}}{\frac{A_1 A_2 \sqrt{2g\frac{\Delta p}{\rho}}}{\sqrt{A_1^2 - A_2^2}}} \quad \text{or} \quad C_d = \frac{\dot{Q}}{\frac{A_2 \sqrt{2g\frac{\Delta p}{\rho}}}{\sqrt{1 - \beta^2}}} \tag{1}$$

History of Orifice

The ISO 5167 series provides standardized guidelines and dimensions for designing orifice flow meters. Following this standard eliminates the need for calibration when measuring fluid flow along an orifice. The concept of applying differential pressure for flow measurement dates back to ancient times. Sextus Julius Frontinus, a former Governor of Britain and Water Commissioner of Rome, first introduced the idea to address the city's water management issues by utilizing pressure differences in fluid flow. In 1738, Daniel Bernoulli, a mathematics professor in St. Petersburg, formulated Bernoulli's theorem, laying the theoretical foundation for flow measurement. Building on this, Giovanni Battista Venturi, a professor at the University of Modena, invented the venturimeter in the 18th century. His experiments demonstrated that fluid velocity increased in the converging section of the device, while a significant drop in static pressure occurred after the throat, highlighting the principle of differential pressure for flow measurement. The orifice meter was first conceptualized by Venturi in 1797 but was practically applied by Clemons Herschel in 1887 at the Holyoke Water Power Company to manage water flow in a 20-inch pipe. Although Herschel was unaware of Venturi's orifice device work, his application became a reference for future studies. In 1890, Forrest M. Towl utilized orifice meters to assess the mass flow rate of natural gas in Columbus, Ohio. Later, around 1904, Thomas R. Weymouth conducted comparison experiments using orifice plates with flange taps and Pitot tubes. Although initially unsuccessful and unpublished, the research laid groundwork for further developments. By 1910, the Alkali Company achieved valuable findings from similar studies. In 1990, Professor S.W. Robinson from Ohio University implemented orifice meters to study gas pressure behavior in Columbus, Ohio. Additionally, experiments in Pennsylvania contributed to the development of flange tap coefficients used in flow analysis. Throughout the 20th century, extensive research by the American Gas Association (AGA) and the American Society of Mechanical Engineers (ASME) culminated in detailed publications such as "History of Orifice Meters" and "The Calibration, Construction, and Operation of Orifices for Metering," offering comprehensive insights into orifice meter development and application.

Types of Orifice

Figure 2(a) and 2(b) demonstrates orifice plates for different flow measurement appliances for different ranges of flow rate. According to the shape sharp edges, concentric, Eccentric, Segmental, Quadrant, orifice are used. Sharp edge type orifice is commonly used for a wide range application because of accuracy of measurement for low viscous fluids such as water, air and gas and high vicious fluids like lubricant, non-Newtonian fluids. Eccentric, Segmental, sector orifice plates are used for non-abrasive, solids and small amounts of liquids flows, like chemical manufacturing, nuclear plants, steel and paper industries. The quadrant edge and conic orifice is also used for viscous fluids. The inlet tip of the conic orifice is bevel type shape and follows a counter-bored passage along the thickness. These types of orifices play an impact on the area rato of the orifice and other applications as per the requirements. The wake created downside of the orifice also impacts on the shape of the orifice on the determination of Δp and the flow coefficient.

Flow through high viscous or sediment laden fluid flow is a social and industrial application to measure and control the flow. The orifice profiles recommended for such type flow are sector, crescent and segmental. The present chapter considers the above type applications. The sector, crescent and segmental orifices are presented in Fig. 2(b). A sector orifice is having the opening like a sector of a circle in the bottom portion. The sector orifice defines the sector angle of the sector opening. The flow rate of the orifice is measured by the difference of the pressure in the upside and downside of the flow. The area ratio of the orifice is directly proportional to the sector angle. The Δp also influences on the flow parameters which related to Re. width ratio is another parameter which affecting pressure drop as per the literature review study. The key parameters effects the flows through orifice are sector angle, Re and thickness ratio. A crescent orifice is another profile of orifice used for high viscous flow. This is a non-axisymmetric shape of opening in the bottom side of the orifice plate. The profile has two radius of curvature inner and outer to shape the orifice shape. The radius of curvatures of the orifice shape provides the orifice opening or area. The shape ratio varies on the variation of the orifice area to pipe area. The pressure drop depends on the shape ratio, width ratio, Re. Segmental Orifice Plate is a specialized flow measurement device designed for fluids containing solid particles, vapors, or water condensates. Featuring a segment-shaped orifice, this plate minimizes the accumulation of solids or gases in front of the orifice, enhancing measurement accuracy. This design is particularly effective in ensuring free flow for solids in liquids or gas bubbles in liquids, making it ideal for various challenging fluid conditions.

The flow through the crescent orifice creates a pressure difference, which is employed to compute the flow rate. The crescent shape helps to better accommodate the complexities of two-phase flows, improving the accuracy of the pressure differential measurement. The flow characteristics, such as the void fraction (the X of the mixture) and liquid velocity, are affected by the crescent orifice. The orifice is a non-axisymmetric type shape depends on two curvatures named inner and outer. Segmental Orifice Plate is a specialized flow measurement device designed for fluids containing solid particles, vapors, or water condensates. Featuring a segment-shaped orifice, this plate minimizes the accumulation of solids or gases in front of the orifice, enhancing measurement accuracy. This design is particularly effective in ensuring free flow for solids in liquids or gas bubbles in liquids, making it ideal for various challenging fluid conditions.

Figure 2. Types of orifice plates according to the shape of orifice.

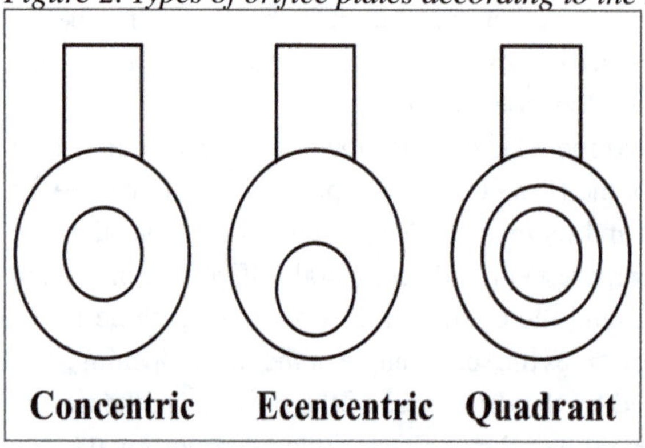

(a)

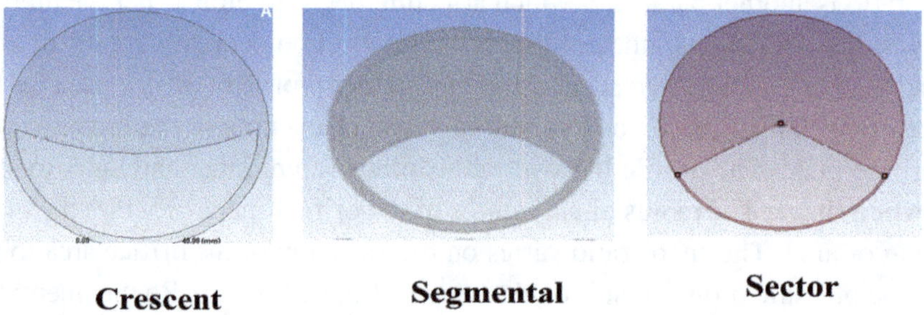

(b)

Figure 3. Types of orifice based on thickness of the orifice.

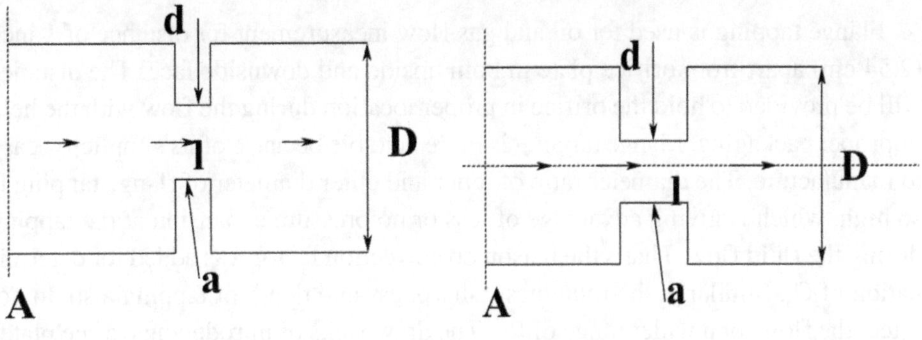

Based on the thickness of the orifice, it is categorized as a thin and thick orifice. Different researchers such as Fossa (2002), Fossa et al. (2006), Roual and Das (2010) classify the orifice based on the geometrical parameters thickness to diameter ratio (s/d). If the s/d ratio is less than 0.5, then the orifice is thin and if the s/d ratio is more than 0.5 then it is thick, as shown in Figures 3(a) and 3(b). The expansion length of orifice is also considered for the above categorization of the orifice. If the orifice length is less than the expansion length of the flow then it is defined as thin orifice, otherwise, the orifice is a thick orifice. This classification plays an important parameter in the stability and accurateness of the determination of the flow rate on two phase flows and highly turbulent flows. The orifice is further classified as per the interaction edge of flowing fluids. The classifications are Square, square with 45°, Sharp, streamlined approach, Slopping and Quadrant edges. The material properties required for orifice plates have good rigidity, non-corrosion, and non-erode against flowing fluid. Erosion can damage the edges of the orifice which affects flow coefficient, pressure drop, smoothness of flow, accuracy in measurement and contraction coefficient. The expansion of fluid in the downside affects the flow expansion which causes loss of pressure energy. To overcome the problems high thermal expansion materials have to be chosen. Different types of materials are used like stainless steel, carbon steel, and titanium, for improving strength Inconel, brass, copper, and Aluminum can be considered two phase flow where heat transfer phenomena take place. Glass, ceramic, different plastic also used as orifice material in visualization, corrosive free reactive fluids.

Orifice Taping or Pressure Tapping

The orifice ISO manual, ISO 5167 reported three types of pressure tapping like flange tapping, D and D/2 tapping and corner tapping

Flange Tapping

Flange tapping is used for oil and gas flow measurement for distance of 1 inch (2.54 cm) apart from orifice plate in both upside and downside face. The distance will be provided to hold the orifice in proper location during the flow with the help of proper packaging. Flange tapping is more suitable because of its simplicity, easy to manufacture. The diameter ratio of inner and outer diameter of flange tapping is so high, which is giving advantage of less or no pressure expansion at the tapping during the fluid flow. That's the reason no correction factor was added for determination of C_d. Similarly, the roundness, sharpness and depth of tapping also do not affect the flow for a wider range of Re. The drawbacks of introducing orifice plates with flange tapping make the flow measuring device more complex. The size of 50 mm or lesser diameter flange tapping affects the measurement of C_d because of pressure energy recovery in the downside zone and vena-contracta region.

D and D/2 Tapping

This type of tapping is positioned 1D distance in the upside and D/2 distance downside apart from the orifice plate. The 1D distance minimizes the effect of flow disturbance that occurs in the upside and downside of orifice because of wake created due to sudden expansion in the flow passage. The fabrication and joining process such as welding creates protuberance of the flange, pipe and orifice disturbed the neighborhood flow velocity. Proper care has to be taken to welding protuberance during fabrication with careful and skillful machining of flange. The flow disturbance can be avoided with correctly placed tapping as mentioned in different databases for a wide range of specifications. The specifications are mentioned as follows: ISO 5167-2: 2003 provides the flange taping diameter is 0.13D and 1.3 cm. Equivalent API standard, informs the tapping diameters should be 0.95 cm (3/8 inch) for 2 inch and 3-inch pipes and 1.27 cm (1/2 inch) for 4-inch pipes and bigger API (2000). Similarly, European data and the American data for tubes of API specification are 0.04D means0.95 cm (3/8 inch). Flange tapping spacing for upside and downside are

(2.54 cm $\pm$ 0.05) cm when $\beta > 0.6$ or $D < 15$ cm

and (2.54 cm $\pm$ 0.1) cm $\beta \leq 0.6$, $\beta > 0.6$ or 15 cm $\leq D \leq$ 100 cm.

Corner Tappings

Corner tappings are another parameter concern for flow along the orifice. Corner tappings mean the location between the orifice plate and the pipe wall. As per the guideline of ISO 5167-2: 2003 single tapping is recommendable for clean fluids and gas flow, but for sewage or high viscous fluids the more rapid corner tapping is recommendable. The tapping size of 2 mm is better for orifice design up to diameter less than 100 mm. The multiple tapping is used for the orifice and flange but each tapping should connect with each other with some arrangement. Fluid flow like hazards, reactive or toxic flows the tapping should give at the top half of the pipe to avoid damages.

LITERATURE REVIEW

A brief literature review presented on the orifice is reported by Panda et al., (2022) and Reader – Harris (2015). Literature in this field predominantly focuses on the geometry and dimensions of orifices, particularly regarding their applicability across various fluids and industrial sectors. When dealing with high-viscosity or sediment-laden flows dominated by gravity, flow measurement presents specific challenges. In such cases, traditional circular orifices are often inadequate. To address these limitations, alternative shapes like crescent, segmental or sector orifices have been introduced to better accommodate sediment flows and assess critical flow constraints such as mass flow rate, Δp, pressure distribution, and discharge coefficient (C_d). Sector orifices, with their openings positioned at the bottom of the pipe, enable the free passage of sediment-laden materials, reducing the risk of upside accumulation. The angle of the sector opening directly influences the flow area and is varied to optimize performance. These orifices are increasingly being adopted as substitutes for conventional concentric plates. Studies have used a pipe with a 40 mm diameter, examining diameter ratios between 0.2 and 0.8 and length-to-diameter ratios from 0.2 to 0.9, Yadav et al., (1995). Sector angles ranging from 72° to 288° were analyzed to determine $C_{d,}$ The study also developed a correlation expressing the discharge coefficient as a function of Re for a given ratio of the orifice area to pipe area as area ratio. In a similar vein, elliptical orifice plates have also been investigated for semi-metallic fluid flows. Experimental studies were carried out using elliptical orifices installed in a 38.1 mm circular pipe, oriented in horizontal, vertical, and angular positions. The diameter ratios explored ranged from 0.2 to 0.8, with water as the working fluid. Key factors such as the expansion length (distance to the vena-contracta), length to diameter ratio (l/d), area rato ratio, and orifice displacement angle were used to evaluate the downside flow characteristics. The sector orifices are studies further studied by Panda (2023) numerically to predict flow charteristics and C_d. The numerical study conducted by considering for θ (30° - 180°), Re (10000 -100000), and s/d (0.1 – 0.9). Air is choosen as a working medium through sector orifice flow. The study was further extended by developing a correlation between Δp and Re, θ and thickness ratio. Panda and Nepak (2023) conducted numerical studies on crescent shaped orifices to optimize their geometric features by analyzing fluid flow behavior. Xue et al. (2021) used crescent orifices as flow meters in gas-stratified flow conditions to examine how irregular geometries affect performance, especially in cases involving high-volume fractions. Segmental orifice is used in a bend of pipe to measure the flow rate. Laser Doppler Annemeter (LDA) and a segmental orifice used for a comparison study with ultrasonic and electromagnetic flow meter for effect of flow rates Straka et al., (2020). The study also varied the segmental orifice area to determine the accuracy of the measurement. Similar, segmental orifice utilized as an ultrasonic meter for a comparison based hybrid analysis Straka et al., (2018). The simulation considered the flow in a turbulent flow on the basis of different turbulent model such as Reynolds averaged Navier Stokes (RANS), k-ω RANS model and large eddy simulations (LES). V-cone orifice flow meter used for measuring wet gas flow for different applications Jazirian (2023), He et al., (2019) and He et al., (2018).

Flow analysis has also been extended to slotted orifices, where the area ratio area ratio was a key input parameter. Experiments and simulations using wet gas (a mixture of air and water) were conducted, incorporating both single phase and multi phase flow models via CFD. These simulations yielded detailed contour maps of static pressure, velocity, and density across the slotted orifice flow domain. Compared to multi hole orifices, slotted orifices demonstrated superior performance in terms of pressure recovery both upside and downside. They also provided a more uniform and stable velocity field, enhancing measurement consistency. Further experimental investigations connected two slotted orifices in series to improve flow measurement accuracy in pipelines carrying gas-liquid mixtures, commonly referred to as wet gas. The findings showed that slotted orifices offer more effective flow regulation and pattern control than conventional orifices, improving their design and function in wet gas applications (Yanfeng, 2006). In practical use, slotted orifices have been applied in the natural gas industry for measuring two-phase flows. These orifices help reduce measurement uncertainties that typically arise in both upside and downside conditions. A comparison based study presents between circular orifice and slotted orifice and found that slotted orifice performs better for a particular application, Morrison et al., (1994). In designs featuring a 0.5 diameter ratio, four larger holes accommodate flange mounting, while four smaller holes serve to minimize gas disturbance at the orifice entrance. Following performance comparisons, slotted orifices were eventually replaced by turbine flow meters and conventional orifice meters to achieve improved measurement control and reduce uncertainty, particularly in dynamic flow conditions (Morrison et al., 2001). The slotted orifice is applied as a flow measuring device to control the flow for natural gas as a turbine flow meter (Gerald et al., 2001). A dual slotted orifice meter developed to measure the flow rate of two phase gas and liquid mixture and determine the pressure and temperature profile, Li et al., (2009). Similarly, the pressure, velocity, and density trend lines and profiles are predicted for a flow regime evaluated numerical Perumal et al., (2011). Athar (2012) examined parameters such as non-dimensional pressure, flow rate, and the limiting discharge coefficient, considering them as functions of various operating conditions specifically, mass flow rate, Re, area rato ratio, l/d ratio, and orifice ellipticity. The pipe diameter considered is 38.1 mm, with an area ratio limit of 0.2-0.84 for two distinct elliptical orifices. The orientations are horizontal and vertical. The study proposed a correlation between C_d and area ratio for the above geometrical ranges.

The energy behavior of fluid flow is influenced by key factors such as velocity and pressure. Conventional fluid dynamics concepts including eddy formation, flow separation, wake formation, cavitations, and frictional losses are typically used to evaluate energy dissipation, particularly in terms of sudden Δp along the flow path. When Δp becomes significant enough to drop the flow pressure below the vapor pressure, vaporization can occur especially in systems using multi square hole orifices. This vaporization leads to hydrodynamic cavitation and further energy losses. In one study, 3-D Particle Image Velocimetry (PIV) and High Speed Photography were employed to analyze fluid flow through multi square hole orifices within a hydrodynamic cavitations reactor. The multi orifice conducted with different arrangement of square orifice to know the hydraulic charteristics for optimum design. These imaging techniques were used to visualize and measure the velocity and pressure distribution for each square orifice configuration, providing valuable insights into flow behavior (Dong, 2013). The concept of combining orifices, such as through the use of two stage orifices, represents a novel configuration aimed at enhancing flow measurement stability. Theoretical, CFD and experimental studies have been conducted to compare two stage orifices with single orifice setups, particularly in terms of C_d and Δp performance. In one such study, mineral oil was used as the working fluid, and experiments were performed across a Re range of 900 to 1700 using an orifice with a 3 mm diameter. Results indicated that the two stage configuration offered a more efficient design for flow measurement than its single stage counterpart. Additionally, this design was shown to be promising for applications involving micro orifices and micro channel flows. Gao and Wu (2019) concluded that, at a given orifice diameter, the two stage setup yields a superior C_d compared to a single orifice.

Further studies on two phase flow examined pressure behavior in systems with sudden expansions and contractions using thick and thin orifice plates. These analyses accounted for both reversible and irreversible losses, with R-113 employed as the working fluid to assess pressure drop characteristics. A volume-averaging method was used to solve the governing flow equations, incorporating variables such as fluid quality, liquid-to-vapor density ratios, and viscosity ratios. The local slip ratio was identified as a key parameter in accurately predicting Δp during numerical analysis Kojasoy et al., (1997). Several flow measurement devices including small orifices and nozzles were experimentally tested under subsonic and critical-flow conditions. The study involved a variety of compressible gases such as air, carbon dioxide (CO_2), argon (Ar), helium (He), and Ar-He mixtures to assess flow characteristics and determine C_d. The orifice configurations tested included knife-edge, straight-bore, rounded-entry, and elliptical-entry types, all with convergent shapes and diameters ranging from 0.9 mm to 1.9 mm. Kayser et al., (1991). Based on these comprehensive experiments, correlations for Δp and C_d were developed across different flow regimes by varying fluid types and orifice edge designs. The findings revealed that perforated plate orifices with a consistent free-area ratio provided better flow coefficients than traditional circular orifices. Moreover, raising the plate thickness to 1.5 times its original value significantly reduced the pressure drop (Gan, 1997).

A comparative correlation-based analysis was conducted on perforated plate orifices and straight-edged orifices, both having the same area ratio, under single-phase flow conditions. This study focused on how variations in Re and geometric parameters influence performance (Shanfang, 2013). Further investigations included both numerical and experimental evaluations of cylindrical, square-edged, and perforated orifices to assess their pressure loss coefficients. Key variables considered in this comparative analysis were plate thickness, area ratio, and Re. The findings suggested that, for a specific area ratio value, the perforated plate orifice demonstrated superior pressure loss characteristics compared to the other configurations (Bullen, 1997). Additionally, the relationship between volumetric flow rate and Δp in perforated orifices was analyzed with respect to structural features such as orifice thickness, porosity, hole arrangement, and upside flow disturbances. Using water as the working fluid in a 2.9 cm diameter pipe, the study found that the perforated orifice plate offered a significant performance advantage, enhancing the C_d by approximately 22.5% to 25.6% over the standard orifice plate (Idelchik, 1994).

An advanced pipe design study was carried out to assess the Δp as a function of pressure loss caused by flow contraction within a pipeline. The experimental investigation was conducted within a Re range of 4×10^4 to 2×10^5, with area ratios varying from 0.13 to 0.7, and different contraction sharpness levels used as input variables. By analyzing the static pressure distribution profiles in both ranges of pipes, the study found that the pressure loss coefficient increases as Re decreases (Morris, 1996). In a related study, the discharge rate and flow coefficient of a safety valve were examined by considering factors such as the valve's diameter ratio and size. Measurements of static and total pressure at inlet of pipe and around the orifice region were used to refine the design of the safety valve. The findings concluded that valve flow stability is influenced by pipe entry losses, diameter ratio, and gravitational head during flow conditions (Dugdale, 1997). Another experiment focused on the flow behavior of viscous fluids specifically molasses through a thin orifice plate inside a fluid chamber. The study employed a large orifice diameter ratio to examine flow dynamics. The orifice had a diameter of 0.1469 m and a thickness of 0.0075 cm, while the chamber's pipe diameter measured 2.85 cm. Viscosity variations in molasses were controlled by adjusting the glycerin to water mass ratio. The Δp across the orifice was calculated using the Hagen-Poiseuille equation, applied over the pipe's length (Chisholm, 1967).

Orifices are typically categorized based on their discharge edge geometry into types such as sharp edged, rounded edged, short tube, and Borda orifices. Among these, rounded edge orifices are known to yield higher flow coefficients, whereas sharp edged orifices are favored for their accuracy in flow measurement applications. A theoretical study focusing on two phase flow through sharp edged orifices was conducted using a gas-liquid mixture. This research aimed to find out the flow coefficient by incorporating interface shear forces into the governing equations. The fluid mixture was modeled as incompressible, with key parameters including phase velocities, Δp, and the weight ratio of the components in the mixture (Lin, 1982). In another investigation, different types of orifices were examined for phase-separated vapor-liquid mixtures, particularly using R-113 refrigerant. The study analyzed variations in Δp and C_d under different vapor-liquid density ratios. A set of correlations for sharp edged orifices was developed based on both experimental and theoretical data. These correlations took into account variables such as mass flow rate, vapor quality, fluid density ratios, and orifice and pipe diameters Akshay et al., (2018). At a macroscopic scale, orifice flow often induces turbulence that results in acoustic phenomena such as sound waves, vibrations, noise, and cavitations. These disturbances reduce energy transfer efficiency across the flow domain. A study investigating the acoustic characteristics of flow through orifices assessed the variation in sound power and frequency spectra. Different orifice shapes circular, square, and rectangular were analyzed using a large eddy simulation turbulence model to understand the flow-induced acoustic behavior (Hasan et al., 2021).

Conical entry orifices are considered superior to sharp edged orifices in terms of flow stability and measurement accuracy. An experimental and computational investigation was conducted to evaluate the performance of conical orifices by varying the conical entry angle across a broad Re range. The study tested conical angles between 15° and 45°, diameter ratios from 0.4 to 0.63, and Re values ranging from 5,000 to 240,000. Water was used as the working fluid. Results revealed that a 30° conical angle delivered optimal performance, with a Δp reduction of approximately 37% to 52% compared to a sharp edged orifice. Additionally, the study analyzed axial flow characteristics, including recirculation zones and flow recovery, using numerical correlation methods (Ho, 1985). In another study, conical entrance orifices were examined under low Re conditions (Re between 100 and 1000) with diameter ratios from 0.247 to 0.448. Both concentric and fully eccentric shapes were tested to predict the C_d. The findings indicated that C_d increased with diameter ratio, while the eccentricity had minimal influence on the outcome (Sanghani, 2016). Further comparative research on various orifice geometries showed that for a fixed Re, sharp edged orifices generate a higher Δp, whereas streamlined orifice designs result in a lower Δp, improving flow efficiency.

CFD techniques have been widely applied to analyze fluid flow through various orifice geometries based on edge design. One such study evaluated six types of orifices, including square, sharp edged, square edged, square edged with a 45° back cut, quadrant edged, streamlined entry, and sloping approach, all tested with water as the working fluid. Among these configurations, the square edged orifice with a 45° back-cut resulted in the highest pressure loss, while the streamlined entry design exhibited the lowest pressure drop for the same flow conditions (Bagade, 2016). Additionally, a comprehensive review conducted by Grace (1951) explored the behavior of multi-hole orifices using various methodologies, such as CFD, Finite Element Method (FEM), numerical simulations, analytical approaches, and experimental techniques. This extensive review provided valuable insights into flow dynamics across orifice plates. Building on this literature, a new chapter is proposed focusing on the design and analysis of different orifice types under various flow conditions. The chapter aims to provide a foundational understanding of orifice design principles and offer practical insights to enhance flow measurement accuracy, efficiency, and reliability across a broad range of industrial applications.

The orifice study also reported in thick and thin orifice for single and two phase applications. The thick orifice plays on improvement in the stability of the flow measurement. The instability of the flow occurs on the downside of the orifice flow because of reverse flow and the negative pressure after orifice. A two phase analysis through thin and thick orifice was conducted for an air-water mixture Fossa and Guglielmini (2002). The pipe diameter is 60 and 40 mm, Re 3×10^4 to 2×10^5, area ratio 0.73 to 0.54 and thickness ratio is 0.025 to 0.59 are considered. The two phase parameter for gas and liquid velocity is 0.3 to 4 m/s and 0.6 to 2 m/s respectively. The study further extended to an experimental analysis for a stratified flow for thin and thick orifice Fossa et al. (2006). A CFD based numerical analysis was carried out with Eulerian - Eulerian multi-phase model to know the effect of individual fluids on an orifice flow. The local Δp, C_d and multi-phase pressure drop are the outcome of the analysis with inlet parameters as velocity and the geometric of the orifice. The study found that multi-phase pressure drop on a single flow analysis contraction coefficient has less effect with Re, but is affected with Re. Based on two-phase results a correlation developed which showed good agreement with existing correlation. The effect of void fraction along the flow evaluated with maximum and minimum in the downside of the orifice flow as slip ratio, Roul and Das (2012). The study was carried out for a diameter of 40 mm, Re (up to 200000), area ratio (0.54 and 0.73). The thickness ratio also varies for thin and thick orifice for a range of 0.025-0.59. The predicted result demonstrates maximum Δp occurs at minimum area ratio and maximum thickness ratio. A velocity and static Δp profile along the orifice gives on the variation of diameter ratio. The static Δp at orifice and increases steadily to get a stable profile in the downside, from which C_d is calculated. Similarly, the velocity profile increases at the orifice and drops to stable condition later in the downside. Similarly type numerical study carried out for thin and thick orifice having area ratio = 0.4 - 0.8, thickness ratio = 0.125 - 0.5, and Re 5000 - 50000 with air, water as single phase fluid and water liquid and water vapour as two phase fluid to know the effects on flow parameters, Patro and Panda, (2021). The studies determine Δp, static pressure, velocity profile and contraction coefficient for the variation of area ratio, thickness ratio, Re and different fluids. Correlation based study conducted for thin and thick orifice conducted to predict the Δp for single phase Panda et al., (2023a) and two phase Panda et al., (2023b) on the effect of area ratio, thickness ratio, Re. The parameters are considered as Re 20000 to 100000, area ratio 0.2 to 0.75. The thickness ratio is 0.1 to 0.5 for thin orifice and 0.6 to 0.9 for thick orifice. The studies are correlated with no. of correlation to develop to determine the Δp. The thin and thick orifice studies further conducted for different compressible and incompressible flow to develop correlation for Δp on the effect of fluid properties density and viscosity, Panda (2023). The fluids are considered with a basic fluid for a compressible flow as air and incompressible as water. The orifice studies considered machine learning based regression studies for large database to predict accurate relation between dependent input and output constraints Khuntia et al., (2023) and Khuntia et al., (2024). Based on the above literature study a chapter proposed to know the optimum profile for the high viscous or sediment flow for the industry applications. The orifices are considered sector, segmental, crescent for these particular applications.

(a) Assumptions

The current numerical analysis is based on the following assumptions:.

- The fluids are assumed to be Newtonian, viscous, and incompressible.
- A steady-state condition is applied throughout the simulation.
- Density and viscosity of fluid are treated as constant across the entire flow domain.
- Only axial components of velocity are taken into account in the flow evaluations.

- The influence of body forces is neglected in the analysis.

(b) Governing equation

The numerical solution is solved by considering the following eq. (1) to eq. (6) as per the flow physics of the numerical problem. The governing equations are presented in the cylindrical coordinate (x, r, θ) form. The orifice models are non–axisymmetric, but the pipe is a axisymmetric. That's the reason the tangential momentum is considered in a short form. The governing equations are taken in account are continuity, Three momentum equations (Axial, Radial and Tangential), Turbulent model k – ε (Turbulent kinetic energy and Dissipation rate of turbulent kinetic energy). The terms U, V, W represent the velocities in the three directions, p is as a pressure term, ν as kinematic viscosity, ν_t as turbulent kinematic viscosity. The effective kinematic viscosity is the summation of viscosities during the turbulent flow.

Continuity

$$\frac{\partial U}{\partial x} + \frac{1}{r}\frac{\partial (rV)}{\partial r} = 0 \tag{1}$$

Axial momentum

$$U\frac{\partial U}{\partial x} + V\frac{\partial U}{\partial r} = \frac{1}{\rho}\frac{-\partial P}{\partial x} + \nu\left(\frac{\partial^2 U}{\partial x^2} + \frac{\partial^2 U}{\partial r^2} + \frac{1}{r}\frac{\partial U}{\partial r}\right) - \left(\frac{\partial \overline{u}^2}{\partial x} + \frac{1}{r}\frac{\partial \overline{uv}}{\partial r}\right) \tag{2}$$

Radial momentum

$$U\frac{\partial V}{\partial x} + V\frac{\partial V}{\partial r} - \frac{W^2}{r}$$

$$= \frac{1}{\rho}\frac{-\partial P}{\partial r} + \nu\left(\frac{\partial^2 V}{\partial x^2} + \frac{\partial^2 V}{\partial r^2} + \frac{1}{r}\frac{\partial V}{\partial r} - \frac{V}{r^2}\right) - \left(\frac{\partial \overline{uv}}{\partial x} + \frac{1}{r}\frac{\partial r\overline{v}^2}{\partial r} - \frac{\overline{w}^2}{r}\right) \tag{3}$$

Tangential momentum

$$U\frac{\partial W}{\partial x} + V\frac{\partial W}{\partial r} - \frac{VW}{r} = \nu\left(\frac{\partial^2 W}{\partial x^2} + \frac{\partial^2 W}{\partial r^2} + \frac{1}{r}\frac{\partial W}{\partial r} - \frac{W}{r^2}\right) - \left(\frac{\partial \overline{uw}}{\partial x} + \frac{\partial \overline{vw}}{\partial r} + \frac{\overline{vw}}{r}\right)$$

. (4)

Turbulence Model

Kinetic energy

$$\frac{\partial(Uk)}{\partial x} + \frac{1}{r}\frac{\partial(rVk)}{\partial r} = \frac{\partial}{\partial x}\left(\nu_t \frac{\partial k}{\partial x}\right) + \frac{1}{r}\frac{\partial}{\partial r}\left(r\nu_t \frac{\partial k}{\partial r}\right)$$

$$+2\nu_t\left[\left[\left(\frac{\partial \overline{u}}{\partial x}\right)^2 + \left(\frac{\partial \overline{v}}{\partial r}\right)^2 + \left(\frac{\overline{v}}{r}\right)^2\right] + \left[r\frac{\partial}{\partial r}\left(\frac{\overline{w}}{r}\right)\right]^2 + \left(\frac{\partial \overline{v}}{\partial x} + \frac{\partial \overline{u}}{\partial r}\right)^2 + \left(\frac{\partial \overline{w}}{\partial x}\right)^2\right] - \beta\rho\omega k \tag{5}$$

Dissipation rate of Turbulent Kinetic energy

$$\frac{\partial(U\varepsilon)}{\partial x} + \frac{1}{r}\frac{\partial(rV\varepsilon)}{\partial r} = \frac{\partial}{\partial x}\left(\nu_t \frac{\partial \varepsilon}{\partial x}\right) + \frac{1}{r}\frac{\partial}{\partial r}\left(r\nu_t \frac{\partial \varepsilon}{\partial r}\right)$$

$$+2\nu_t\left[\left[\left(\frac{\partial \overline{u}}{\partial x}\right)^2 + \left(\frac{\partial \overline{v}}{\partial r}\right)^2 + \left(\frac{\overline{v}}{r}\right)^2\right] + \left[r\frac{\partial}{\partial r}\left(\frac{\overline{w}}{r}\right)\right]^2 + \left(\frac{\partial \overline{v}}{\partial x} + \frac{\partial \overline{u}}{\partial r}\right)^2 + \left(\frac{\partial \overline{w}}{\partial x}\right)^2\right] - \beta\rho\omega\varepsilon \tag{6}$$

$\nu_{eff} = \nu + \nu_t$, $\nu_t = C_\mu \frac{k}{\omega}$, $\beta = 0.09$, $C_\mu=1$
ν_t= eddy viscosity and

(c) Mesh Independent Study

Determining the result accurately a proper grid size (mesh) is essential. The appropriate grid sizes are accomplished by a grid independence test, which is carried out for all three geometry of orifices such as sector, segmental and crescent. The individual grid independence test provides good appropriate results and out of these the segmental orifice results with details are as follows as below. The segmental orifice shape depends on the curvature r radius (R) to define the opening of the area of the flow. The grid study was carried out with an area ratio 0.3; thickness ratio 0.3, with flow coefficient Re 56000 and size of the pipe is 4 inch (101.6 mm). The Δp is considered as the testing output flow constraint. The region of orifice is taken into account for the Δp prediction. The grid size of the particular area generates more densely to implement more grids. Numbers of grid sizes are taken into account such as 12.56, 23.02, 74.64, 136.23 million grids. The increase in the grid size from 74.64 million to 136.23 million predicted significant improvement but after 152.36 million gives a less improvement in the *Δp* results. The numerical study considered 152.36 million size meshes to carry out the numerical simulation from the above grid independent study.

(d) Numerical Solution

The current numerical studies are workings with a single-phase problem, air as single phase fluid. Velocity inlet boundary condition is considered at the inlet, pressure outlet at the outlet, no-slip and stationary wall as the wall of the orifice pipe. The fluid flow analysis is conducted by solving the fundamental governing equations, which include the conservation of mass, momentum, turbulent kinetic energy, and the dissipation rate of turbulence. The turbulent model used in this study is the standard k-ε turbulence model. This approach is applied in a single-phase flow framework. To numerically solve these equations, the Finite Volume Method (FVM) is employed, particularly within the control volume surrounding the orifice geometry, as implemented through the ANSYS Workbench platform (developed by Ansys Inc.). To enhance the accuracy and reliability of the simulation results, a high-quality computational mesh is generated, ensuring optimal aspect ratios throughout the domain. The simulation utilizes the viscous flow model, with the k-ε turbulence model chosen to effectively capture the behavior of turbulent flow structures. Appropriate boundary conditions are carefully applied at the inlet, outlet, and wall boundaries, as defined in the simulation setup, to ensure the fidelity of the numerical solution. The interaction between pressure and velocity fields is handled using the SIMPLE (Semi-Implicit Method for Pressure-Linked Equations) algorithm, which is widely recognized for its stability and efficiency in solving pressure-based flow problems. For spatial discretization, the least-squares cell-based method is adopted to handle gradient calculations. The standard discretization scheme is used for the pressure field, while a second-order upwind scheme is applied to the momentum equations, as well as to the turbulent kinetic energy and its dissipation rate equations, to improve solution accuracy and minimize numerical diffusion. Under-relaxation factors (URFs) are adjusted in accordance with convergence requirements and are applied to variables such as pressure, density, body forces, momentum, turbulent viscosity, turbulent kinetic energy, and turbulent dissipation rate. To monitor the progress of convergence, residual values are evaluated, with thresholds set at 10^{-3} for continuity and 10^{-5} for momentum and turbulence-related equations. The convergence history is visually tracked using residual plots. Before initiating the numerical solution, standard initialization procedures are performed to ensure that all variables begin from physically reasonable starting values, thus facilitating a smooth and stable solution process.

RESULT ANALYSIS

A numerical study proposed to find the optimum profile of orifice for high viscous or semimetal flow. Among the available orifice profile sector, crescent and segmental orifice are chosen for the study. The different profiles of orifice such as sector, crescent and segmental are validated with the related works of the orifice profile. The sector orifice validates with the available literature of Yadav et al., (1995) with the consideration of diameter of the pipe 40mm, area ratio area ratio is 0.5, sector angle 72°, Re 50000. The current work with the validation study found an 8% deviation from the prediction of C_d as compared to the previous d reported results. The crescent orifice validates with the results of Nepak and Panda (2023) to predict a results of 10% deviation. The validation parameter choosen for crescent orifice is pipe diameter 102.6 mm, area ratio is 0.48 thickness ratio 0.3 with Re 60000. Similarly, the segmental orifice model predicted a deviation of 10% with the comparison of result of Golijanek-Jędrzejczyk et al, (2021). The input parameters for the segmental orifice are pipe diameter 50 mm, diameter ratio 0.5, Re 12000. The validation study is followed by the parametric analysis. The studies determine the pressure drop and Cdas predicted constraints. The input constraints considered for the studies are Re, area ratio and thickness ratio. The ranges of parameter are Re upto 100000, thickness ratio upto 0.5 and area ratio 0.6.

The Fig. 4(a) shows the comparison static pressure profile of the for the three considered orifice profiles. The profiles found close trends with a deviation of less than 1%. The sector orifice produces higher value and segmental orifice shows the lower value. The static pressure increases suddenly after the flow passes through the orifice. The profile gets a stable profile after 0.5D distance. Similarly, Fig. 4 (b) shows the comparison velocity magnitude profile for the sector, crescent and segmental orifice. The velocity magnitude profile is the reciprocal of the static pressure profile. The velocity magnitude profile rapidly increases after the flow through the orifices and the sector orifice produces the highest value on a particular location. The deviation in the orifice profile for the velocity magnitude is less than 0.5%.

The Fig. 5 depicts that Δp and C_d on the effects of Re for different profiles of the orifice for a constant area ratio. The figs also compared with the additional profiles of orifice as circular along to the considered profiles. The profiles found that sector orifice provides more Δp and circular profiles have less. The C_d also get the higher value for circular profile and lowest for the sector orifice. The profiles found such a trend line because of circular profile centric to the pipe flow, but the other three profiles more affected by the wall effects of the pipe flow. The sector orifice gets more Δp and less C_d because of straight edges of orifice compared to the curvature edges of crescent and segmental orifices.

Figure 4. Flow profile of orifice along the flow length

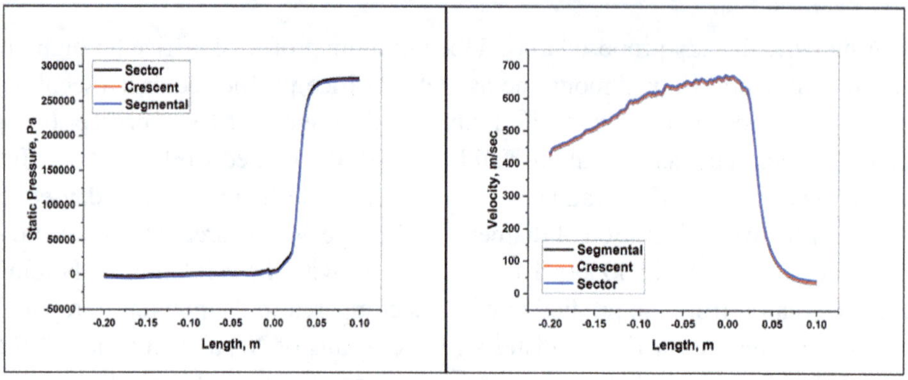

Figure 5. Δp and C_d profile on the effect of Re

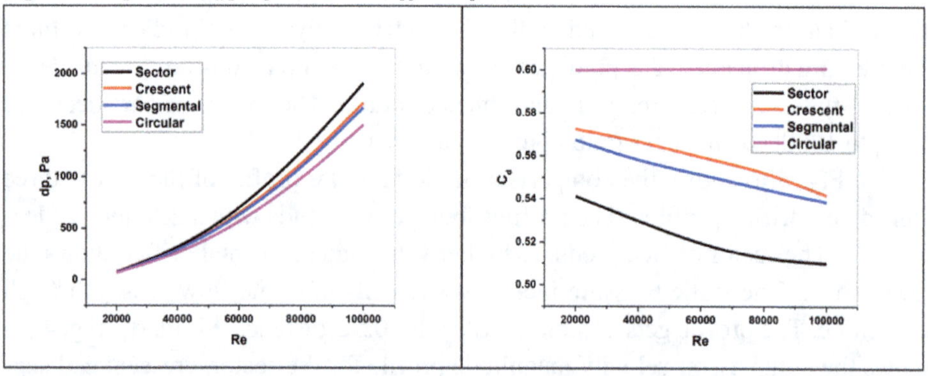

The Fig. 6(a) and Fig. 6(b) depict the Δp profile for the effect of area and thickness ratio respectively. The Δp profile decreases with increase in the area ratio. The lower area ratio produces narrow passages at the orifice (due to sudden contraction of the area) which guides more Δp. The increases in the area ratio open up more area which allows more fluid to flow and guides less Δp. The sector orifice produces more Δp among other orifice profiles. The Δp profile is also the same in case of thickness ratio, as the thickness ratio increases the pressure drop decreases. The reason for this phenomenon is that a thicker profile produces more stable length along the orifices which guides less Δp for all the orifice profiles. Sector orifice has the better profiles among other profiles for a constant thickness ratio.

Figure 6. Δp on the effect of area ratio and thickness ratio

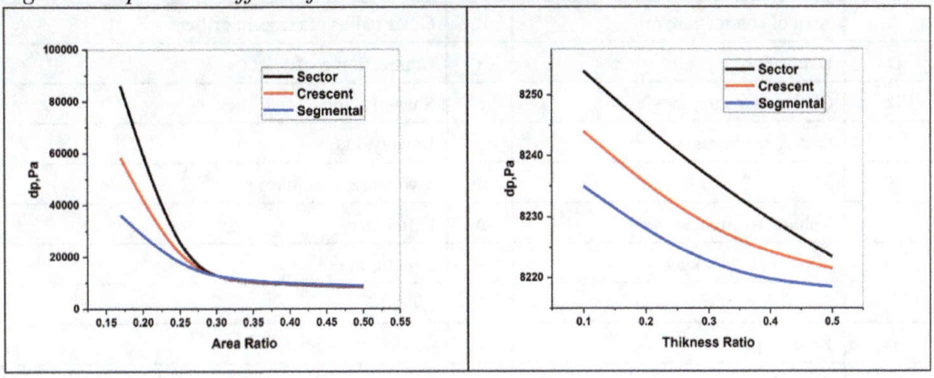

CONCLUSIONS

The present chapter gives brief details about an orifice as a flow measurement device for high viscous flow. The chapter has discussed different types of orifices according to the shape, thickness and edges of their application. Among the various shapes of an orifice crescent, sector, segmental, are considered for the high viscous type of flow. The orifice flow is studied with single and multi-phase types to determine the Δp, flow phenomenon and C_d. The input parameters are considered area ratio, thickness ratio and Re. The conclusions of the present works are that Δp is directly proportional to the Re, static pressure and velocity increases suddenly after the orifice. The pressure drop decreases with the increases in the area ratio and thickness ratio. The main conclusions of the studies for the high viscous or semimetal flow sector orifice have better profile compared to the crescent and segmental orifice. The present chapter gives a broad range of orifice in spite of the future design, better understanding, accurate in the measurement, stability of flow, minimize error, utmost life time for the flow measurement point of view.

Table 1. Abbreviations

P	Pressure, kPa	w	Width Factor
U, V	Velocity, m/s	x	Manometric height difference, m
A_1	Pipe area, m^2	θ	Sector angle
A_2	Orifice area, m^2	R_1	Inner radius of crescent orifice

continued on following page

Table 1. Continued

A_c	Area of contraction, m^2	R_2	Outer radius of crescent orifice
D_1	Diameter of pipe, m	σ_c	Contraction coefficient
D_2	Orifice diameter, m	σ	Shape Factor
t	Orifice thickness, m	ρ	Density, kg/m^3
S	Slip	Φ	Two phase multiplexer
X	Volume fraction, kg/kg	Δ	Difference
m	Mass flow rate, kg/s	z	Length, m
C_d	Coefficient of discharge	l	Liquid
p	Primary phase	g	Gas
q	Secondary phase	β	Area ratio (a/A)
Q	Volume flow rate, m^3/s		

REFERENCES

Athar, M. (2003). ISH M. A. Ansari M. & Khan M. A., "Flow characteristics of elliptical orifice plates,". *ISH Journal of Hydraulic Engineering*, *9*(2), 22–35.

Bagade, V. S., Suryawanshi, P. M., & Nalavade, S. M. (2016). A review of multi-hole orifice plate. *International Journal for Research in Applied Science and Engineering Technology*, *7*(4), 3197–3208.

Bullen, P. R., Cheeseman, D. J., Hussain, L. A., & Ruffel, A. E. (1987). The Determination of Pipe Contraction Coefficients for Incompressible Turbulent Flow. *International Journal of Heat and Fluid Flow*, *8*, 111–118.

Chisholm, D. (1967). "Flow of incompressible two-phase mixtures through sharped edged orifices," *Mechanical Engine metering. Science*, *9*, 72–78.

Dong, Z., Yang, Y. G., Chen, Q., & Bin, S. (2013). A study of hydraulic characteristics of multi-square-hole orifice plates. *Applied Mechanics and Materials*, *256*, 2470–2473.

Dugdale, D. S. (1997). Viscous flow through a sharp-edged orifice. *International Journal of Engineering Science*, *35*(8), 725–729.

Emerson, "*Fundamentals of orifice meter measurement,* White paper," https://www.emerson.com/documents/automation/white-paper-fundamentals-of-orifice-meter-measurement-daniel-en-us-188730.pdf (2017).

Fossa, M., & Guglielmini, G. (2002). Pressure Drop and Void Fraction Profiles during Horizontal Flow through Thin and Thick Orifices. *Experimental Thermal and Fluid Science*, *26*, 513–523.

Fossa, M., Guglielmini, G., & Marchitto, A. (2006). Two-Phase Flow Structure Close to Orifice Contractions during Horizontal Intermittent Flows. *International Communications in Heat and Mass Transfer*, *33*, 698–708.

Gan, G., & Riffat Saffa, B. (1997). Pressure loss characteristics of orifice and perforated plates. *Experimental Thermal and Fluid Science*, *14*, 160–165.

Gao, J., & Wu, F. (2019). Investigation of flow through the two-stage orifice. *Engineering Applications of Computational Fluid Mechanics*, *13*(1), 117–127.

Gassan, H., Abdul-Majeed, A., & Maha, R. A. (1991). Correlations developed to predict two phase flow through wellhead chokes. *Journal of Canadian Petroleum Technology*, *30*, 47–55.

Gerald, L. M., Dwayne, T., Carl, B., & Hall, K. R. (2001). Response of a slotted orifice flow meter to an air/water mixture. *Flow Measurement and Instrumentation*, *12*, 175–180. DOI: 10.1016/S0955-5986(01)00018-8

Golijanek-Jędrzejczyk, A., Mrowiec, A., Hanus, R., Zych, M., Heronimczak, M., & Świsulski, D. (2021). The assessment of metrological properties of segmental orifice based on simulations and experiments. *Measurement*. Advance online publication. DOI: 10.1016/j.measurement.2021.109601

Grace, H. P., & Lapple, C. E. (1951). Discharge Coefficient of Small Diameter Orifices and Flow Nozzles. *Trans. ASME*, *73*, 639–647.

Hasan, D. (2021). Effect of conical angle on the hydraulic properties of plate flows: A numerical approach. *Flow Measurement and Instrumentation*, *81*, 102026.

He, D. H., Chen, S. L., & Bai, B. F. (2018). New model for measuring stratified gas-liquid flow by Chisholm-model-based V-cone flowmeter. *CIESC Journal*, *69*, 3428–3435. DOI: 10.11949/j.issn.0438-1157.20180136

He, D. H., Chen, S. L., & Bai, B. F. (2019). Numerical simulation of stratified gas-liquid flow inside V cone throttle device. *Zhongguo Shiyou Daxue Xuebao. Ziran Kexue Ban*, *43*, 151–158. DOI: 10.1615/InterJFluidMechRes.2023047456

Ho, Y. S., & Leung, T. P. (1985). Performance of conical entrance orifice plates at low Reynolds numbers. *International Journal of Heat and Fluid Flow*, *6*(2), 122–125.

Idelchik, I. E., Malyavskaya, G. R., Martynenko, O. G., & Fried, E. (1994). *Handbook of Hydraulic Resistances*. CRC Press.

Jazirian, H., Jafarkazemi, F., & Rabieefar, H. (2023). A numerical model for simulating separated gas-liquid two-phase flow with low GVF in a V-cone flow meter. *Flow Measurement and Instrumentation*, *90*, 102329.

Kadam, A. R., & Chaudhuri, M. B. (2018). Acoustic study and behaviour of flow through orifices. *Procedia Manufacturing*, *20*, 154–159.

Kayser, J. C., & Shambaugh, R. L. (1991). Discharge coefficients for compressible flow through small-diameter orifices and convergent nozzles. *Chemical Engineering Science*, *46*(7), 1697–1711.

Khuntia, A. K., Misra, S., Alok, P., & Panda, S. K. (2023). Determination of pressure drop through a circular orifice for two-phase flow with Machine learning technique. *Flow Measurement and Instrumentation*, *94*, 102470.

Khuntia, A. K., Panda, S. K., & Misra, S. (2024). Two-phase flow measurement through orifice-meter with regression analysis. *International Journal of Fluid Mechanics Research, 52*(2), 41–54.

Kojasoy, G., Kwame, M. P., & Chang, C. T. (1997). Two-Phase Pressure drop in Multiple Thick and Thin Orifices Plates. *Experimental Thermal and Fluid Science, 15*, 347–358.

Li, Y., Wang, X. J., & Geng, Y. F. (2009). Study on wet gas online flow rate measurement based on dual slotted orifice plate. *Flow Measurement and Instrumentation, 20*, 168–173. DOI: 10.1016/j.flowmeasinst.2009.04.002

Lin, Z. H. (1982). Two Phase Flow Measurement with Sharp Edge Orifices. *International Journal of Multiphase Flow, 8*, 683–693.

Morris, S. D. (1996). Liquid flow through safety valves: Diameter ratio effects on discharge coefficients, sizing and stability. *Journal of Loss Prevention in the Process Industries, 9*(3), 217–224.

Morrison, G. L., Dwayne, T., Carl, B., & Hall, K. R. (2001). Response of a slotted orifice flow meter to an air/water mixture. *Flow Measurement and Instrumentation, 12*, 175–180.

Morrison, G. L., Hall, K. R., Holste, J. C., Macek, M. L., Ihfe, L. M., & De Otte, R. E.Jr. (1994). Comparison of orifice and slotted plate flow meters. *Flow Measurement and Instrumentation, 5*, 71–77. DOI: 10.1016/0955-5986(94)90039-6

Panda, S. K. (2023). An orifice flow analysis on the basis of density and viscosity effects of fluids. *WSEAS Transactions on Heat and Mass Transfer, 18*, 140–146.

Panda, S. K. (2024, February). Sector orifice analysis with a correlation to determine flow coefficient. *Journal of Engineering and Thermal Sciences, 4*(1), 1–9. DOI: 10.21595/jets.2024.23641

Panda, S. K., Choudhury, B. K., & Rath, K. C. (2022). A Literature Review on Orifice as a Flow Measuring Device. *ECS Transactions, 107*(1), 815. DOI: 10.1149/10701.0815ecst

Panda, S. K., & Nepak, S. (2023). Optimization of geometrical profile of a crescent orifice. *International Journal of Fluid Mechanics Research*. Advance online publication. DOI: 10.1615/InterJFluidMechRes.2023047456

Panda, S. K., & Patra, A. (2021). Determination of Coefficient of Contraction of Orifice with Variation of Geometrical Parameter. In Palanisamy, M., Ramalingam, V., & Sivalingam, M. (Eds.), *Theoretical, Computational, and Experimental Solutions to Thermo-Fluid Systems. Lecture Notes in Mechanical Engineering*. Springer., DOI: 10.1007/978-981-33-4165-4_38

Panda, S. K., Rath, K. C., & Choudhury, B. K. (2023). A two phase analysis of orifice based on a non-dimensional number. *Flow Measurement and Instrumentation, 94*(102470).

Panda, S. K., Rath, K. C., & Choudhury, B. K. (2023). Determining the flow correlation for an orifice with a non-dimensional number. *Flow Measurement and Instrumentation, 90*, 102338. DOI: 10.1016/j.flowmeasinst.2023.102338

Perumal, K., Michael, W., & Ming, B. (2011) A CFD study of low pressure wet gas metering using slotted orifice meters, *Flow Measurement and Instrumentation*, 22, 33-42, ISSN 0955-5986, DOI: 10.1016/j.flowmeasinst.2010.12.002

Reader-Harris, M. (2015). *Experimental fluid mechanics, "Orifice plates and venture tubes*. Springer., DOI: 10.1007/978-3-319-16880-7

Roul, M. K., & Dash, S. K. (2012). Single-Phase and Two-Phase Flow through Thin and Thick Orifices in Horizontal Pipes. *Journal of Fluids Engineering, ASME, 134*, 091301–1.

Sanghani, C. R., Jayani, D. C., Dobariya, C. R., Sabhadiya, G. K., & Gohil, J. R. (2016). Comparative Analysis of different profile geometrics for pressure drop. *International Journal of Science Technology & Engineering*, *2*(10), 494–498.

Shanfang, H., Taiyi, M., Dong, W., & Zhong, L. (2013). Study on discharge coefficient of perforated orifices as a new kind of flow meter. *Experimental Thermal and Fluid Science*, *46*, 74–83.

Straka, M., Fiebach, A., Eichler, T., & Koglin, C. (2018). Hybrid simulation of a segmental orifice plate. *Flow Measurement and Instrumentation*, *60*, 124–133. DOI: 10.1016/j.flowmeasinst.2018.02.006

. Straka M., Koglin C. Eichler T., (2020) Segmental orifice plates and the emulation of the 90°-bend, *tm - Technisches Messen*, 87, 18-31. .DOI: 10.1515/teme-2019-0120

Xue, Z. L., Xu, Y., Zhang, T., Zhang, J. L., & Chen, Z. H. (2021) Research on numerical simulation of stratified gas-liquid flow through crescent orifice plate, *2021 IEEE International Instrumentation and Measurement Technology Conference (I2MTC)*, 2021,1-6, doi: DOI: 10.1109/I2MTC50364.2021.9459904

Yadav, H. S. (1995). ISH M., Athar M., ISH M., Kaushal D. R., Flow Characteristics of Sector Orifice Plates. *ISH Journal of Hydraulic Engineering*, *1*(2), 84–94.

Yanfeng, G., Zheng, J., & Shi, T. (2006). Study on the metering characteristics of a slotted orifice for wet gas flow. *Flow Measurement and Instrumentation*, *17*, 123–128.

Chapter 10
Comparative Analysis and Optimization of Fin Material and Convection Effects for EV Battery Thermal Management

S. K. Mohammad Shareef
https://orcid.org/0000-0002-2634-9273
CVR College of Engineering, Hyderabad, India

G. Amba Prasad Rao
National Institute of Technology, Warangal, India

ABSTRACT

There are several things that can power with lithium-ion batteries thanks to their high energy density and long cycle life. However, they face several challenges, including thermal management issues, which can lead to performance degradation and safety concerns. This can result in capacity loss, reduced lifespan, and potential safety hazards. The right thermal management prevents overheating and maintains optimal operating temperatures. Active and passive cooling are used to enhance thermal management of batteries. Passive cooling methods, such as phase change materials (PCMs), are commonly hampered by low thermal conductivity, making them less efficient at dissipating heat quickly. To optimize the thermal management of Li-ion batteries, a numerical simulation was performed using ANSYS-FLUENT with different materials for fins and various fin numbers at a discharge rate of 8C.

DOI: 10.4018/979-8-3373-3621-3.ch010

The results of this study show that the temperature reduction achieved by installing copper fins with fewer numbers is comparable to the temperature reduction achieved by using a higher number of aluminum fins.

1. INTRODUCTION

There are many pollutants that are released by diesel-powered engines, such as nitrogen oxides and particulates that contribute to air pollution and respiratory problems. Additionally, diesel engines produce higher levels of carbon dioxide than some alternative fuels, exacerbating climate change. Efforts to reduce these emissions include the development of cleaner technologies and stricter regulations on diesel vehicles. To reduce diesel emissions, individuals can choose to take public transportation or carpool instead of driving diesel-powered vehicles. Additionally, regular maintenance of diesel engines can improve their efficiency and reduce emissions. Lastly, considering the switch to electric or hybrid vehicles can significantly lower one's carbon footprint. There are also many advantages to using electric vehicles, as they can use a more significant percentage of the energy that is generated by electrical power from the grid to power the wheels, as opposed to internal combustion engines. Moreover, as the energy grid becomes increasingly powered by renewable sources, the overall carbon footprint of electric vehicles continues to decrease. An electric vehicle's main issue is thermal runway (Li et al., 2020; Liu, Gan, Chen et al, 2023). Thermal runway occurs when the battery (Akula, Minnikanti, & Balaji, 2024; Mane et al., 2024) of an electric vehicle overheats, leading to a decrease in its efficiency and capacity. This can cause the vehicle to slow down or even stop, putting the driver, passengers, and others in danger. Therefore, thermal runway needs to be addressed to ensure the safety of electric vehicle drivers. Enhanced thermal management (Fan et al., 2021; Fu et al., 2019; Liu, Jin, Li et al, 2023; Panchal et al., 2017) systems help maintain optimal battery temperatures, which in turn ensures consistent performance and extends the lifespan of the battery(Ping, Peng, Kong, Chen, & Wen, 2018). By preventing overheating, these systems allow electric vehicles to operate efficiently, maintaining their range and power output.

Electric mobility has accelerated demand for robust battery thermal management systems (BTMS) because lithium-ion cells operate safely and efficiently only within a narrow temperature band (≈20–40 °C). Deviations degrade performance and shorten life: at elevated temperatures, parasitic reactions and non-uniform heat generation hasten capacity fade and can escalate to thermal runaway; at low temperatures, sluggish ion transport and plating risks undermine power delivery and safety. These challenges are magnified during fast charging and high C-rate discharge typical of modern EV duty cycles, making precise thermal control and spatial uniformity across cells a primary design requirement for pack-level reliability and safety (Hwang, 2024; Liu, Zhang, Zhou, Xiong, & Wang, 2022).

A broad spectrum of BTMS strategies has been explored. Air cooling is simple and low-cost but struggles at high heat flux; liquid cold plates deliver strong heat removal but add mass, packaging complexity, and cost. Passive phase-change-material (PCM) concepts efficiently buffer transients yet suffer from low thermal conductivity, prompting fin assistance or composite enhancements. Recent work has therefore advanced hybrid approaches—PCM combined with fins, heat pipes, liquid cooling, or thermoelectric elements—to improve peak-temperature suppression and temperature uniformity under aggressive loads and climates (Choudhari, Dhoble, & Panchal, 2020; Hwang, 2024; Li, Bai, Zhou, Wu, Wei et al, 2024; Rao & Kumar, 2024; Wang, Gao, Yan, & Wang, 2020; Xu et al., 2023). Parallel efforts target special functions such as integrated preheating/cooling for all-climate operability and immersion/boiling schemes for highly uniform heat extraction in compact modules (Akula, Minnikanti, & Balaji, 2024; Xu et al., 2023). Despite this progress, designs often trade thermal performance against weight, volume, and manufacturability, and many studies remain confined to lab-scale demonstrations without clear pathways to scalable EV pack integration.

The rapid electrification of the transportation sector has not only increased the demand for efficient lithium-ion batteries but also highlighted the importance of ensuring their safety, reliability, and thermal stability. According to the International Energy Agency, global EV sales surpassed 14 million in 2023 and are projected to continue growing exponentially over the next decade. This sharp rise emphasizes the need for innovative battery designs that can sustain high power densities, support ultra-fast charging, and maintain safety under extreme climatic conditions. Within this framework, BTMS development has emerged as one of the most critical enabling technologies for achieving large-scale EV adoption.

The complexity of thermal management arises from the non-uniform heat generation in battery modules, where localized hot spots can accelerate side reactions, degrade electrode materials, and cause uneven aging across cells. These thermal inconsistencies often dictate the service life of the entire pack rather than individual cells, underscoring the necessity for highly uniform cooling strategies. Moreover, as battery capacities scale up and EV manufacturers push for thinner, lighter, and more compact modules, passive safety margins are becoming insufficient. This transition calls for BTMS designs that are not only efficient but also compact, lightweight, and cost-effective.

Another critical factor influencing BTMS design is the push toward sustainable and recyclable materials. While copper offers superior thermal conductivity, its higher cost, density, and environmental extraction footprint may pose long-term challenges. Aluminum, on the other hand, is abundant, lighter, and more recyclable, making it appealing for large-scale EV applications. Thus, evaluating trade-offs between conductivity, weight, and manufacturability is essential for developing practical BTMS configurations.

In addition, extreme operating conditions such as fast charging at high ambient temperatures or cold-weather operation impose stringent requirements on BTMS. For example, during fast charging, localized heating can raise cell temperatures by more than 15 °C within minutes, making cooling system responsiveness vital. Conversely, in sub-zero climates, inadequate thermal control may hinder charging acceptance, forcing EVs to rely heavily on auxiliary heaters that reduce driving range. Consequently, BTMS must provide dual functionality: effective cooling during aggressive discharge/charge events and reliable preheating during cold starts.

Recent advancements in numerical simulation tools such as ANSYS FLUENT and COMSOL Multiphysics have enabled researchers to model multi-physics interactions involving heat generation, conduction, and convection in complex battery pack geometries. Such modeling not only accelerates design optimization but also helps in identifying practical trade-offs between fin materials, fin counts, and convective boundary conditions before physical prototyping. Despite these advances, limited studies have bridged the gap between laboratory-scale demonstrations and EV-scale implementation, particularly in quantifying how different fin materials influence cooling performance under realistic convective environments.

This highlights the need for comprehensive investigations, like the present study, which explicitly quantify material-specific efficiency under varying airflow conditions. Such efforts will guide the design of lightweight, compact, and thermally robust BTMS architectures that directly support the scalability and sustainability of future EV technologies.

In addition to performance and durability considerations, thermal safety has become one of the most pressing issues in battery research. Thermal runaway, which is typically triggered when local hotspots exceed the decomposition threshold of electrolyte and electrode materials, can escalate rapidly through exothermic chain reactions. Once initiated, thermal runaway may propagate across adjacent cells, causing catastrophic failure of entire modules. Studies have shown that the onset temperature for thermal runaway in $LiFePO_4$ cells lies between 200–250 °C, but uneven heating or poor cooling uniformity can accelerate onset even at lower surface temperatures. Hence, minimizing both absolute peak temperatures and spatial temperature gradients is critical for preventing localized degradation and ensuring safe operation under real-world duty cycles.

From an engineering design perspective, the optimization of BTMS involves striking a balance between multiple competing factors:

- achieving high heat dissipation without excessive system mass,

- ensuring temperature uniformity across cells without occupying large packaging volume,
- maintaining low cost while incorporating high-performance materials, and
- enabling scalability from single-cell laboratory prototypes to large EV battery packs.

These competing requirements explain why hybrid approaches combining active and passive cooling methods are gaining significant traction. For instance, air-cooled systems are lightweight and inexpensive but insufficient at high C-rates; liquid-cooling plates ensure superior heat removal but demand pumps, tubing, and sealing mechanisms that increase both cost and complexity. PCM-based systems mitigate transient thermal spikes but require conductivity enhancers such as fins or graphite foams to remain effective during continuous operation. Within this context, fin-assisted PCM systems represent a particularly promising compromise, offering both transient buffering and enhanced conduction pathways.

The choice of fin material remains underexplored despite its pivotal influence on BTMS performance. Copper, with its high thermal conductivity (~400 W/m·K), enables efficient lateral heat spreading and reduces the number of fins required to achieve a given cooling target. Aluminum, while less conductive (~205 W/m·K), is lighter and more cost-effective, making it attractive for mass production. Beyond these conventional choices, emerging alternatives such as graphite sheets and carbon-based composites offer anisotropic conductivity that could outperform metals in specific orientations while further reducing weight. Yet, the practical trade-offs between copper and aluminum fins under realistic convective heat transfer environments have not been quantified systematically, leaving a key gap in EV-focused BTMS research.

Another dimension to consider is the impact of operating environments on thermal management requirements. For example, EVs deployed in tropical climates must withstand prolonged exposure to ambient temperatures exceeding 40 °C, which places significant stress on passive air cooling. Conversely, vehicles in colder regions face challenges of low temperature charging efficiency, necessitating integrated heating functions. Designing a BTMS that can flexibly adapt to such diverse climatic conditions without incurring excessive energy overhead remains a crucial challenge for manufacturers. Consequently, optimization strategies must address not only material and geometry but also external convective conditions that realistically represent natural convection, fan-assisted flows, or ducted air circulation in EV battery enclosures.

Recent industry trends further underscore the importance of compact and lightweight BTMS designs. Automakers are increasingly pursuing higher energy density packs by stacking thinner cells more closely together, leaving limited space for auxiliary cooling structures. At the same time, consumer demand for extended driving range requires minimizing the parasitic weight of non-energy-storing components. In this scenario, identifying material–geometry–convection trade-offs become vital to maximize thermal safety while maintaining energy density and vehicle range.

Against this backdrop, an important gap persists: the community has not systematically quantified how fin material selection (e.g., copper vs. aluminum) and convective boundary conditions (h) jointly govern cooling effectiveness, temperature uniformity, and the practical trade-off between fin count/size and system mass/volume in fin-assisted PCM/air configurations. Prior studies frequently optimize geometry in isolation or report temperature reductions without connecting material choice to pack-level constraints such as compactness and weight (Li, Bai, Zhou, Wu, Wei et al, 2024). The present study addresses this gap by (i) validating a high-fidelity electro-thermal model against established data (Liu, Zhang, Zhou, Xiong, & Wang, 2022); and (ii) comparing copper and aluminum fins across realistic ranges of heat-transfer coefficients to expose material-specific efficiency and the possibility of achieving equivalent cooling with fewer, higher-conductivity fins—a result directly relevant to scalable, lightweight BTMS for EV modules (Chu et al., 2019).

The rapid adoption of electric vehicles is not only a technological shift but also a policy-driven movement. Governments across the globe are implementing stricter emission standards and incentivizing zero-emission vehicles through subsidies, tax exemptions, and carbon credit schemes. These initiatives directly accelerate the deployment of large-scale lithium-ion battery packs in automotive fleets. However, with this surge, there is increasing regulatory pressure to ensure battery safety and thermal stability. International standards such as UN 38.3 and ISO 12405 explicitly demand that EV battery packs maintain thermal uniformity and demonstrate resilience against thermal runaway propagation. Compliance with these standards places additional emphasis on the development of robust BTMS designs capable of addressing real-world safety scenarios.

Beyond safety, sustainability and recyclability are emerging as critical concerns in BTMS design. Copper extraction and refining are energy-intensive and environmentally taxing, whereas aluminum, while abundant and easier to recycle, provides lower conductivity and requires more material to achieve equivalent performance. This trade-off extends the discussion from purely technical metrics to life-cycle assessment (LCA) and environmental impact. As EV adoption scales, choosing fin materials that balance performance with sustainability will influence not only engineering outcomes but also the long-term viability of supply chains and recycling infrastructures.

The integration of next-generation materials offers another promising direction for BTMS development. Graphene-coated fins, carbon nanotube composites, and metal foams have demonstrated superior thermal conductivity-to-weight ratios compared to conventional metals. These materials not only enhance lateral heat spreading but also reduce mass penalties, making them particularly suitable for compact EV modules. However, high costs, manufacturing complexity, and limited scalability remain major obstacles. Therefore, a comparative evaluation of readily available metals such as copper and aluminum under realistic convective boundary conditions remains essential for bridging laboratory innovations with industry-scale applications.

In addition, the shift toward fast-charging infrastructure worldwide imposes stringent requirements on BTMS performance. Ultra-fast chargers (≥350 kW) can replenish up to 80% of a vehicle's capacity in less than 20 minutes, but such aggressive charging profiles generate intense, non-uniform heat within cells. Without effective cooling, surface temperatures can spike beyond recommended limits within minutes, risking accelerated aging or safety failures. Thus, optimizing fin-assisted cooling systems under varying convective conditions is not merely an academic exercise but a practical necessity to support the global fast-charging ecosystem.

Another dimension that deserves attention is the thermal–electrochemical coupling in batteries. Elevated temperatures accelerate undesirable side reactions such as solid electrolyte interphase (SEI) decomposition, gas generation, and electrolyte breakdown, while excessive cooling at low temperatures may reduce ionic transport and increase impedance. These coupled phenomena directly link BTMS efficiency with electrochemical performance and cycle life. Consequently, studies that isolate and quantify the impact of design variables—such as fin material, number, and convection intensity—provide valuable input for integrating thermal and electrochemical models into holistic EV battery management systems.

Finally, the commercial scalability of BTMS hinges on manufacturability, reliability, and cost-effectiveness. While academic studies often report laboratory-scale improvements, industrial adoption requires solutions that can be mass-produced, integrated into compact vehicle platforms, and maintained over long service lifetimes. For this reason, comparative studies focusing on widely used fin materials such as copper and aluminum, combined with realistic convective conditions, are directly relevant to bridging the gap between theory and practice. This approach not only highlights material-driven performance optimization but also provides insights into weight reduction and compactness, which are critical drivers in next-generation EV pack design.

The optimization in this study is defined as the trade-off between fin material selection, number of fins, and convective boundary conditions. By quantifying the performance of copper versus aluminium fins, we demonstrate that equivalent cooling can be achieved with fewer copper fins, thereby optimizing pack weight and volume without compromising safety.

2. LITERATURE REVIEW

Battery thermal management has been extensively studied through multiple strategies, each with distinct strengths and limitations. Early works on air cooling demonstrated simplicity and low cost but consistently reported inadequate heat removal under high C-rate discharges, limiting their application in modern EV duty cycles (Liu, Zhang, Zhou, Xiong, & Wang, 2022). To overcome this, liquid cooling systems emerged, offering strong temperature suppression and uniformity, though at the expense of added mass, packaging complexity, and cost (Rao & Kumar, 2024; Xu et al., 2023). Parallel studies explored PCM-based approaches, which effectively buffer transient heat spikes but suffer from low intrinsic conductivity, necessitating structural enhancements through fins, foams, or composites (Dagdevir & Ding, 2024; Li, Bai, Zhou, Wu, Wei et al, 2024). To balance these trade-offs, hybrid systems—such as PCM combined with fins, thermoelectric coolers, or liquid plates—were introduced, showing improved uniformity and stability (Akula, Minnikanti, & Balaji, 2024; Choudhari, Dhoble, & Panchal, 2020; Chu et al., 2019).

While these categories reflect substantial innovation, most studies remain design-specific and demonstrate limited scalability beyond laboratory modules. Furthermore, the majority of efforts optimize fin geometry or PCM enhancement, while the role of fin material selection under different convective boundary conditions remains largely unquantified. This gap is critical because fin material choice directly influences both conductivity and pack weight, which are decisive factors for compact, lightweight EV battery packs. The present work therefore contributes by systematically comparing copper and aluminium fins under realistic convective environments, offering material-specific insights relevant for practical EV-scale BTMS designs

(Panchal et al., 2017) developed a transient electrochemical-thermal model to predict the temperature distribution in a large-sized 20 Ah $LiFePO_4$/graphite prismatic battery under varying discharge rates. The authors implemented a Newman pseudo two-dimensional (P2D) framework, incorporating coupled mass, charge, and energy conservation equations to simulate electrochemical and thermal behaviours within the battery. Their model was validated against experimental data using thermocouple measurements and IR thermal imaging, showing strong agreement with the observed results. By capturing detailed temperature gradients and identifying hot spots, especially near the current collector tabs, their model provides critical insights for battery thermal management and design improvements in high-performance applications such as electric vehicles.

(Liu, Zhang, Zhou, Xiong, & Wang, 2022) proposed a structurally enhanced hybrid battery thermal management system (BTMS) combining thermoelectric coolers (TECs) and phase change material (PCM) supported by copper fins to improve heat transfer. The study analysed the effects of fin thickness and TEC input current on thermal performance using a three-dimensional simulation model. It was found that increasing fin thickness extended the temperature control time and improved the coefficient of performance (COP), although it also caused a rise in temperature difference during PCM phase transition. The optimized design identified 4 mm thick fins and a TEC input current of 3A as providing a balance between cooling efficiency, uniformity, and energy consumption.

(Li, Bai, Zhou, Wu, Wei et al, 2024) proposed a liquid immersion cooling (LIC) system for prismatic lithium-ion battery packs using a fluorinated liquid (FS49) to enhance thermal management. The system utilized the boiling of FS49 directly on the battery surface, enabling efficient and uniform heat removal. By optimizing parameters like cell spacing and liquid filling ratio and introducing a wick structure to reduce coolant volume they improved both volumetric and gravimetric energy density without sacrificing cooling performance. Experimental and numerical analyses, including high-speed imaging and VOF-based simulations, confirmed the LIC system's ability to significantly reduce temperature differences and state-of-charge imbalances across cells, ensuring thermal stability even under high discharge rates.

(Xu et al., 2023) proposed a versatile battery thermal management system (BTMS) for prismatic battery modules that integrates a biomimetic cephalofoil fin structure and film heaters to perform both cooling and preheating. Using a combination of computational fluid dynamics and experimental validation, they analyzed various filler materials, fin geometries, and optimization strategies. The inclusion of 1 mm-thick PCM sheets and cephalofoil fins reduced the maximum temperature by 6.6 °C and improved temperature uniformity. A many-objective optimization and local refinements (e.g., adjusting fin height and air flow path) further enhanced cooling performance. For preheating, film heaters enabled the module temperature to rise from −5 °C to 20 °C within 10 minutes while maintaining a uniform temperature distribution, demonstrating effective thermal control across a wide range of environmental conditions.

(Akula, Minnikanti, & Balaji, 2024) developed a PCM-based pin fin composite heat sink to enhance the thermal management of cylindrical 18650 lithium-ion batteries under high discharge rates. The study introduced square-section pin fins into paraffin wax PCM and compared them against traditional plate-fin configurations using both numerical simulations and in-house experiments. The best configuration, 260 pin fins of 1 mm sizes, showed significantly lower battery surface temperatures and better temperature uniformity. The authors conducted extensive parametric studies, including variations in fin size, number, PCM type, and ambient conditions, and validated the model experimentally. Their design outperformed plate-fin systems by 3.3 °C in temperature regulation and maintained the maximum surface temperature difference under 2.2 °C, providing an effective and lightweight passive cooling solution for EV applications.

(Dagdevir & Ding, 2024) numerically investigated the thermal performance of a hybrid battery thermal management system that integrates a novel helical fin geometry with composite phase change materials (CPCMs) based on paraffin and expanded graphite (EG). The study analyzed how variations in fin width, number of helical turns, and EG mass fraction affect the safe operating time and temperature uniformity of a cylindrical Li-ion battery. Results showed that increasing fin width and using an optimal EG fraction (12%) substantially improved heat dissipation by breaking thermal resistance paths and enhancing PCM conductivity. The best configuration (Fin-W6-NR3 with 12% EG) extended safe operation time by 244% over the no-fin, pure PCM case, indicating the design's potential for stationary energy storage, albeit with increased mass and volume.

(Chu et al., 2019) experimentally and numerically evaluated the thermal performance of a novel passive heat sink incorporating triangular fins with alternating layout to enhance both radiation and natural convection. Five configurations were tested including rectangular fins and triangular fins with various angles and layouts—under varying pressures and heat inputs. Results showed that alternating triangular fins significantly increased the effective heat transfer coefficient (h_{eff}), particularly under atmospheric conditions and vertical orientation. At optimal fin angles (e.g., 150°), heat sinks achieved up to 16% higher heff compared to the rectangular fin baseline. CFD simulations confirmed the performance gains by revealing improved airflow entrainment and larger view factors. This design offers a cost-effective, passive solution for applications where fan-less heat dissipation is critical.

(Weng, Ouyang, Yang, Chen, Zhang, & Wang, 2020) experimentally optimized internal fins within a phase-change-material (PCM) module for lithium-ion battery thermal management by evaluating different fin shapes and quantities. The study compared rectangular, triangular, and circular fins, and found that longitudinal fins improved air convection at the upper part of the battery, while circular fins enhanced heat conduction at the bottom. Although increasing fin count improved performance up to a point, four rectangular fins yielded the best trade-off between cooling and PCM volume. An optimized configuration combining two circular fins at the bottom and four longitudinal fins at the top lowered the maximum battery temperature to 29.1 °C during 1C charging a 5.5% drop over the rectangular-only fin design, demonstrating better thermal balance and smaller temperature gradients. The results highlight the practical viability of hybrid fin designs for enhancing PCM-based passive battery cooling.

(Choudhari, Dhoble, & Panchal, 2020) numerically optimized the thermal performance of a phase-change material (PCM) module for lithium-ion battery cooling by analyzing the effects of various fin shapes, numbers, and external convection.

Using paraffin RT-42 as PCM, five fin geometries—rectangular, triangular, trapezoidal, I-shape, and T-shape were evaluated under 1C, 2C, and 3C discharge rates. Results showed that fin structures significantly reduced battery temperature, with rectangular and I-shaped fins performing best due to greater heat conduction area. An optimum configuration four I-shaped aluminum fins combined with an external convection coefficient of 15 W/m^2 K reduced battery temperature by 2.38% at 2C and 9.28% at 3C over a finless PCM module. While increasing fin count improved cooling initially, excessive fins reduced PCM volume and thermal storage capacity. The study concluded that optimized hybrid PCM-fin-convection configurations effectively enhanced passive battery thermal management under high heat generation scenarios.

(Wang, Gao, Yan, & Wang, 2020) numerically investigated the thermal performance of a lithium-ion battery module using graphite sheet fins and liquid cold plates to improve heat dissipation and temperature uniformity. A model consisting of 70 prismatic battery cells and six cold plates was developed to evaluate the effects of discharge rate, coolant flow, and fin material and thickness. Results showed that graphite sheets due to their anisotropic thermal conductivity offered better surface temperature uniformity and reduced peak temperatures compared to aluminum alloy fins, particularly at higher discharge rates. Additionally, the study found that increasing fin thickness significantly enhanced heat transfer during early cooling but had diminishing returns later. An optimal balance between fin thickness, material, and liquid temperature was suggested to improve cooling efficiency without increasing the module's gravimetric density, highlighting the promise of graphite sheet fins for lightweight, high-efficiency EV battery thermal management systems.

Amba Prasad Rao and Kumar (Rao & Kumar, 2024) conducted a comprehensive review on integrated battery thermal management systems (BTMS) for lithium-ion batteries (LIBs) used in electric vehicles (EVs), emphasizing the critical need for hybrid cooling solutions to mitigate thermal runaway (TR) risks. The study highlighted various combinations of passive (phase change materials—PCMs, composite PCMs with metal foams, graphite, nanofluids) and active (air, liquid, and heat pipe) cooling methods. Results across numerous configurations showed that hybrid systems, especially those combining PCM with liquid cooling or heat pipes, superior temperature uniformity and control during high charge/discharge cycles, fast charging, and extreme climates. Innovations such as fin-enhanced PCM, delayed liquid cooling, and internal/external cold plate designs reduced peak temperatures and temperature differences significantly, while optimizing space, weight, and energy consumption. The review further discussed advancements in machine learning for state-of-health prediction and reinforced the need for safe, compact, and economical BTMS architectures to ensure the reliability and sustainability of EV battery systems.

The lithium-ion battery (LIB) is used in many industries, including electric vehicles (EVs), energy storage power stations, electronics equipment industry, and so on, due to its high energy efficiency, longer cycle life, and lower self-discharge rate compared to other types of batteries. LIBs are greatly affected by temperature during their working process and during their safety performance. Temperatures between 20 and 40 degrees Celsius are generally recommended for LIBs. It is possible for LIBs to run away from their thermal equilibrium when the temperature exceeds this range. A temperature below this range results in reduced energy efficiency because lithium ions diffuse at a slower rate and lithium ions deposit on active materials and electrolytes because of the slower diffusion rate. To guarantee battery performance, a thermal management system in the battery module must be designed and implemented. The Lib's cooling process is based on air-cooling, liquid cooling, PCM cooling, and composite cooling systems. Aside from being more efficient, cheaper to maintain, and more flexible, air-cooling systems are less expensive to install. A few electric vehicles (EVs) have air conditioning, such as the Toyota Prius, Ford C-Max, Mitsubishi i-MiEV Minicab, and Volkswagen e-Golf. Since air has a poorer heat conductivity and capacity, air-cooling systems applied to EVs may fail under high-power or high-temperature conditions. It is therefore common for storage power stations and energy storage companies to conduct thermal management of energy storage battery modules (ESBM) using air-cooled systems due to their lower power consumption and safer operation scenarios in comparison with electric vehicles, particularly portable ones.

Due to the widespread use of EVs (Electric Vehicles), vehicle power batteries have become increasingly important. Due to the battery's ability to function in extreme environments such as 253.15 K or 323.15 K, excessive temperatures will cause the battery to dissolve or even explode. In contrast, too low temperatures will damage it irreparably. In batteries, the temperature difference between individual cells should not be too large; otherwise, it will affect the health of individual cells or even the battery's overall output. BTMS (Battery Thermal Management System) technology therefore determines EV's future. BTMS-related technologies have recently been the subject of a large number of research. BTMS can be active or passive, depending on whether they require external energy to drive. A major focus of active BTMS research is the improvement of structural parameters and the adjusting of performance parameters.

While extensive research has been carried out on battery thermal management systems, most existing studies primarily emphasize reducing surface temperature or enhancing cooling performance using PCM, fins, or hybrid systems. However, there is limited work that systematically evaluates the role of fin material selection (copper vs. aluminum) under varying convective conditions, particularly in terms of efficiency, weight, and scalability for real EV applications. Additionally, few studies integrate numerical simulations with a focus on optimizing fin count versus thermal conductivity trade-offs, which is essential for achieving compact and lightweight BTMS designs. These limitations restrict the applicability of prior solutions to full-scale EV modules (Hwang, 2024).

To address this gap, the present study numerically investigates the comparative thermal performance of copper and aluminum fins at different heat transfer coefficients, providing insights into material-specific efficiency, cooling effectiveness, and design optimization for scalable EV battery modules.

System-Level Trade-Offs in BTMS

Recent reviews emphasize that battery thermal management systems (BTMS) must satisfy multiple, often competing requirements: suppress peak temperature, minimize temperature gradients, reduce weight/volume, and remain cost-effective. Air cooling is attractive due to simplicity but is typically inadequate at high discharge rates, whereas liquid cold plates offer strong thermal control at the expense of packaging complexity and mass. Phase change material (PCM)-based systems buffer transient loads efficiently but require conductivity enhancement for sustained operation. Hybrid solutions that combine PCM with fins, liquid plates, or thermoelectric elements increasingly dominate research, reflecting a shift toward multi-objective optimization rather than single-parameter improvement ().

Fin Material and Anisotropic Conductors

While much prior work focused on fin geometry, emerging studies show that fin material selection significantly influences pack compactness and mass. Graphite fins, due to anisotropic conductivity, have been used to improve lateral heat spreading while reducing weight compared to metals. (Wang, Gao, Yan, & Wang, 2020) demonstrated that graphite sheet (Li, 2025) fins coupled with liquid plates improve uniformity in large prismatic modules. More recently, Li et al. (Wahab, 2025) used pyrolytic graphite honeycomb fins to enhance cooling in liquid-assisted designs, highlighting material choice as a determinant of both thermal effectiveness and pack weight. These findings support the notion that conductivity per unit mass, not just geometry, defines optimal fin configurations.

Immersion and Boiling-Assisted Cooling

Immersion cooling using dielectric liquids has attracted growing attention due to its superior temperature uniformity and transient performance. Wahab et al. (Wahab, 2025) reviewed immersion technologies, noting significant gains in ΔT reduction at high C-rates but challenges with fluid stability and serviceability. Chandrasekaran et al. (Chandrasekaran, 2025) experimentally studied droplet and boiling-based immersion modes, confirming high heat flux handling but also highlighting packaging and cost barriers. Compared to immersion systems, fin-assisted air or PCM cooling remains lighter and more practical for compact EV battery modules

Hybrid PCM–Fin–Heat Pipe Systems

Several researchers explored hybrids that combine PCM with fins and/or heat pipes to extend safe operating time and enhance temperature uniformity. Yu et al. (Li, 2025) summarized how fin placement governs PCM utilization before saturation. Cai et al. (Cai, 2024) applied multi-objective optimization to PCM–graphite–liquid systems, showing that material distribution must be co-optimized with convection for practical performance. These studies collectively underline that conductivity enhancement and convection must be tuned together, motivating the current work's focus on material × convection trade-offs.

Thermal Runaway and Safety Limits

Beyond efficiency, safety remains a critical BTMS driver. Several recent works () highlight that thermal runaway onset is highly sensitive to absolute temperature and cell-to-cell ΔT, reinforcing the need for uniform cooling. For prismatic cells, recommended safe operation remains in the 20–40 °C band, with ΔT typically constrained to <3–5 °C to prevent localized degradation. This justifies why your study emphasizes both maximum temperature suppression and uniformity, not only absolute cooling.

Positioning the Research Gap

Although some studies compared copper and aluminum fins, most simply report that copper performs better due to higher conductivity (). However, very few quantify the fin count reduction achievable with copper at specified convection levels. Others show aluminum can be preferable when mass dominates the optimization under low convection, whereas copper and graphite outperform under higher h values where conductivity can be fully utilized. The present study fills this gap by systematically comparing copper and aluminum fins under $h = 5–25\ W/m^2K$, demonstrating that 15 copper fins achieve equivalent performance to 20 aluminum fins, thereby quantifying a practical optimization path for lightweight BTMS.

Battery thermal management systems (BTMS) are critical for ensuring the safe and efficient operation of lithium-ion batteries in electric vehicles (EVs). Maintaining the battery temperature within a narrow range (20–40 °C) and minimizing the temperature gradient across cells ($\Delta T \leq 3$–5 °C) are essential to prevent performance degradation and thermal runaway.

Early studies on passive systems explored phase change materials (PCM), which absorb latent heat but suffer from low thermal conductivity. To overcome this, fin-enhanced PCM modules were developed. Numerical analyses revealed that rectangular and I-shaped fins provide greater conduction pathways than triangular or trapezoidal fins, although excessive fin counts reduce PCM volume and storage capacity (Weng, Ouyang, Yang, Chen, Zhang, & Wang, 2020). Similarly, hybrid geometries combining circular fins at the cell base with longitudinal fins at the top achieved better temperature uniformity without significantly increasing mass or volume (Wang, Gao, Yan, & Wang, 2020).

The choice of fin material plays a pivotal role in BTMS performance. Copper fins, due to their superior thermal conductivity, enable effective lateral heat spreading and allow equivalent cooling with fewer fins, thus reducing pack weight and size. Aluminum, while less conductive, is lighter and cheaper, making it attractive for large-scale manufacturing. Comparative studies have demonstrated these trade-offs clearly, showing that copper reduces peak temperature more effectively while aluminum remains advantageous from a mass-production standpoint (Alghassab, 2024; Li, 2025). More recently, anisotropic materials such as pyrolytic graphite honeycomb fins have shown promise in balancing conductivity with lightweight design in prismatic modules (Ping, Peng, Kong, Chen, & Wen, 2018).

External convection (h) strongly influences these systems. Under natural convection ($\approx 5\ W{\cdot}m^{-2}{\cdot}K^{-1}$), both copper and aluminum fins show limited benefit. However, with forced convection in the range of 15–25 $W{\cdot}m^{-2}{\cdot}K^{-1}$, copper's higher conductivity becomes much more effective. Hybrid PCM-fin systems have demonstrated reduced maximum temperature (T_max) and improved uniformity when coupled with moderate airflow (Liu, Zhang, Zhou, Xiong, & Wang, 2022). Other studies combined PCM with thermoelectric coolers, showing that fin thickness and TEC current directly affect temperature control and system efficiency (Li, Bai, Zhou, Wu, & Wei, 2024).

Beyond fin-assisted PCM systems, immersion cooling has emerged as a benchmark due to its excellent uniformity. Experiments with fluorinated liquids reported ΔT suppression below 2 °C during high C-rate cycling, though practical issues remain, including sealing, coolant cost, and long-term reliability (Wahab, 2025). Reviews further highlight the growing interest in immersion strategies but note challenges related to integration and maintenance in large EV packs (Elshaer, 2024).

Innovative material-based enhancements are also under investigation. Open-cell copper foams integrated with PCM have been shown to outperform conventional plate fins by accelerating melt front propagation and enhancing effective conductivity

(Schöberl, 2024). Such porous structures offer superior thermal networks compared to flat fins, especially under transient high loads.

The safety dimension of BTMS cannot be overlooked. Large-scale automotive tests confirmed that thermal runaway propagation is highly sensitive to ΔT between cells. Even when average surface temperatures remain within safe limits, localized hotspots can trigger cascading failures (Schaeffler, 2024). In-situ measurements of internal thermal gradients further confirm that uniform cooling is essential to delay thermal runaway onset (Rahmani, 2024).

Recent reviews synthesize these findings and converge on several design imperatives: (i) keep cell operating range within 20–40 °C, (ii) limit ΔT to ≤3–5 °C, (iii) integrate conduction enhancers such as fins, foams, or graphite sheets, and (iv) co-optimize with convection for sustained performance (). However, these reviews also stress that the combined effect of fin material selection and external convection has not been systematically quantified. The present study addresses this gap by comparing copper and aluminum fins under practical convective conditions, thereby offering insights into lightweight, compact, and safe BTMS designs for scalable EV applications.

Battery thermal management systems (BTMS) are increasingly critical for electric vehicle (EV) applications due to the growing demand for higher energy density and fast charging. Several recent reviews emphasize that achieving effective BTMS requires balancing multiple criteria: temperature suppression, cell-to-cell uniformity, weight reduction, and cost-effectiveness ().

(Amer et al., 2024) classified BTMS approaches into internal cooling, external cooling, tab cooling, and hybrid methods, highlighting the importance of reducing internal thermal gradients in large-format cells. Similarly, (Liu, 2025) categorized systems into air, liquid, PCM, and advanced hybrid methods, noting that while liquid-based cooling is dominant, PCM-based and hybrid approaches offer advantages in transient control and weight. Alawi et al. (Alawi, 2025) further reviewed active, passive, and hybrid systems and emphasized the integration of intelligent control strategies such as machine learning for predictive BTMS management. (Murugan, 2025) also pointed out that scalability, packaging constraints, and non-uniform heating remain major challenges.

Special focus has been given to PCM-based hybrid systems. Awasthi et al. (Awasthi, 2025) reviewed hybrid PCM designs for prismatic LIBs, demonstrating that conduction enhancers (such as fins and graphite) and active coupling (liquid or air) significantly improve performance over PCM alone. (Nasiri, 2025) surveyed PCM systems and highlighted the role of conductive fillers and fin structures in overcoming low intrinsic conductivity. Numerical and experimental works confirm these findings: (Alma'asfa et al., 2025) showed that combining PCM with fins in cylindrical cells reduces both peak temperature and ΔT under high C rates, while

(Narkhede, 2023) optimized fin geometry and PCM thickness for transient thermal loads. Similarly, (Cui et al., 2023) developed a transverse fin–casing composite structure, demonstrating enhanced PCM melting uniformity. (Mane, 2024) extended this approach by showing that long fins embedded in PCM improve overall heat dissipation capacity.

Convection also plays a decisive role. (Zhao, 2024) noted that external convection is often the limiting factor in PCM-based BTMS, where airflow or coolant flow must be optimized alongside conduction pathways. (Argade & De, 2024) optimized Z-type airflow arrangements and reported that higher inlet velocities and tapered manifolds lowered Tmax and ΔT. Karimi Kisomi et al. (Karimi Kisomi, 2025) compared PCM, air, and PCM+fin systems in cylindrical cells and found that the hybrid approach produced the lowest Tmax and highest uniformity, with ΔT below 1 K.

Hybrid cooling concepts that combine multiple mechanisms are also being investigated. (Lyu, Su, Li, Yan, & Chen, 2024) proposed a compact hybrid BTMS incorporating nanofluid cooling, PCM embedded with aluminum foam, and pulsed flow, achieving substantial temperature reductions with minimal parasitic power penalties. At pack scale, (Zhang, 2025) analyzed airflow and PCM integration in a 16-cell pack and confirmed the influence of manifold design and PCM timing on thermal stability. (Balasubramanian, 2025) studied cylindrical modules with honeycomb PCM and cooling plates, concluding that hybridization enhances both Tmax control and ΔT uniformity.

Recent research highlights the potential of hybrid BTMS that combine active and passive strategies. (Lyu, Su, Li, Yan, & Chen, 2024) demonstrated that coupling nanofluid microchannels with PCM and pulsed flow improved cooling efficiency while minimizing power overhead. Similarly, spiral-fin PCM structures were shown to outperform conventional fin arrangements in enhancing thermal uniformity (Liu et al., 2022) (Liu, Zhang, Zhou, Xiong, & Wang, 2022). Experimental work on pin-fin PCM composites confirmed surface temperature reductions of ~3.3 °C compared with plate-fin systems (Akula et al., 2024) (Akula, Minnikanti, & Balaji, 2024). Composite PCM with fins or fillers further enhances thermal conductivity, as shown in studies on fin–PCM heat sink analogues (Elshaer et al., 2023) (Elshaer et al., 2023). To contextualize these innovations, Rahmani et al. (2024) (Rahmani et al., 2024) reviewed recent developments and emphasized hybridization, material optimization, and scalability as major future directions.

These findings converge on a common gap: while fin geometry, PCM enhancement, and hybrid coupling are well studied, the combined influence of fin material choice (copper vs. aluminum) and realistic convective conditions has not been systematically addressed. The present work directly quantifies this interaction, providing design-level insights for lightweight and scalable EV BTMS.

3. MODEL FOR ANALYSIS

Figure 1. Dimensions of physical model (Shareef & Rao, 2025)

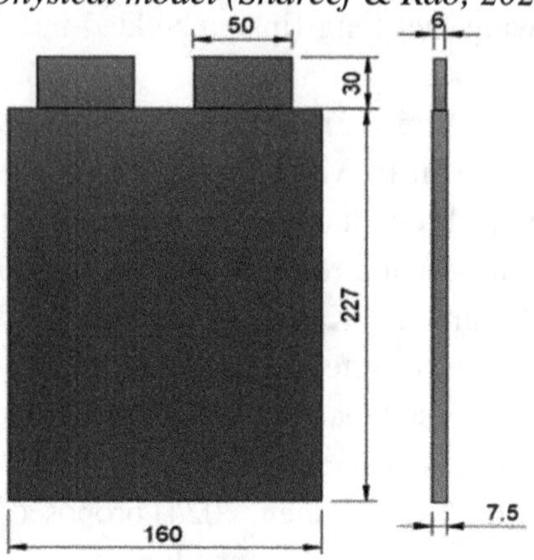

The battery model used for numerical analysis is shown in Fig.1. In the design phase, Fusion 360 is used, and in the analysis phase, ANSYS-FLUENT is used.

Table 1. Physical Properties of the battery (Shareef & Rao, 2025)

Nominal capacity	20 Ah
Minimum potential	3.2 V
Maximum potential	4 V
Cathode	LiFePO4
Anode	Graphite
C-Rate	8C
Dimensions	160 mm × 220 mm × 7.5 mm

All parameters in Table 1 were taken from manufacturer datasheets and validated literature sources (Li, Bai, Zhou, Wu, Wei et al, 2024; Liu, Zhang, Zhou, Xiong, & Wang, 2022; Wang, Gao, Yan, & Wang, 2020).

4. NUMERICAL METHODOLOGY

The numerical analysis was conducted in ANSYS-FLUENT using a three-dimensional model of a 20 Ah $LiFePO_4$ prismatic battery cell with dimensions of $160 \times 220 \times 7.5\ mm^3$. External fins were designed on the major surfaces of the cell using Fusion 360, and the model was simplified by assuming isotropic material properties and uniform internal heat generation. The volumetric heat generation (Qg) was estimated from the discharge C-rate and internal resistance of the cell, following the approach of (Liu, Zhang, Zhou, Xiong, & Wang, 2022).

The transient heat conduction equation with volumetric heat generation was solved:

$$\rho c_p \, \partial T/\partial t = k_b \nabla^2 T + Q_g \quad (1)$$

Where ρ represents density of the material (kg/m^3), cp represents specific heat capacity at constant pressure (J/(kg·K)), T represents temperature (K or °C,) k_b represents thermal conductivity of the battery material (W/(m·K)) and Q_g represents heat generation per unit volume (W/m^3).

$$\rho_{\text{fin}} c_{p,\text{fin}} \, \partial T/\partial t = \nabla \cdot (k_{\text{fin}} \nabla T) \quad (2)$$

Where ρ_{fin} is density of the "fin" material, $c_{p,fin}$ is specific heat capacity of the "fin" material and k_{fin}- "Fin" material thermal conductivity.

The computational domain was meshed with unstructured tetrahedral elements, with local refinement near the fin–air interface. A grid independence study was performed, increasing the element count from 12,764 to 2.3 million. The temperature response stabilized beyond 308,560 elements, which was adopted for all simulations as a compromise between accuracy (<2% deviation) and computational time. A pressure-based transient solver with a time step of 0.01 s and convergence criterion of 10^{-6} was used for energy equations.

Model validation was carried out using experimental results reported by (Liu, Zhang, Zhou, Xiong, & Wang, 2022) for a single cell at 2C discharge and 300 K ambient. The simulated peak temperature rise differed by less than 3% from measured values, demonstrating that the numerical model accurately captures thermal behavior under high discharge conditions. This validated model was then applied to parametric investigations comparing copper and aluminum fins across different convective environments.

Boundary Conditions

Convective boundary condition: External heat transfer from fins to air was modelled using a uniform convective boundary condition on all exposed fin and cell surfaces with ambient air temperature $T\infty = 300$ K. To represent practical operating envelopes from passive to fan-assisted cooling, we prescribed h = 5, 10, 15, 20, and 25 $W \cdot m^{-2} \cdot K^{-1}$. The lower bound (≈5 $W \cdot m^{-2} \cdot K^{-1}$) corresponds to natural convection in still air inside an enclosure; intermediate values (10–15 $W \cdot m^{-2} \cdot K^{-1}$) represent low to moderate forced convection (low-flow fans or leaky plenums); and higher values (20–25 $W \cdot m^{-2} \cdot K^{-1}$) approximate compact, fan-assisted ducted flows achievable in BTMS modules. This span aligns with commonly reported ranges for air-cooled battery modules and allows us to quantify how material choice (Cu vs.

Al) interacts with external convection to determine peak temperature and uniformity. The convective heat flux was imposed in ANSYS Fluent as

q^"=h(T_"surface" -T_∞) (3)

The convective heat-transfer coefficient (h) was imposed as a boundary condition at the fin–air interface, representing the external airflow environment. This value was kept identical for both copper and aluminum cases (e.g., h = 25 W/m^2K) to ensure that the comparison isolates the influence of fin material conductivity. In this approach, copper and aluminum affect only conduction within the fins, while the convective environment is applied uniformly. Typical ranges used for air-side convection in compact electronics/BTMS geometries shown in Table.2.

Table 2. Typical ranges used for air-side convection in compact electronics/BTMS geometries

Scenario (air side)	Representative h ($W \cdot m^{-2} \cdot K^{-1}$)
Natural convection in still air (enclosure)	5
Gentle fan / low face velocity	10
Moderate forced convection	15
Fan-assisted flow in a simple plenum	20
Compact ducted flow in BTMS enclosure	25

Grid Independent Check

Various grid elements, including 12764 to 2323200, were tested on single-cell LiFePO4 prismatic batteries. As shown in Fig.2. With increasing grid elements, the cell's temperature decreases, but not much at 308560. Numerical analysis is then performed on grid number 308560.

Figure 2. Grid independent check (Shareef & Rao, 2025)

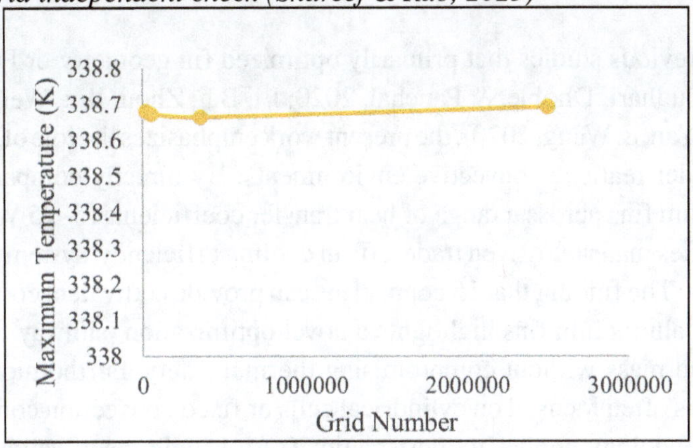

Validation of the Results

(Panchal et al., 2017). provided experimental data for a single cell at 2C at 300K (27°C) for validation. Simulated results are very close to experimental results. This shows that the simulations are accurate and reliable. Fig. 3. Compares simulation results with experimental data.

Figure 3. Validation of simulation results against experimental data reported by (Shareef & Rao, 2025).

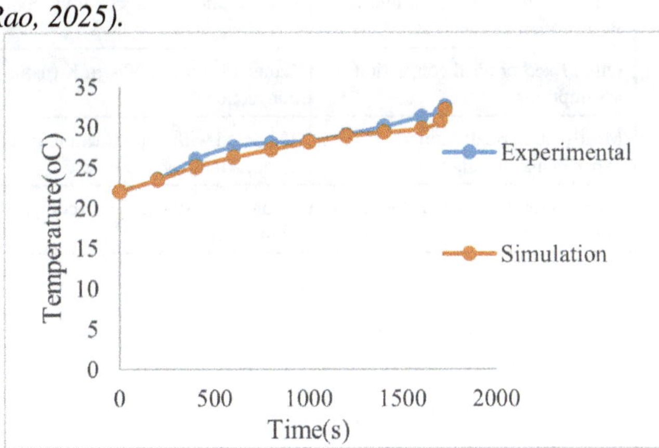

5. RESULTS AND DISCUSSIONS

Unlike previous studies that primarily optimized fin geometry or PCM combinations (Choudhari, Dhoble, & Panchal, 2020; Li, Bai, Zhou, Wu, Wei et al, 2024; Wang, Gao, Yan, & Wang, 2020), the present work emphasizes the role of fin material selection under realistic convective environments. By directly comparing copper and aluminium fins across a range of heat transfer coefficients (5–25 W/m²K), this study identifies material-driven trade-offs in cooling efficiency, system weight, and compactness. The finding that 15 copper fins can provide equivalent cooling performance to 20 aluminium fins highlights a novel optimization pathway for reducing pack size and mass without compromising thermal safety. Furthermore, whereas earlier studies often focused on cylindrical cells or fixed convection conditions, the present methodology applies to a prismatic 20 Ah $LiFePO_4$ cell validated against experimental data (Liu, Zhang, Zhou, Xiong, & Wang, 2022), thereby enhancing reliability and applicability to full-scale EV modules. The comparison of previous studies vs. present study is shown in Table.3

Table 3. Comparison of Previous Studies vs. Present Study

Aspect	Previous Studies	Present Study
Cell Type Studied	Mostly cylindrical cells or simplified module	20 Ah prismatic $LiFePO_4$ cell
Focus Parameter	Fin geometry (rectangular, triangular, helical)	Fin material choice (Copper vs. Aluminum)
Fin/PCM Optimization	Increased fin numbers or hybrid PCM combinations	Optimization of fin count vs. conductivity
Convective Conditions	Often fixed or ideal convection assumptions	Range of h = 5–25 W/m²K (natural → forced convection)
Validation Approach	Mostly numerical; limited experimental validation	Validated with experimental data ((Liu, Zhang, Zhou, Xiong, & Wang, 2022))
Practical Contribution	Focused on lab-scale performance only	Scalable design insights (compactness & weight reduction)

Figure 4. Temperature contours of the model

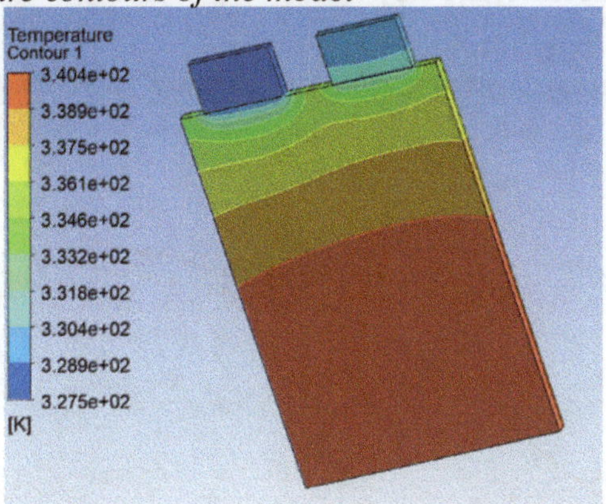

Figure 5. battery with 20 Aluminium fins

Fig.4 shows that the maximum temperature of the battery at 300K atmosphere temperature is 339.5K (~66.3 °C) operating at discharge rate of 8C and 20 verticals fins were added to external surface of the battery as shown in Fig.5.

Simulations are carried out at different fin (Ping, Peng, Kong, Chen, & Wen, 2018) material such as aluminium and copper by changing heat transfer coefficient (h) values.

Aluminium Material with 20Fins

Figure 6. Battery temperatures at various "h" (W/m^{2}K) values for 20-Aluminium fins

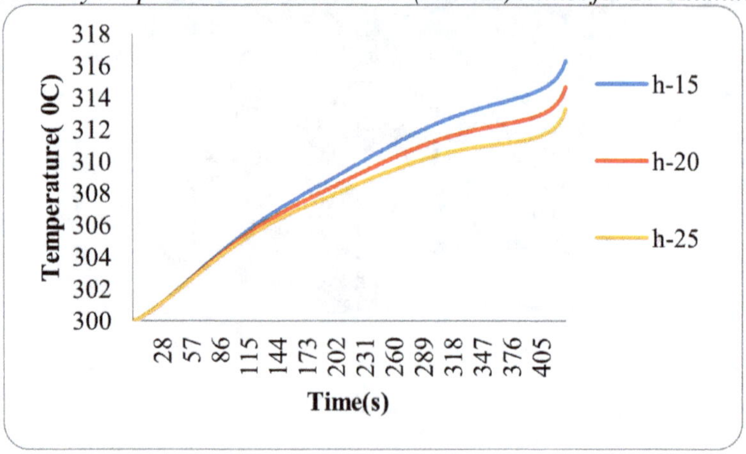

Fig.6 shows the battery temperature variation at various "h" values for aluminium fins. It is observed that the battery temperature is reduced from 339.5K (~66.3 °C) to 313.24K (~40.1 °C) by using 20 fins at "h" value of 25 W/m^{2}K. The decreased temperature is because of aluminium material having high thermal conductivity and increasing convective heat transfer values also helps to decrease the battery temperature.

Figure 7. Battery temperature contours at different "h" (W/m2K) values for 20-aluminum fins

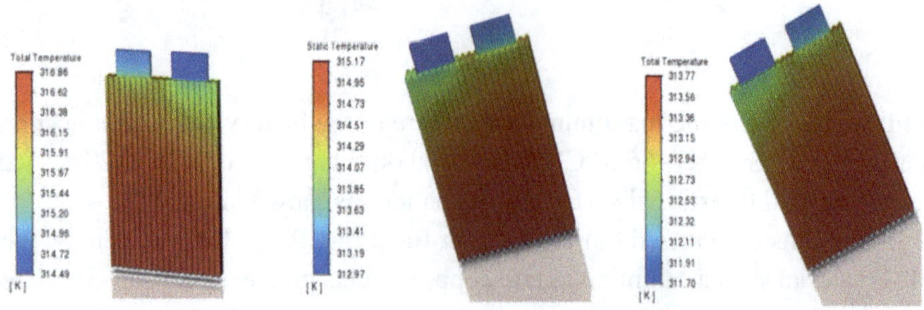

a) At h-15 b) At h-20 c) At h-25

Fig.7 shows the Battery temperature contours at different "h" values for 20-aluminum fins

Copper Material with 20 Fins

Figure 8. Battery temperatures at different "h" (W/m^2K) values for 20-Copper fins

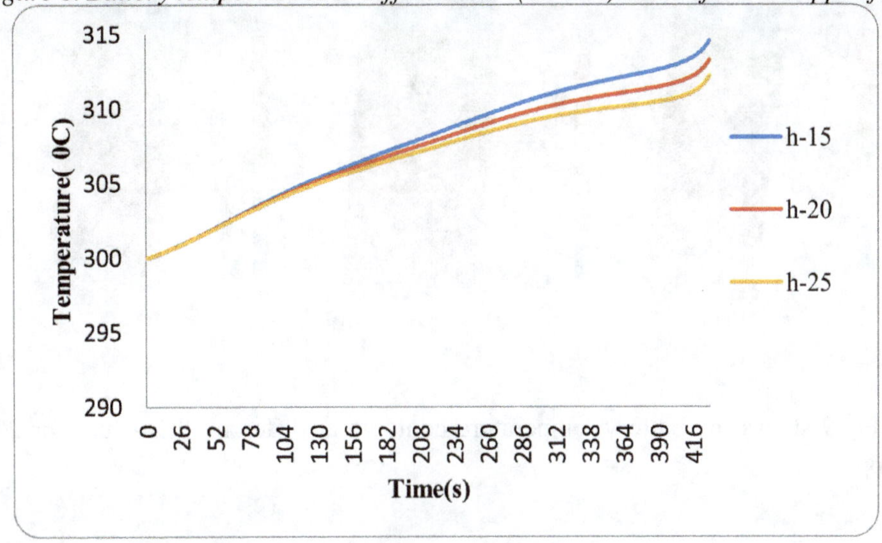

Fig.8 shows the variation in temperature of battery at different "h" values for copper fins. It is observed that the temperature of the battery is reduced from 339.5K (~66.3 °C) to 312.12K (~39.0 °C) by using 20 fins at "h" value of 25 W/m^2K. The decreased temperature is because of copper material having high thermal conductivity than that of aluminium and increasing convective heat transfer values also helps to decrease the battery temperature.

Figure 8 confirms the expected physical trend: as the convective heat transfer coefficient decreases, the battery temperature increases due to weaker external cooling. This behaviour was consistently observed for both copper and aluminium fins.

Although the absolute peak temperature differences between copper and aluminum fins are relatively small (~1–2 °C), the design implication is significant: copper fins achieve equivalent cooling with 25% fewer fins compared to aluminium. This reduction in fin count translates to potential weight and volume savings, which are critical for pack-level battery thermal management optimization.

Figure 9. Battery temperature contours at different "h" (W/m²K) values for 20-Copper fins

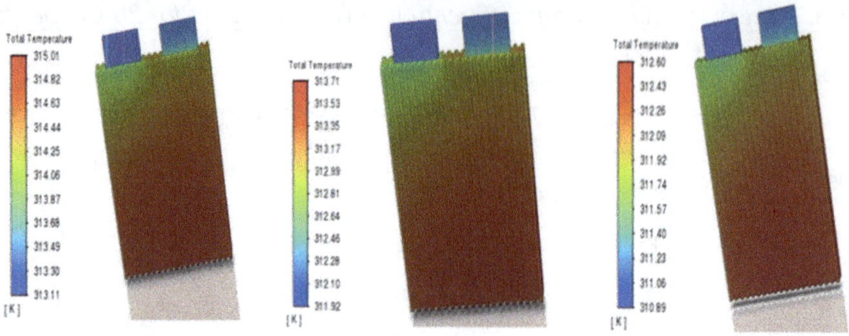

a) At h-15 b) At h-20 c) At h-25

Fig.9 shows the battery temperature contours at different "h" values for 20-Copper fins

Copper Material with 15 Fins

Figure 10. Battery temperatures at different "h" (W/m²K) values for 15-Copper fins

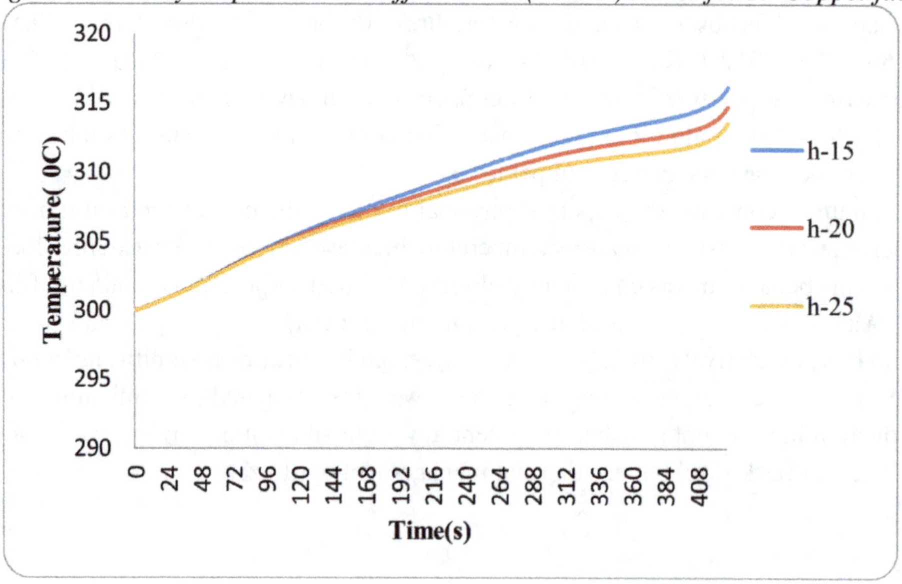

Fig.10 shows the variation in temperature of the battery with 15-copper fins. From the Fig.11. It is concluded that the reduction temperature by 20-alumimun fins

with "h" of 25 W/m2K is same as that of copper 15-fins as a result of the copper fins' high thermal conductivity.

Figure 11. Temperatures of the battery at different "h" values of Copper and Aluminium fins

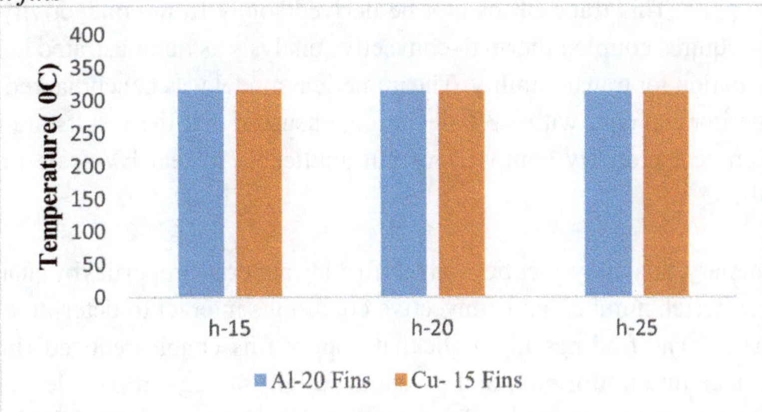

6. CONCLUSIONS

This study numerically evaluated the comparative thermal performance of copper and aluminum fins for a prismatic 20 Ah $LiFePO_4$ cell under varying convective conditions. While the superior conductivity of copper is well established, the novelty of this work lies in quantifying its impact under realistic EV-relevant environments and identifying material–geometry–convection trade-offs that inform scalable BTMS design. The major conclusions are:

- Quantified material advantage: Copper fins consistently achieved lower peak temperatures than aluminum fins. At h = 25 W/m^2K, copper fins reduced the maximum cell temperature to ~312.1 K compared to ~313.2 K with aluminum. More importantly, simulations showed that 15 copper fins can achieve equivalent cooling performance to 20 aluminum fins, representing a 25% reduction in fin count. This demonstrates that material selection directly enables more compact, lighter BTMS designs despite copper's higher density.
- Interaction with convection: Raising the convective heat transfer coefficient from 5 to 25 W/m^2K lowered cell temperature by nearly 26 K, underscoring that external airflow enhancement is as critical as fin material choice. Notably, at low h-values (natural convection), both Al and Cu fins showed

limited improvements, whereas under forced convection, copper fins delivered superior temperature uniformity (<2.0 K vs. ~3.3 K for aluminum).

- Design trade-offs for pack integration: Although copper is heavier per unit volume, its higher conductivity allows achieving equivalent cooling with fewer fins, potentially resulting in an overall lighter and more compact battery pack. This trade-off cannot be derived solely from conductivity values but requires coupled thermal–convective analysis as demonstrated here.
- Validation for transferability: The numerical model was benchmarked against experimental data with <3% deviation, ensuring that the results are not just theoretical property comparisons but applicable to real EV-scale prismatic cells.

In summary, this study goes beyond reaffirming material properties by quantifying how fin material, number, and convective conditions interact to determine BTMS performance. The findings highlight that copper fins enable reduced fin count, better temperature uniformity, and potential weight savings in pack-level design. These insights are critical for developing lightweight, compact, and scalable battery thermal management architectures for electric vehicles. Future work will extend this methodology to hybrid PCM–fin systems, multi-cell modules, and fast-changing conditions to further bridge laboratory research with practical EV pack integration.

This study highlights an optimization pathway where material selection (copper vs. aluminum), fin number, and convection jointly influence BTMS performance. Specifically, copper fins allow equivalent cooling with 25% fewer fins, reducing pack size and weight, which constitutes a practical design optimization for scalable EV modules.

REFERENCES

Akula, R., Minnikanti, A., & Balaji, C. (2024). Pin fin–PCM composite heat sink solution for thermal management of cylindrical Li-ion battery. *Applied Thermal Engineering*, *248*, 123146. DOI: 10.1016/j.applthermaleng.2024.123146

Akula, R., Minnikanti, A., & Balaji, C. (2024). Pin fin–PCM composite heat sink solution for thermal management of cylindrical Li-ion battery. *Applied Thermal Engineering*, *248*, 123146. DOI: 10.1016/j.applthermaleng.2024.123146

Akula, R., Minnikanti, A., & Balaji, C. (2024). Pin fin–PCM composite heat sink solution for thermal management of cylindrical Li-ion batteries. *Applied Thermal Engineering*, *248*, 123146. DOI: 10.1016/j.applthermaleng.2024.123146

Alawi, A.. (2025). Comprehensive review of thermal management systems for lithium-ion batteries. *Batteries*, *11*(3), 221.

Alghassab, M. A.. (2024). Comparative study of aluminum and copper fins in PCM modules. *Journal of Energy Storage*, *75*, 109245. DOI: 10.1016/j.est.2024.109245

Alghassab, M. A.. (2024). Comparative study of aluminum and copper fins in PCM-enhanced battery modules. *Journal of Energy Storage*, *75*, 109245.

Alma'asfa, S. I., Abdul Aziz, M. S., Khor, C. Y., & Fraige, F. Y. (2025). Thermal management of cylindrical lithium-ion batteries with fin and PCM hybrid systems. *Journal of Thermal Science*, *34*(6), 1789–1804.

Amer, M. M., Shouman, M. A., Salem, M. S., Kannan, A. M., & Hamed, A. M. (2024). Advances in thermal management systems for Li-ion batteries: A review. *Thermal Science and Engineering Progress*, *53*, 102714.

Argade, S., & De, A. (2024). Optimization of Z-type airflow cooling for lithium-ion battery packs. *Applied Thermal Engineering*, *229*, 120834.

Awasthi, A.. (2025). Phase change material-based hybrid BTMS for prismatic lithium-ion batteries: A review. *Journal of Thermal Engineering*, *11*(2), 155–172.

Balasubramanian, D.. (2025). Hybrid battery thermal management using honeycomb PCM and cooling plates. *Scientific Reports*, *15*, 1651.

Cai, Y.. (2024). Multi-objective optimization of PCM–graphite–liquid BTMS. *Energy*, *276*, 127456. DOI: 10.1016/j.energy.2024.127456

Chandrasekaran, M.. (2025). Experimental study on immersion and boiling-assisted cooling of Li-ion cells. *Journal of Energy Storage, 110*, 112945. DOI: 10.1016/j.est.2025.112945

Chen, Z.. (2024). Fin geometry and material optimization in PCM-based battery modules. *Journal of Energy Storage, 74*, 108921. DOI: 10.1016/j.est.2024.108921

Choudhari, V. G., Dhoble, A. S., & Panchal, S. (2020). Numerical analysis of different fin structures in phase change material module for battery thermal management system and its optimization. *International Journal of Heat and Mass Transfer, 163*, 120434. DOI: 10.1016/j.ijheatmasstransfer.2020.120434

Choudhari, V. G., Dhoble, A. S., & Panchal, S. (2020). Numerical analysis of different fin structures in phase change material modules for battery thermal management and optimization. *International Journal of Heat and Mass Transfer, 163*, 120434.

Chu, W. X., Lin, Y. C., Chen, C. Y., & Wang, C. C. (2019). Experimental and numerical study on the performance of passive heat sink having alternating layout. *International Journal of Heat and Mass Transfer, 135*, 822–836. DOI: 10.1016/j.ijheatmasstransfer.2019.02.034

Cui, M., Zhu, Z., Yang, D., He, Z., & Liu, Y. (2023). Numerical study of battery thermal management with transverse fin–casing composite structures. *Journal of Energy Storage, 65*, 107462.

Dagdevir, T., & Ding, Y. (2024). Numerical investigation of battery thermal management by using helical fin and composite phase change material. *Journal of Energy Storage, 75*, 109674. DOI: 10.1016/j.est.2023.109674

Elshaer, A. M.. (2024). Comparative performance of copper foam and plate fins for PCM-based battery thermal management. *International Journal of Heat and Mass Transfer, 212*, 124789.

Elshaer, A. M., Abdalrahman, T., & Alghassab, M. A. (2023). Effect of fin and composite PCM integration on thermal regulation in heat sinks. *Scientific Reports, 13*, 11572. DOI: 10.1038/s41598-023-37639-3 PMID: 37414796

Fan, R., Zheng, N., & Sun, Z. (2021). Evaluation of fin intensified phase change material systems for thermal management of Li-ion battery modules. *International Journal of Heat and Mass Transfer, 166*, 120753. DOI: 10.1016/j.ijheatmasstransfer.2020.120753

Fu, J., Xu, X., & Li, R. (2019). Battery module thermal management based on liquid cold plate with heat transfer enhanced fin. *International Journal of Energy Research, 43*(9), 4312–4321. DOI: 10.1002/er.4556

Hwang, F. S.. (2024). Review of battery thermal management systems in electric vehicles. *Renewable & Sustainable Energy Reviews, 182*, 113298. DOI: 10.1016/j.rser.2023.113298

Karimi Kisomi, M.. (2025). Comparative analysis of PCM, air, and hybrid cooling systems in lithium-ion batteries. *Journal of Energy Storage, 95*, 112509.

Li, H.. (2025). Prismatic module with pyrolytic graphite honeycomb fins and liquid plates. *International Journal of Heat and Mass Transfer, 223*, 125005. DOI: 10.1016/j.ijheatmasstransfer.2025.125005

Li, H.. (2025). Prismatic battery thermal management using pyrolytic graphite honeycomb fins and liquid plates. *International Journal of Heat and Mass Transfer, 223*, 125005.

Li, W., Jishnu, A. K., Garg, A., Xiao, M., Peng, X., & Gao, L. (2020). Heat transfer efficiency enhancement of lithium-ion battery packs by using novel design of herringbone fins. *Journal of Heat Transfer, 142*(2), 021401. DOI: 10.1115/1.4046160

Li, Y., Bai, M., Zhou, Z., Wu, W.-T., & Wei, L.. (2024). Immersion cooling of prismatic lithium-ion batteries using fluorinated liquid. *Applied Thermal Engineering, 257*, 124453.

Li, Y., Bai, M., Zhou, Z., Wu, W.-T., Wei, L., Hu, C., Liu, X., Gao, S., Li, Y., & Song, Y. (2024). Thermal management for the prismatic lithium-ion battery pack by immersion cooling with fluorinated liquid. *Applied Thermal Engineering, 257*, 124453. DOI: 10.1016/j.applthermaleng.2024.124453

Liu, F., Gan, S., Chen, Y., Qin, W., & Li, J. (2023). Thermal characteristics of Li-ion battery based on phase change material–aluminum plate-fin composite heat dissipation. *Energy Science & Engineering, 11*(1), 62–78. DOI: 10.1002/ese3.1357

Liu, H.. (2025). A review of lithium-ion battery thermal management strategies. *Energies, 18*(4), 1654.

Liu, H., Jin, C., Li, H., & Ji, Y. (2023). A numerical study of PCM battery thermal management performance enhancement with fin structures. *Energy Reports, 9*, 1793–1802. DOI: 10.1016/j.egyr.2023.04.214

Liu, X., Zhang, C., Zhou, J., Xiong, X., & Wang, Y. (2022). Thermal performance of hybrid TEC–PCM battery thermal management systems with fin enhancement. *Applied Energy, 322*, 119503.

Liu, X., Zhang, C., Zhou, J., Xiong, X., & Wang, Y. (2022). Thermal performance of spiral-fin PCM structures for cylindrical lithium-ion batteries. *Energies*, *15*(23), 8847. DOI: 10.3390/en15238847

Liu, X., Zhang, C. F., Zhou, J. G., Xiong, X., & Wang, Y. P. (2022). Thermal performance of battery thermal management system using fins to enhance the combination of thermoelectric cooler and phase change material. *Applied Energy*, *322*, 119503. DOI: 10.1016/j.apenergy.2022.119503

Lyu, Z., Su, J., Li, Z., Yan, H., & Chen, L. (2024). Compact hybrid BTMS with nanofluid, PCM, and pulsed flow strategies. *Energy Conversion and Management*, *301*, 118213.

Lyu, Z., Su, J., Li, Z., Yan, H., & Chen, L. (2024). A compact hybrid battery thermal management system using nanofluid cooling, PCM, and pulsed flow strategies. *Energy Conversion and Management*, *301*, 118213. DOI: 10.1016/j.enconman.2024.118213

Mane, N. S.. (2024). Thermal control of electronics using composite PCM with long fins. *Energy Storage Technology*, *12*(6), 105014.

Mane, N. S., Kodancha, P., Hemadri, V., & Tripathi, S. (2024). Investigation on cooling performance of composite PCM and graphite fin for battery thermal management system of electric vehicles. *Energy Storage*, *6*(6), e70024. DOI: 10.1002/est2.70024

Murugan, M.. (2025). Recent developments in battery thermal management: Challenges and opportunities. *Energy Science & Engineering*, *13*(1), 105–132.

Narkhede, S.. (2023). Design and thermal analysis of fin–PCM integrated battery thermal management systems. *International Journal of Applied Mechanics and Engineering*, *28*(4), 755–770.

Nasiri, M.. (2025). Phase change material systems for lithium-ion battery thermal management: Advances and perspectives. *Journal of Energy Storage*, *91*, 112345.

Panchal, S., Dincer, I., Agelin-Chaab, M., Fraser, R., & Fowler, M. (2017). Transient electrochemical heat transfer modeling and experimental validation of a large sized $LiFePO_4$/graphite battery. *International Journal of Heat and Mass Transfer*, *109*, 1239–1251. DOI: 10.1016/j.ijheatmasstransfer.2017.03.005

Ping, P., Peng, R., Kong, D., Chen, G., & Wen, J. (2018). Investigation on thermal management performance of PCM-fin structure for Li-ion battery module in high-temperature environment. *Energy Conversion and Management*, *176*, 131–146. DOI: 10.1016/j.enconman.2018.09.025

Ping, P., Peng, R., Kong, D., Chen, G., & Wen, J. (2018). Thermal management performance of PCM-fin structures for Li-ion modules in high-temperature environments. *Energy Conversion and Management*, *176*, 131–146.

Rahmani, A.. (2024). Recent advancements in battery thermal management for electric vehicles. *Batteries*, *10*(2), 112.

Rahmani, A., Sharma, S., & Kumar, R. (2024). Recent advancements in battery thermal management systems for electric vehicles. *Batteries*, *10*(8), 265. DOI: 10.3390/batteries10080265

Rao, A. P., & Kumar, S. R. S. (2024). A review of integrated battery thermal management systems for lithium-ion batteries of electric vehicles. *E-Prime: Advances in Electrical Engineering. Electronics and Energy*, *8*, 100526. DOI: 10.1016/j.prime.2024.100526

Rasool, G.. (2024). Recent advancements in battery thermal management for EVs. *Frontiers in Energy Research*, *12*, 1402932. DOI: 10.3389/fenrg.2024.1402932

Roy, A.. (2024). Reliability analysis of prismatic lithium-ion batteries under varying thermal environments. *Philosophical Magazine*, *104*(12), 2135–2154. DOI: 10.1080/14786435.2024.1234567

Saber, N.. (2025). Battery thermal management strategies for prismatic cells: Challenges and opportunities. *Energies*, *18*(5), 2154.

Schaeffler, S.. (2024). In-situ measurement of internal thermal gradients and their effect on thermal runaway onset. *Journal of Power Sources*, *589*, 233523.

Schöberl, J.. (2024). Thermal runaway propagation in large-format automotive Li-ion modules. *Journal of Energy Storage*, *72*, 108619.

Shareef, S. M., & Rao, G. A. P. (2025). Comparative Evaluation of Phase Change Materials and Fins in Battery Thermal Management During High Discharge. *Energy Storage*, *7*(7), e70271. DOI: 10.1002/est2.70271

Sun, X.. (2023). Critical review of thermal runaway prediction and mitigation. *Energy Materials Advances*, *2*, 100008. DOI: 10.34133/energymatadv.000008

Wahab, A.. (2025). Immersion cooling for EV batteries: Recent innovations and challenges. *Renewable & Sustainable Energy Reviews*, *195*, 113281. DOI: 10.1016/j.rser.2025.113281

Wahab, A.. (2025). Advances and challenges in immersion cooling for electric vehicle batteries. *Renewable & Sustainable Energy Reviews*, *195*, 113281.

Wang, G., Gao, Q., Yan, Y., & Wang, Y. (2020). Thermal management optimization of a lithium-ion battery module with graphite sheet fins and liquid cold plates. *Automotive Innovation*, *3*(4), 336–346. DOI: 10.1007/s42154-020-00121-1

Wang, G., Gao, Q., Yan, Y., & Wang, Y. (2020). Thermal management optimization of a lithium-ion battery module with graphite sheet fins and liquid cold plates. *Automotive Innovation*, *3*(4), 336–346. DOI: 10.1007/s42154-020-00121-1

Wang, G., Gao, Q., Yan, Y., & Wang, Y. (2020). Thermal management optimization of lithium-ion battery modules with graphite sheet and aluminum fins. *Automotive Innovation*, *3*(4), 336–346.

Wang, P.. (2025). Thermal runaway modeling of large 280 Ah LFP prismatic cells. *Energy*, *280*, 127988. DOI: 10.1016/j.energy.2025.127988

Weng, J., Ouyang, D., Yang, X., Chen, M., Zhang, G., & Wang, J. (2020). Optimization of the internal fin in a phase-change-material module for battery thermal management. *Applied Thermal Engineering*, *167*, 114698. DOI: 10.1016/j.applthermaleng.2019.114698

Weng, J., Ouyang, D., Yang, X., Chen, M., Zhang, G., & Wang, J. (2020). Optimization of internal fins in PCM modules for improved battery thermal management. *Applied Thermal Engineering*, *167*, 114698.

Worwood, D.. (2019). Experimental benchmarking of fin materials for large pouch cells. *Energies*, *12*(23), 4495. DOI: 10.3390/en12234495

Xu, L.. (2024). Thermal management and heat transfer in lithium-ion batteries: A comprehensive review. *Energies*, *17*(16), 3873.

Xu, X., Chen, X., Shen, J., Kong, J., Zhang, H., & Zhou, F. (2023). Thermal management system for prismatic battery module with biomimetic cephalofoil fin and film heater. *Applied Thermal Engineering*, *227*, 120379. DOI: 10.1016/j.applthermaleng.2023.120379

Yu, Z.. (2023). BTMS with heat-pipe/PCM hybrids: A review. *Journal of Energy Storage*, *68*, 107532. DOI: 10.1016/j.est.2023.107532

Zhang, X.. (2025). Comprehensive analysis of airflow and PCM coupling in Li-ion battery packs. *Energy Reports*, *11*(2), 10921–10935.

Zhao, Y.. (2024). Review of PCM-based battery thermal management systems: Convection and conduction effects. *Journal of Energy Storage*, *73*, 108920.

Chapter 11
Performance Analysis of a Low-Charge NH3 System With CO2-Based Natural Circulation Loop for Cold Storage Applications

Yogesh Arun Katu
https://orcid.org/0009-0007-5416-2960
Indian Institute of Technology, Kharagpur, India

Krushna Mohan Das
https://orcid.org/0000-0002-1409-3362
Indian Institute of Technology, Kharagpur, India

Maddali Ramgopal
Indian Institute of Technology, Kharagpur, India

P. Srinivasa Rao
Indian Institute of Technology, Kharagpur, India

ABSTRACT

In light of increasing environmental and regulatory concerns, there is a growing emphasis on the use of natural refrigerants. In the present study theoretical performance of a cold storage that uses a NH3 based VCRS with a CO2 based natural circulation loop (NCL) is analyzed. The secondary loop is a rectangular, two-phase, buoyancy driven, NCL with adiabatic riser and downcomer. As addition of secondary loop introduces, energy and exergy analyses of system are carried out

DOI: 10.4018/979-8-3373-3621-3.ch011

to estimate the performance. The key operating parameters such as evaporating and condensing temperatures, temperature differences in evaporator, condenser and intermediate heat exchanger are found to have a significant effect on COP, second law efficiency and exergy destruction per unit refrigeration capacity of the system. Based on the data generated, suitable correlations are developed between the operating and performance parameters. Results show that despite the minor penalty over single-stage ammonia refrigeration system, indirect system provides a safer working space by reducing charge of ammonia.

1. INTRODUCTION

Approximately one-third of global food production is lost due to spoilage, with developing countries like India experiencing nearly 40% post-harvest losses. India requires an estimated 61 million tonnes of cold storage capacity, but current facilities provide only 26.85 million tonnes, resulting in substantial food and resource wastage. Developing efficient cold chain infrastructure is therefore critical to reduce post-harvest losses and ensure year-round availability of perishable products (Parfitt et al., 2010; Arvindraj et al., 2020).

The proper selection of refrigeration system type and working fluids is crucial for achieving energy-efficient and environmentally friendly cold storages. Based on cooling method, refrigeration systems can be classified as direct, indirect, or cascade systems. Refrigerants are broadly categorized as CFCs, HCFCs, HFCs, HFOs, hydrocarbons (HCs), and inorganic fluids. Due to their environmental impacts, CFCs, HCFCs, and HFCs are being phased out and replaced with eco-friendly alternatives (UNEP, 2014). Hydrocarbons (e.g., propane R-290, isobutane R-600a) and inorganic refrigerants (e.g., ammonia R-717, carbon dioxide R-744) offer environmentally safe options and can be applied across most refrigeration and air-conditioning sectors (Evans et al., 2019; ISHRAE, 2015).

Ammonia stands out in the history of refrigerants as the only natural refrigerant that has maintained its prominence due to its exceptional thermophysical and transport properties, which result in superior system performance. It is widely available, cost-effective, and environmentally friendly, with zero ozone depletion potential (ODP) and global warming potential (GWP). Additionally, ammonia's distinctive pungent odour allows for easy detection of leaks. These advantages make ammonia a popular choice for large-scale refrigeration applications. But ammonia is toxic in nature and is also mildly flammable. As a result, use of large quantities of ammonia in refrigerated space poses fire and other hazards. However, it can be used as primary refrigerant with some non-toxic, non-flammable secondary refrigerant so that it can be confined to plant rooms away from the cold storage. Also ammonia charge is reduced when it is used as a primary refrigerant in hybrid systems (International Institute of Ammonia Refrigeration [IIAR], 2019). In hybrid systems, ammonia based refrigeration system is coupled with another system that uses a different working fluid. This can be achieved by using cascade refrigeration systems or secondary/indirect refrigeration systems.

Similarly, carbon dioxide has a critical point of 31.1 °C, which is close to the ambient temperature, making it less suitable as a primary refrigerant for high heat sink temperatures. However, these same characteristics make CO_2 an excellent secondary refrigerant, particularly in natural circulation loop (NCL) systems. CO_2 offers several advantages: its high coefficient of thermal expansion enhances buoyancy forces, thereby strengthening natural circulation; its low viscosity minimizes frictional losses in the loop; and its relatively high operating pressure combined with a low critical temperature ensures strong driving forces for circulation, even at moderate temperature lifts.

Compared to refrigerants such as R-717 (ammonia), R-22, or R-134a, the loop diameters required for CO_2 are smaller for the same heat input and loop length, which translates into more compact and cost-effective system designs. CO_2 also exhibits the smallest temperature difference across heat exchangers in both liquid and vapour phases, ensuring better thermodynamic performance. When compared to water in NCLs under identical heat transfer and loop height conditions, CO_2 consistently enables higher circulation rates and more compact designs (Kumar & Ramgopal, 2009).

From a practical standpoint, CO_2 avoids common issues associated with brines such as corrosion, material incompatibility, and toxicity concerns. It is also inexpensive, abundant, and environmentally benign, with the added benefit that large-scale utilization of captured CO_2 in closed refrigeration systems can contribute to climate change mitigation. Taken together, these thermal, hydraulic, environmental, and economic advantages provide a strong justification for selecting CO_2 as the secondary fluid in the NH_3– CO_2 cascade cold storage system.

Unlike FCL, the NCLs do not have any moving components (e.g. pumps), hence NCLs are simpler in construction and are inherently more reliable than forced circulation system. Hence for the present study, CO_2 based secondary natural circulation system is proposed with ammonia as a primary refrigerant for cold storage for storing vegetables.

Cascade systems, which employ ammonia as the high-temperature refrigerant and CO_2 as the low-temperature refrigerant, are becoming increasingly popular due to their excellent overall performance. Several studies are carried out on NH_3/CO_2 cascade refrigeration systems (Lee et al., 2006; Getu and Bansal, 2008; Dopazo and Seara, 2011; Rezayan and Behbahaninia, 2011; Mehdi et al., 2014; Patel et al., 2019; Laguri et al., 2021). Additionally, recent work has focused on CO_2 based natural circulation loops. For example, a homogeneous-flow model was developed to study the design and performance of two-phase CO_2 NCLs, highlighting the importance of optimizing loop size and mass flow for standstill pressure management (Das et al., 2025). Furthermore, a quantitative assessment of low-GWP zeotropic CO_2 mixtures demonstrated their potential to reduce high standstill and operating pressures while maintaining favourable performance (Das et al., 2025).

Despite these advances, there remains limited work on comprehensive performance evaluation of hybrid systems combining a primary ammonia circuit with a secondary CO_2 NCL, particularly with detailed component-wise energy and exergy analysis and comparative assessment against conventional single-stage ammonia and cascade refrigeration systems. The addition of a secondary loop introduces additional irreversibility due to the extra heat transfer resistance between the primary and secondary refrigerants. To assess the extent of this on the system performance, energy and exergy analyses are performed.

In conventional refrigeration systems, ammonia is typically utilized either in pure ammonia systems or in cascade refrigeration systems in conjunction with another suitable refrigerant. Carbon dioxide, propane, propylene, isobutane, ethane, and ethylene are reported to be suitable with ammonia in cascade systems (Sarkar et al., 2013). As cascade systems also offer a solution to reduce the ammonia charge similar to the NCL system, in the present study a comparison is made among the proposed NH_3 refrigeration system with an indirect CO_2 NCL, conventional single stage NH_3 system and the cascade refrigeration systems with ammonia in high temperature cycle and other suitable natural refrigerants in low temperature cycle.

2. SYSTEM DESCRIPTION AND THERMODYNAMIC ANALYSIS:

The system proposed comprises of two sub-systems, namely CO_2 NCL and NH_3 vapor compression refrigeration system (VCRS). Figure 1 shows the schematic of the NH_3/CO_2 indirect refrigeration system considered in the present study.

Figure 1. Schematic of the NH_3/CO_2 Indirect Refrigeration System

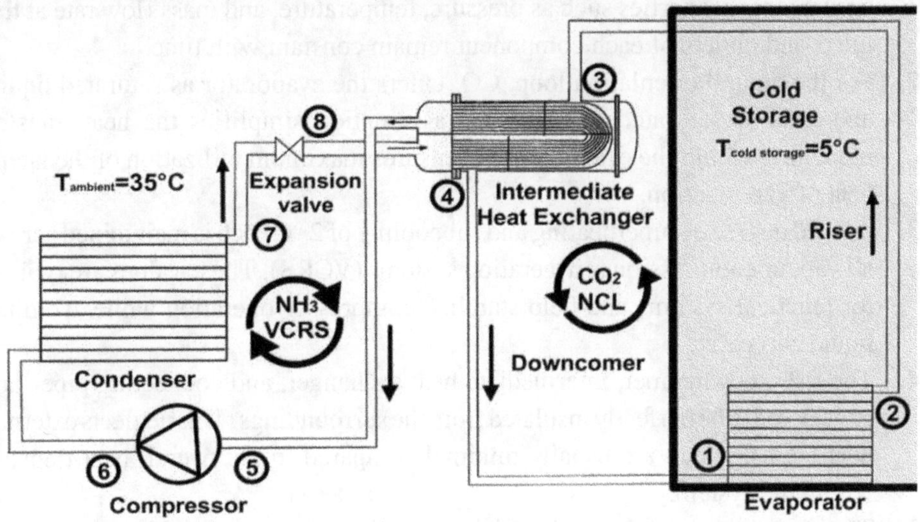

The primary refrigeration system working with ammonia as refrigerant, consists of compressor, condenser, expansion valve and evaporator. Secondary NCL refrigeration system which is to be used to maintain temperature inside cold storage comprises of the intermediate heat exchanger, CO_2 evaporator, riser and downcomer. The evaporator of secondary CO_2 system is inside cold storage as shown in figure 1. Air in the cold room is circulated over the evaporator by using fans which reduces the temperature and moisture content of the air. This cold and dry air is now circulated throughout the room to maintain the required conditions inside the cold storage. The cold storage temperature is assumed to be 5 °C as it is suitable for most of the vegetables and fruits.

For the sake of simplicity, the conventional ammonia system considered here is a single stage ammonia system. For the sake of comparison, the performance of the proposed system is compared with that of a single stage ammonia VCRS and cascade systems with ammonia on high temperature side with carbon dioxide, propane, propylene, isobutane, ethane, and ethylene on low temperature side (Sarkar et al., 2013).

A thermodynamic model for each system is developed and solved using EES (Engineering Equation Solver). The following assumptions are made to simplify the analysis:

1. All components are assumed to operate under steady-state conditions. This ensures that properties such as pressure, temperature, and mass flow rate at the inlets and outlets of each component remain constant with time.
2. For the natural circulation loop, CO_2 enters the evaporator as saturated liquid and exits as saturated vapour. This assumption simplifies the heat transfer modelling within the evaporator and ensures maximum utilization of the latent heat of vaporization.
3. A fixed degree of superheating and subcooling of 2 °C each is maintained across all vapour compression refrigeration systems (VCRS). These values are typical for practical systems and help stabilize compressor operation while avoiding liquid carryover.
4. The riser, downcomer, intermediate heat exchanger, and connecting pipes are considered to be perfectly insulated from the surroundings. This neglects external heat losses, which are usually minimal compared to the overall heat transfer within the system.
5. The isentropic and volumetric efficiencies of compressors for refrigerants other than NH_3 and CO_2 are taken as 0.7 and 0.85, respectively, which are standard values reported in literature (Lee et al., 2006; Sarkar et al., 2013).
6. All throttling processes are assumed to be isenthalpic, which is a standard thermodynamic assumption for expansion valves and capillaries where no work interaction or significant heat transfer occurs.
7. Changes in kinetic and potential energy across system components are neglected, as these are insignificant compared to enthalpy changes in refrigeration processes.
8. The power consumption of the NH_3 condenser fan and the CO_2 evaporator fan is neglected. Their contribution to the overall energy balance is comparatively small, and excluding them simplifies the analysis without significantly affecting accuracy.

The input parameters used for the analysis of single stage ammonia, the NH_3/CO_2 indirect refrigeration system and cascade systems are summarized in Table 1. These values are based on the design of a small 5 MT cold storage for storage of fresh food.

Table 1. Input Parameters for the Energy and Exergy Analysis

Parameter	Value
Cooling capacity	1.5 TR (≈5.25 kW)
Refrigerated (cold storage) space temperature	5 °C
Ambient temperature	35 °C
Temperature difference for heat transfer: Between refrigerant & air (in condenser) Between secondary fluid (CO_2) & air (in evaporator) Between refrigerants (in IHX or cascade heat exchanger)	 7 K 7 K 3 K
Degree of superheating and subcooling in all VCRS	2 K
Pressure drop on secondary fluid (CO_2) side (Das et al., 2024) Evaporator Downcomer Riser	 11.11 kPa 17.99 kPa 4.09 kPa

For each component in the NH_3/CO_2 indirect refrigeration system, mass, energy, and exergy balances are applied under steady-state conditions, as depicted in Figure 1. The exergy balance quantifies the maximum useful work that can be obtained from the energy in the system and identifies where irreversibilities reduce this potential. In this system, exergy is lost due to several mechanisms:

- Evaporator (1–2): losses occur as CO_2 absorbs heat from the cold storage, with entropy generation in the heat transfer process.
- Riser (2–3) and Downcomer (4–1): frictional losses and mixing reduce the exergy carried by the CO_2 loop.
- Intermediate Heat Exchanger (3–4): heat exchange between NH_3 and CO_2 results in exergy destruction due to temperature differences.
- Compressor (5–6) and Condenser (6–7): irreversibilities from compression and heat rejection to the ambient decrease the exergy of NH_3.
- Expansion Valve (7–8): throttling generates exergy loss due to isenthalpic expansion.

While total energy is conserved throughout the system, exergy identifies where the energy has reduced ability to perform useful work. For each of the components depicted in Figure 1, the mass balance [eq. (1)], energy balance [eq. (2)] and exergy balance [eq. (3)] under steady state are as shown below.

$$\sum_{in} \dot{m} = \sum_{out} \dot{m} \quad (1)$$

$$\dot{Q} - \dot{W} + \sum_{in} \dot{m}h - \sum_{out} \dot{m}h = 0 \quad (2)$$

$$\dot{E}x_{D,k} = \left(1 - \left(T_{amb}/T_r\right)\right) \cdot \dot{Q}_k + \sum \dot{m}_R \cdot \left(h_{in} - h_{out} - T_{amb} \cdot \left(s_{in} - s_{out}\right)\right) \tag{3}$$

Where, $\dot{Q}_k$ is the rate of heat transfer through the kth component at temperature T_r and $\dot{E}x_{D,k}$ is the exergy destruction rate in W.

Isentropic efficiencies (η_{isen}) of the ammonia and CO_2 compressors are found by using eq. (4) (IIAR, 2008) and eq. (5) (Lee et al., 2006), respectively.

$$\eta_{isen,NH_3} = 1.26 - 0.56R_p + 0.22R_p^2 - 0.043R_p^3 + 0.0042R_p^4 \tag{4}$$

Where, R_p is the compression ratio and eq. (4) is valid for the range of R_p from 2 to 14.

$$\eta_{isen,CO_2} = 0.89810 - 0.923R_p - 0.00416R_p^2 \tag{5}$$

Volumetric efficiencies (η_{vol}) of the ammonia and CO_2 compressors are found using eq. (6) & (7), respectively (Lee et al., 2006).

$$\eta_{vol,NH_3} = 1.03231 - 0.0508R_p - 0.00076R_p^2 \tag{6}$$

$$\eta_{vol,CO_2} = 1.13413 - 0.15293R_p - 0.00816R_p^2 \tag{7}$$

Due to the loss of efficiency in compressor, exergy is destructed in the compressor. But when using eq. (3) for calculating the exergy destruction in compressor and condenser, destruction of the exergy due to inefficiencies in the compressor appears in the exergy destruction of the condenser. Hence, to account for the contribution of compressor to exergy destruction correctly, the exergy destruction in the condenser is found with 100% isentropic efficiency first. This is treated as the actual exergy destruction in the condenser. Now, the difference in exergy destructions for 100% efficiency and actual efficiency of compressor is found and is added to the exergy destruction of the compressor calculated with actual efficiency. Total exergy destruction of the complete system is obtained by adding exergy destruction of each component. First law efficiency or coefficient of the performance (COP) of the system is given by eq. (8) and Carnot COP by eq. (9).

$$COP = \frac{\dot{Q}_{Evaporator}}{\dot{W}_{Compressor}} \tag{8}$$

$$COP_{Carnot} = \frac{T_{Cold\_Storage}}{T_{amb} - T_{Cold\_Storage}} \tag{9}$$

Second law efficiency or exergy efficiency is given by eq. (10).

$$\eta_{II} = \frac{COP}{COP_{Carnot}} \tag{10}$$

Ideal work is given by eq. (11) and lost work by eq. (12).

$$\dot{W}_{ideal} = \frac{\dot{Q}_{Evaporator}}{COP_{Carnot}} \tag{11}$$

$$\dot{W}_{lost} = \dot{W}_{actual} - \dot{W}_{ideal} \tag{12}$$

For cascade system, coefficient of the performance (COP) of the system is given by eq. (13),

$$COP = \frac{COP_{LTC} \cdot COP_{HTC}}{1 + COP_{LTC} + COP_{HTC}} \tag{13}$$

For comparison of different systems, the single stage R-717 system is considered as the base system. COP of any other system relative to single stage R-717 system is given by eq. (14)

$$COP_{Relative} = \frac{COP_{system}}{COP_{SSR-717}} \tag{14}$$

The volumetric refrigeration capacity of the system which is an indication of the compactness of the compressor is given by eq. (15) and volumetric refrigeration capacity relative to single stage R-717 system is given by eq. (16)

$$Vc = \frac{\dot{Q}_{Evaporator}}{\sum \dot{m} \cdot v_{CompressorIn}} \tag{15}$$

$$Vc_{Relative} = \frac{Vc_{System}}{Vc_{SSR-717}} \tag{16}$$

Figures 2 & 3 show the pressure (P)-enthalpy (h) and temperature (T) – entropy (s) plots of the NH_3 VCR system with CO_2 in the secondary loop.

Figure 2. P-h Diagram of the System

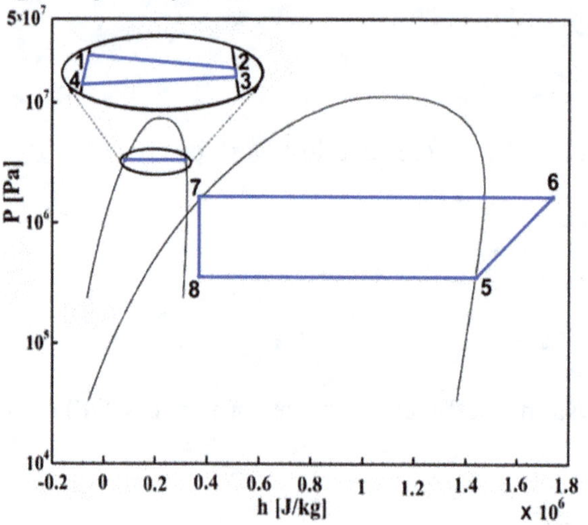

Figure 3. T-s Diagram of the System

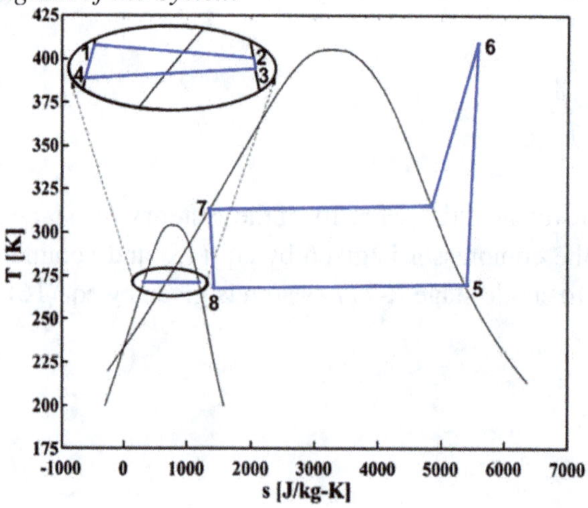

In the figures 2 and 3, 1-2-3-4 is CO_2 natural circulation loop (NCL) and 5-6-7-8 is ammonia VCRS. 1-2 is CO_2 evaporation in evaporator, 2-3 is pressure drop in riser, 3-4 is heat exchange in intermediate heat exchanger and 4-1 is pressure rise downcomer. As shown in the figures, in the natural circulation loop (NCL) secondary system, the pressure drop and temperature difference in the source and sink are very small, causing the process to appear as a nearly straight lines on P-h and T-s plots. Hence enlarged views of NCL are shown in these figures to indicate typical pressure and temperature variations.

3. RESULTS AND DISCUSSION

Using the developed mathematical models, the thermodynamic analyses are performed. A performance comparison is made between single stage ammonia VCRS, R-717/R-744 system with NCL and ammonia cascade system with different refrigerants in low temperature circuit. Component-wise exergy destruction and effect of different operating paerameters on system performance is studied for the NH_3 refrigeration system with an indirect CO_2 NCL.

3.1 Thermodynamic Analysis of the NH_3 Refrigeration System With an Indirect CO_2 NCL

Using input parameters listed in table 1, the results obtained are presented in table 2.

Table 2. Thermodynamic Analysis Results of the NH_3 Refrigeration System with CO_2 NCL

Parameter	NH_3 refrigeration system with CO_2 NCL
COP	3.634
COP_{carnot}	9.272
ηl	0.392
$\dot{W}_{actual}$	1452 W
$\dot{W}_{ideal}$	569 W
$\dot{W}_{lost}$	882.8 W
$\dot{E}x_{D,Total}$	882.8 W

From table 2, it can be observed that the lost work and total exergy destruction in the system are equal, indicating that the model accurately accounts for exergy destruction in all the components. Figure 4 below shows component-wise exergy destruction in the system.

Figure 4. Component-wise Exergy Destruction in the System

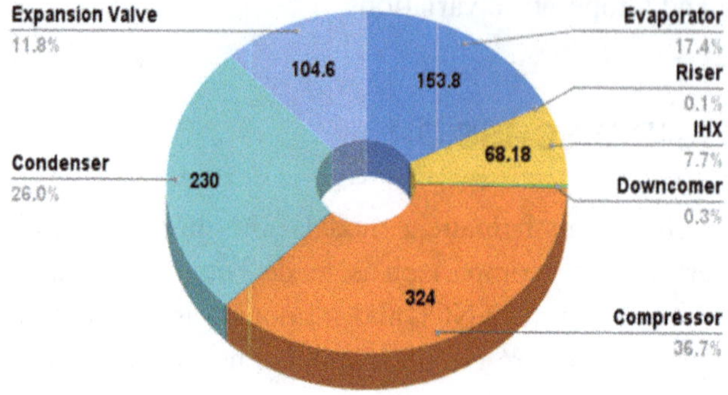

From figure 4 it can be seen that the maximum exergy destruction is in the compressor followed by condenser and CO_2 evaporator. For ammonia, compressor discharge temperature is very high compared to other refrigerants. Hence, condensers of ammonia generally have more exergy destruction compared to other refrigerants. For the indirect system, minimum exergy destruction takes place in riser and downcomer. As these components are assumed to be perfectly insulated, pressure drop is the source for the exergy destruction in both these components.

With the developed thermodynamic model, the effects of various parametrs on COP, second law efficiency and total exergy destruction are analysed. These parameters include pressure drop in NCL, degree of subcooling & degree of superheating for ammonia, temperature difference between refrigerant and air inside cold storage in evaporator, temperature difference between refrigerant and ambient air in condenser, evaporating temperature and condensing temperature. To examine the effect of pressure drop in NCL, pressure drop is varied from 0 to twice the design pressure drop value. With this variation, the COP decreases by 1.63%, the second law efficiency decreases by 1.7%, and the total exergy destruction increases by 2.9%. Though pressure drop in NCL plays crucial role in designing the NCL, it does not have significant effect on performance parameters of the complete system.

Similarly, the degree of superheating and subcooling are varied from 0 to 7 K. For the degree of superheating, when varied from 0 to 7 K, changes in COP, second law efficiency, and exergy destruction per unit refrigeration capacity are 1.61%, 1.6%, and 2.7%, respectively. For the degree of subcooling, when varied from 0 to 7 K, changes in COP, second law efficiency, and exergy destruction per unit refrigeration capacity are 3.12%, 3.22%, and 5.16%, respectively.

It is observed from the analysis that changes in evaporating temperature, condensing temperature, temperature difference between the refrigerant and air inside the cold storage in the evaporator, temperature difference between the refrigerant and ambient air in the condenser, and the temperature difference between ammonia and CO_2 in the intermediate heat exchanger are having significant effect of performance parameters. Regression analysis is conducted to establish the relationship between these independent parameters and each of the three dependent parameters. Table 3 shows the range of variation in the independent parameters considered for the regression analysis.

Table 3. Independent Parameter Ranges for Regression Analysis

Independent parameter	Minimum value (°C)	Maximum value (°C)
Temperature difference between air and refrigerant in condenser (ΔT_{Cond})	1	10
Temperature difference between air and refrigerant in evaporator (ΔT_{Evap})	1	10
Condensing temperature (T_{Cond})	32	52
Evaporating temperature (T_{Evap})	-17	3
Temperature difference between ammonia and CO_2 in intermediate heat exchanger	1	10

Equations (17), (18) & (19) given below show the relation between the above independent parameters and COP, second law efficiency and total exergy destruction per unit refrigeration capacity of the system, respectively.

$$COP = 12.028 - 0.115 \cdot \Delta T_{Evap} + 0.0012 \cdot \Delta T^2_{Evap} - 0.105 \cdot \Delta T_{Cond} + 0.0013 \cdot \Delta T^2_{Cond} - 0.219 \cdot T_{Cond} + 0.0015 \cdot T^2_{Cond} + 0.1002 \cdot T_{Evap} + 0.0003 \cdot T^2_{Evap} - 0.099 \cdot \Delta T_{IHX} + 0.00034 \cdot \Delta T^2_{IHX} \quad (17)$$

$$\eta_{II} = 0.071 - 0.0123 \cdot \Delta T_{Evap} + 0.00012 \cdot \Delta T^2_{Evap} - 0.0112 \cdot \Delta T_{Cond} + 0.00013 \cdot \Delta T^2_{Cond} + 0.0199 \cdot T_{Cond} - 0.00019 \cdot T^2_{Cond} - 0.0049 \cdot T_{Evap} - 0.0003 \cdot T^2_{Evap} - 0.0106 \cdot \Delta T_{IHX} + 0.000034 \cdot \Delta T^2_{IHX} \quad (18)$$

$$\overline{Ex_D} = 0.0328 + 0.0057 \cdot \Delta T_{Evap} + 0.000123 \cdot \Delta T^2_{Evap} + 0.00548 \cdot \Delta T_{Cond} + 0.000082 \cdot \Delta T^2_{Cond} - 0.0022 \cdot T_{Cond} + 0.0000637 \cdot T^2_{Cond} - 0.0024 \cdot T_{Evap} + 0.00026 \cdot T^2_{Evap} + 0.006 \cdot \Delta T_{IHX} + 0.0002 \cdot \Delta T^2_{IHX} \quad (19)$$

In equations (17), (18) & (19), all independent parameters are in °C and all dependent parameters are dimensionless. Table 4 shows the root mean square (RMS) value and R^2 value for eqns. (17), (18) & (19).

Table 4. Root Mean Square (RMS) Value and R^2 Value for Developed Correlations

Equation	RMS	R^2 (%)
Eq. (17)	0.0183	99.83
Eq. (18)	0.00192	99.47
Eq. (19)	0.00153	99.69

RMS and R^2 values for the above equations are well within the range to predict the accurate values. Hence eq. (17), (18) & (19) can be used to predict the COP, second law efficiency and total exergy destruction per unit refrigeration capacity in the system.

3.2 Comparison With Conventional Systems

In this work, the conventional systems refer to the single stage ammonia VCRS and cascade refrigeration systems with ammonia in high temperature circuit and selected low GWP refrigerants in low temperature circuit. Carbon dioxide (R-744), propane (R-290), isobutane (R-600a), propylene (R-1270), ethane (R-170), nitrous oxide (R-744a) and fluoromethane (R-41) are considered to be well-suited for use in cascade refrigeration systems with ammonia (R-717) (Sarkar et al., 2013). For a cold storage application designed to store vegetables at a temperature of 5°C, the performance of these cascade refrigeration systems is analysed in terms of COP. To compare these systems, it is necessary to determine the optimum cascade temperature that maximizes the COP for the selected application. Figure 5 shows the variation in COP of these cascade systems with the cascade condensing temperature (T_{CC}).

Figure 5. Variation in COP with Cascade Condensing Temperature

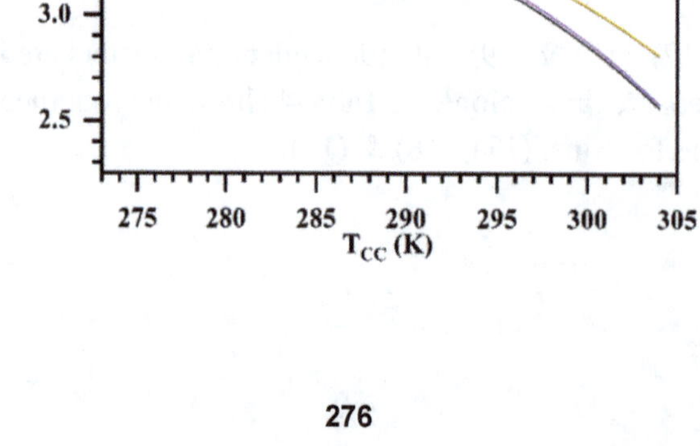

From figure 5, it is observed that for the cascade refrigeration system with R-290, R-600a, and R-1270, the optimum COP can be achieved within the operating temperature range of the selected application, even though the variation is very small. For other refrigerants, there is no optimum cascade condensing temperature in the given range. The optimum cascade condensing temperatures for R-290, R-600a, and R-1270 are 292 K, 295 K, and 292 K respectively. Therefore, for comparison, R-290, R-600a, and R-1270 are chosen as refrigerants in the cascade systems with ammonia in the high-temperature circuit. Figure 6 shows the variation of COP with ambient temperature for proposed system and selected cascade refrigeration systems.

Figure 6. Variation in COP with Ambient Temperature

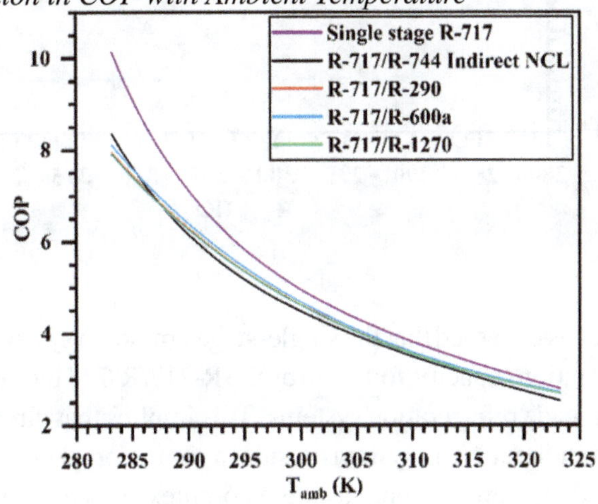

From figure 6, it is observed that the single-stage ammonia system exhibits the highest COP compared to the other systems. This is expected as the other systems must overcome an additional temperature difference in the cascade or intermediate heat exchanger. At low ambient temperatures, the COP of the R-717/R-744 system with indirect NCL performs better than the cascade systems. However, as the ambient temperature increases, its COP becomes marginally lower compared to the cascade systems, with a maximum difference of 6.4% observed within the operating range of ambient temperatures. Figure 7 shows the variation in volumetric refrigeration capacities of the systems with ambient temperature.

Figure 7. Variation in Volumetric Refrigeration Capacities with Ambient Temperature

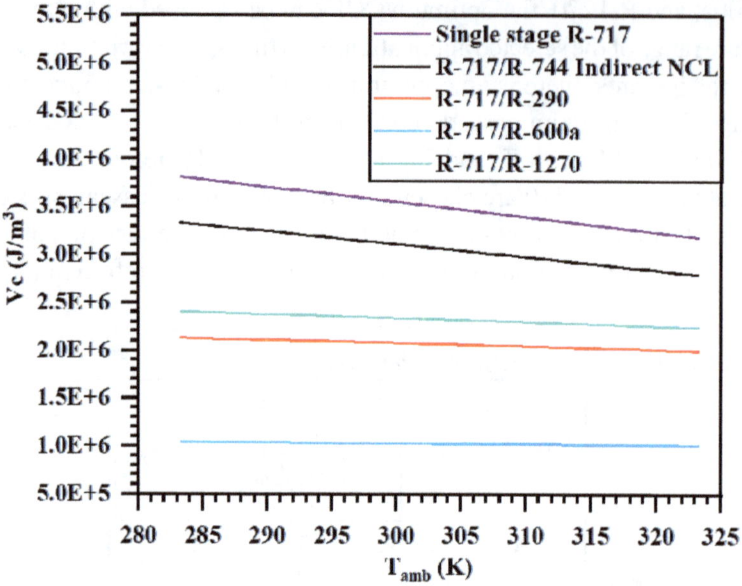

From figure 7, it is observed that the single-stage ammonia system has the highest volumetric refrigeration capacity, followed by the R-717/R-744 indirect NCL system, and then all the cascade refrigeration systems. This implies that direct R-717 system and R-717/R-744 indirect NCL systems result in more compact compressor compared to the cascade systems, as only a single compressor is used in these systems.

3.3 Design Considerations and Practical Implications

While the theoretical study shows thc potential of the proposed NH_3/CO_2 system, there are several practical challenges in implementing it. Availability of small-capacity ammonia compressors, such as 1.5 TR, is limited, which makes selecting the right compressor challenging. Handling ammonia requires strict safety precautions since it is toxic and flammable; the system must ensure proper leakage detection and ventilation. For the natural circulation loop (NCL), careful design of the downcomer and its height is critical to ensure proper circulation. Additionally, the design should minimize pressure drops across all components to reduce the required height of the NCL. CO_2 presents challenges due to its high standstill pressure at ambient temperature, requiring all components to be rated for high pressure. The initial CO_2 charging also requires care: it is charged to cylinder pressure, then cooled by the primary system to reduce pressure, and this process is repeated until the desired pressure is achieved. Addressing these challenges is essential for safe operation and for achieving the predicted system performance.

In spite of these challenges, the proposed NH_3/CO_2 system offers several practical advantages. The secondary CO_2 loop allows retrofitting into existing ammonia plants with minimal modifications. Reduced ammonia charge improves safety in occupied spaces and simplifies compliance with regulations. High volumetric refrigeration capacity enable compact compressors, reducing the system footprint and potentially lowering both energy consumption and maintenance costs. Furthermore, the use of natural refrigerants ensures environmental compliance while maintaining reliable operation in cold storage applications.

4. CONCLUSIONS

This study presents a comprehensive theoretical performance analysis of an ammonia-based vapor compression refrigeration system integrated with a carbon dioxide (CO_2) natural circulation loop serving as a secondary circuit. The analysis focused on component-wise exergy destruction, parametric influences, regression model development, and comparative evaluation with conventional refrigeration systems.

Key findings from the exergy and parametric analysis are as follows:

- The highest exergy destruction occurs in the ammonia compressor (36.7%), followed by the ammonia condenser (26%) and CO_2 evaporator (17.4%). Losses in the riser (0.1%) and downcomer (0.3%) are negligible.
- Performance metrics such as COP, second law efficiency, and total exergy destruction are strongly influenced by evaporating temperature, condensing temperature, and heat exchanger temperature differences.
- Parameters such as pressure drop in the CO_2 NCL, degree of subcooling, and superheating of NH_3 show relatively minor impacts within the studied range.
- Using the simulation results, empirical correlations were developed to relate performance parameters with the governing variables, with regression models yielding excellent predictive accuracy ($R^2 \approx 99\%$) across the considered range.

Additionally, the performance of the proposed indirect system was compared with two conventional refrigeration configurations: a single-stage ammonia VCRS and ammonia-based cascade refrigeration systems employing various low-temperature refrigerants (R-290, R-600a, R-1270, among others). The analysis revealed following comparative findings:

- The single-stage NH_3 system achieved the highest COP and volumetric refrigeration capacity.
- The proposed NH_3/CO_2 system demonstrated COP comparable to cascade systems.
- The proposed system outperformed cascade configurations in volumetric refrigeration capacity, indicating potential for enhanced compressor compactness and reduced equipment size.
- Single-stage ammonia operation offers superior efficiency under lower ambient temperatures, cascade and indirect NCL systems become comparatively competitive at higher ambient conditions.

Despite a slightly lower COP compared to the single-stage ammonia system, the proposed NH_3/CO_2 system offers significant advantages in terms of safety and environmental containment. By substantially reducing the ammonia charge and limiting its presence to the plant room, the system minimizes the risk associated with ammonia leakage in occupied spaces. Therefore, the proposed configuration strikes a favourable balance between thermodynamic performance and operational safety, making it a promising solution for cold storage and similar refrigeration applications.

Nomenclature

COP- coefficient of performance
$\dot{E}x_{D,k}$- rate of exergy destruction in kth component (W)
$\overline{Ex_D}$- rate of exergy destruction per unit refrigeration capacity
HTC- high temperature circuit
h- specific enthalpy (J/kg)
LTC- low temperature circuit
$\dot{m}$- mass flow rate (kg/s)
P- pressure (Pa)
$\dot{Q}$- heat transfer rate (W)
Rp- compression ratio
s- specific entropy (J/kg-K)
T- temperature (K)
T_{CC}- cascade condensing temperature (K)
VCRS- vapor compression refrigeration system
Vc- volumetric refrigeration capacity (J/m^3)
v- specific volume (m^3/kg)
$\dot{W}$- work (W)

Subscripts

amb- ambient
Cond- condenser

Evap- evaporator
IHX- intermediate heat exchanger
R- refrigerant
Greek symbol
η_l- second law efficiency
η_{isen}- isentropic efficiency
η_{vol}- volumetric efficiency

ACKNOWLEDGEMENTS

The authors would like to acknowledge the technical and financial support provided by Science & Engineering Research Board (SERB) and Indian Society of Heating, Refrigerating and Air Conditioning Engineers (ISHRAE).

REFERENCES

Das, K. M., Guha, A., & Ramgopal, M. (2024). Studies on CO_2 based two-phase natural circulation loop. In *Proceedings of the 27th National and 5th International ISHMT-ASTFE Heat and Mass Transfer Conference.*

Das, K. M., Guha, A., & Ramgopal, M. (2025). Quantitative assessment of various low-GWP zeotropic mixtures of CO_2 as the working fluid in two-phase natural circulation loops. *Applied Thermal Engineering*, *266*, 125512.

Das, K. M., Ramgopal, M., & Katu, Y. A. (2025). Performance of a refrigeration system with a two-phase CO_2-based natural circulation loop for cold storage applications. *Thermal Science and Engineering Progress*, *58*, 103184.

Dopazo, J. A., & Seara, J. F. (2011). Experimental evaluation of a cascade refrigeration system prototype with CO_2 and NH_3 for freezing process applications. *International Journal of Refrigeration*, *34*, 257–267.

Evans, J., Curlin, J., & Clark, E. (2019). IIR UN Environment brief on cold storages and refrigerated warehouse. *UN Environment and IIR* (2019).

Getu, H. M., & Bansal, P. K. (2008). Thermodynamic analysis of an R744-R717 cascade refrigeration system. *International Journal of Refrigeration*, *31*, 45–54.

Indian Society of Heating, Refrigeration and Air Conditioning Engineers. (2015). *ISHRAE position document on refrigerants for Indian refrigeration & air conditioning industries - Challenges & responsibilities.*

Indian Society of Heating. Refrigeration and Air Conditioning Engineers. International Institute of Ammonia Refrigeration. (2019). *The history of ammonia refrigeration* International Institute of Ammonia Refrigeration.

International Institute of Ammonia Refrigeration. (2008). *Ammonia Data Book: The Profile of a Sustainable Refrigerant.* International Institute of Ammonia Refrigeration.

Kumar, K., & Ramgopal, M. (2009). Carbon dioxide as a secondary Fluid in natural circulation loops. *Proceedings of the Institution of Mechanical Engineers. Part E, Journal of Process Mechanical Engineering, 223*(3), 189–194.

Laguri, V., Kumar, K., & Kumar, P. (2021). Performance analysis of CO_2/natural refrigerants for cascade refrigeration system. In *Sixth International Conference on polygeneration, IISc.*

Lee, T. S., Liu, C. H., & Chen, T. W. (2006). Thermodynamic analysis of optimal condensing temperature of cascade condenser in CO_2/NH_3 cascade refrigeration systems. *International Journal of Refrigeration, 29*, 1100–1108.

Mehdi, A., Najaf, B., & Rinaldi, A. F. (2014). Exergetic, economic and environmental (3E) analyses, and multi-objective optimization of a CO_2/NH_3 cascade refrigeration system. *Applied Thermal Engineering, 65*, 42–50.

Parfitt, P., Barthel, M., & Macnaughton, S. (2010). Food waste within food supply chains: quantification and potential for change to 2050. *Philosophical Transactions of Royal Society*, 3065-3081. Arvindraj, K., Chinna, A. R., & Paul, J. (2020). A review: Present scenario of cold chain storage facilities in India. *AIP Conference Proceedings*, •••, 2207–2215.

Patel, V., Panchal, D., Prajapati, A., Mudgal, A., & Davies, P. (2019). An efficient optimization and comparative analysis of cascade refrigeration system using NH_3/CO_2 and C_3H_8/CO_2 refrigerant pairs. *International Journal of Refrigeration, 102*, 62–76.

Rezayan, O., & Behbahaninia, A. (2011). Thermodynamic optimization and exergy analysis of CO_2/NH_3 cascade refrigeration systems. *Energy, 36*, 888–895.

Sarkar, J., Bhattacharya, S., & Lal, A. (2013). Selection of suitable natural refrigerants pairs for cascade refrigeration system. *Institution of Mechanical Engineers, Part A: Journal of Power and Energy, 227*(5), 612–622.

United Nations Environment Programme. (2014). *Environmental effects of ozone depletion and its interactions with climate change: 2014 assessment*. United Nations Environment Programme.

Chapter 12
Renewable Energy Applications in Thermo-Fluid Systems

Aditya Shrivastav
https://orcid.org/0009-0000-8461-8171
Ajeenkya D.Y. Patil University, India

Sunil Sankathala
https://orcid.org/0009-0004-7809-7647
Ajeenkya D.Y. Patil University, India

Kashvi Chaturvedi
https://orcid.org/0009-0005-3192-3400
Ajeenkya D.Y. Patil University, India

Susanta Das
https://orcid.org/0000-0002-9314-3988
Ajeenkya D.Y. Patil University, India

ABSTRACT

The rising interest of renewable resources due to the increasing demand of sustainable energy solutions world-wide has heightened the development of thermo-fluid systems. These systems are very significant in facilitating fluid conversion, storage and distribution of energy in adherence to the principles of thermodynamics and fluid mechanics. Due to the increasing rate of climate change and exhaustion of materials, renewable sources have become viable to achieve such as solar, wind, biomass and geothermal sources which have been encouraged by advances in computational fluid dynamics (CFD), and material science, as well as hybrid systems countering intermittency and increasing reliability. In recent research, the combination of mul-

DOI: 10.4018/979-8-3373-3621-3.ch012

tiple sources of renewable energy is emphasized, with solutions providing a move toward operational resilience and system efficiency. This chapter offers a wholesome investigation of the applications of renewable energy within thermo-fluid systems with an emphasis on principles, development and hybrid methods.

1. INTRODUCTION

The world energy industry is experiencing a radical shift, which is dictated by the need to curb climate change, depletion of resources, and sustainable development. The increasing energy demand across the globe due to population growth, urbanization, and economic growth has enhanced the constraints and environmental impacts of the fossil-based energy systems (Nazir et al., 2025; Khan et al., 2017). The use of fossil fuels is directly connected to greenhouse gases, air and water pollution, and the increase in global warming, which is why there is a necessity to change the paradigm of energy production and usage. This awareness has created a rising agreement among policy-makers, industry leaders, and researchers on the need to move towards low-carbon options (Nazir et al., 2025; Khan et al., 2017; Novozhilov, 2001).

Renewable energy technologies have therefore become the mainstay of modern-day energy policies, which provide renewable and clean energy solutions that cut carbon footprints by large margins as opposed to traditional sources of energy. Their installations are even more enabled by limitless climate commitments that align therewith, including people of the Paris Agreement, and robust national and international policies to encourage decarbonization and energy security (Nazir et al., 2025). In this transition, thermo-fluid systems play a central role, since they control the underlying processes of heat transfer, fluid flow, energy conversion, and thermal management, in a broad spectrum of renewable applications. In solar thermal collectors, the flow of winds, bio-mass and geothermal heat-exchangers, the efficiency, reliability and scalability of renewable technologies cannot be debated free of the nature of thermo-fluid optimization (Nazir et al., 2025; Khan et al., 2017). This renewable energy system integration has been traced by improvements in the field of materials science, computational and systems engineering which enable a clear way of modeling renewable energy systems and optimizing them. Computational fluid dynamics (CFD), specifically, can already provide an effective means of simulating the complex flow and heat transfer that could be used to inform the design of high-efficiency heat exchangers, new flow streams, and hybrids. These advancements support resource utilization, reduce intermittency and energy storage capacity, which is utilized to make the thermo-fluid centrality in the new global renewable power industry (Nazir et al., 2025; Khan et al., 2017; Novozhilov, 2001).

Solar energy is another good example, as thermo-fluid principles are the basis of solar thermal collectors, concentrated solar power systems, photovoltaic/thermal hybrids, and thermal storage. The optimization of fluid flows, heat exchange areas, and thermal management plans control the performance of the system, and CFD-based designs lead to the efficiency and stability of the system (Khan et al., 2017; Novozhilov, 2001). Likewise, wind power plants are the perfect example of the combination of aerodynamics and thermodynamics, and aerodynamics of the blades, rotor design, turbulence, and temperature control are key factors of the energy output and the lifetime of the turbine. The recent work on blade materials, active flow control, and cooling strategies demonstrates the current innovation in the field (Cheng et al., 2025; Chen et al., 2025; Ji et al., 2024). In biomass systems, thermo-fluid dynamics determines the efficiency of thermochemical conversion processes like gasification, pyrolysis and combustion. Fluidized bed reactors, which are commonly used in biomass conversion, rely on the interaction of fluid flow, temperature distribution and reaction kinetics to maximize yields and minimize emissions. The development of new catalysts, feedstock pretreatment, and system design integrations is an emerging field of research that demonstrates the increasing role of biomass in the context of circular economies and net-zero emission plans (Zhang, 2025; Shahzad et al., 2024; Wang & Wu, 2023). Geothermal systems also rely on the circulation of fluids and the transfer of heat in both the subsurface reservoirs and engineered systems. Enhanced geothermal systems (EGS) enhance the use of resources by stimulating fractured reservoirs, but there are issues related to scaling, fouling, and management of the reservoir. New developments, such as nanofluid-based heat exchangers and deep-well, are expected to enhance the efficiency and feasibility of geothermal energy by a great extent (Ahmadi et al., 2018).

Hybrid renewable energy systems are the frontiers of sustainable energy innovation, where thermo-fluid synergies among two or more resources, such as solar-geothermal, solar-biomass, and wind-solar combinations, can be used to provide continuous supply, increased resource utilization, and resilience in operations (Kariuki et al., 2024). These systems share common infrastructure like hybrid heat exchangers and thermal storage tanks and improvements in control algorithms and system-level optimization improve the technical and economic performance of these systems. The growing interest in hybridization indicates that there is a greater need to perform thermo-fluid modelling and experimentation to ensure dependability and scalability across multiple conditions (Kariuki et al., 2024). A combination of all these trends proves that thermo-fluid systems are critical in the renewed form of energy technology development. They underpin the science and engineering foundations of efficiency, flexibility, and sustainability and as such, they lie at the centre of the world transitioning to low carbon energy futures (Nazir et al., 2025).

The thermofluids are crucial in renewable energy since it enhances the performance of such devices as Solar Air Heaters (SAHs) (Sutar et al., 2024a; Sutar et al., 2024b). In such systems, specially shaped ribs or turbulators may be added to the plate absorber to increase flow and heat transfer of air in the system. The rib shape, geometries include quadratic and triangular shapes, produce turbulence in airflow thereby enhancing heat absorption and thermal performance. With the help of computational software such as Ansys Fluent, researchers consider the factors of rib size, pitch, airflow velocity, etc. to maximize energy transfer. Findings indicate that the thermal enhancement ratio can also be increased substantially by optimised rib designs, and solar-based heating systems are better at use in areas such as drying crops, food processing, space heating and ventilation (Sutar et al., 2024a; Sutar et al., 2024b). This shows that thermo-fluid systems provide a direct benefit to renewable energy technologies by striving to make the best use of the solar power in real-life scenarios (Sutar et al., 2024a; Sutar et al., 2024b). This chapter presents the detail account of renewable energy applications in the thermo-fluid systems, and the significance of renewable energy solutions in the development of sustainable energy solutions. It looks into the hybridization between systems, and also considers the questions and prospects of optimal energy conversion and management.

2. FUNDAMENTALS OF THERMO-FLUID SYSTEMS IN ENERGY ENGINEERING

Energy engineering is a science that is heavily anchored on the laws of thermodynamics and fluid mechanics, which are the laws governing the changes of energy being transformed, stored and utilized in renewable systems. The first law of thermodynamics or the principle of energy conservation means that energy may not be created and can only be changed. This law is the key background needed in the measurement of the input and output energies of renewable energy systems, such as solar thermal collectors and geothermal, and therefore, the design and the evaluation of the effective conversion cycles (Oyedepo et al., 2025; Gorgulu, 2023). In addition to this, the second law of thermodynamics also brings in the notion of entropy, which defines the direction of energy transformations and the irreversibility of practical processes. This efficiency constraint is especially applicable to renewable technologies, where entropy generation is a key factor in determining exergy performance, as well as system-level performance. Exergy analysis, which takes into account both the quality and the quantity of energy, has become a necessity in optimizing renewable systems, including photovoltaic/thermal collectors, wind turbines, and biomass reactors (Nabwey et al., 2024; Gorgulu, 2023).

Meanwhile, the conservation laws of the fluid mechanics: mass, momentum, energy, are the analytical foundations of a thermo-fluid system design. A mass conservation law is the continuity equation, the application of which is oriented at predicting the processes of flow and phase changes, as well as contacts of the airflow with wind turbines or the dynamics of working fluids in geothermal loops (Gorgulu, 2023). The Navier Stokes equations capture momentum conservation, the basis of analyses of aerodynamic lift and drag on wind turbines and flow distribution in solar collectors and biomass reactors. The equations of thermodynamics are related to fluid mechanics via the energy conservation equations, and model the heat transfer, fluid heating and cooling, and the pressure work interactions. The interplay of these principles with advanced CFD enables the simulation of multiphase flows and turbulent regimes of flow, which is an effective design instrument in the design of efficient solar collectors, turbine rotor, and fluidized bed reactors in order to change biomass (Nabwey et al., 2024).

The above fundamental laws also form the basis of heat transfer, energy storage and conversion processes that determine the performance of renewable technologies. Conduction, convection, and radiation are the three modes of heat transfer that are very important in the determination of the efficiency of solar thermal plants, concentrated solar power (CSP) systems, and thermal energy storage devices. The parameters that are crucial in the process of design and energy performance optimization include heat transfer coefficients, thermal conductivity, the characterization of the flow regime, and the dynamics of the phase-change (Saha & Rupam, 2023). For example, thermal battery and latent heat reservoirs are both types of energy storage systems that are based on well-considered thermo-fluid systems capable of balancing the rate of charge and discharge, reducing energy losses and increasing responsiveness. Storage density and flexibility of the renewable systems have also been improved by the new materials possessing high thermal conductivity and novel flow designs (Yu et al., 2025; Saha & Rupam, 2023).

It is particularly noticeable in energy conversion systems in which thermo-fluid processes play a role. In solar systems, flat-plate and evacuated tube collectors are optimized to operate at maximum flow regimes and surface heat transfer coefficient, and nanofluids and molten salts are employed as working fluids to enhance the performance and the stability of these collectors (Nayak et al., 2024; Nkele et al., 2023). CSP plants require highly accurate fluid dynamic and thermal management, with concentrated solar energy needing to be efficiently captured and circulated by high-temperature fluids through thermodynamic cycles such as Rankine or Brayton, and storage facilities regulating the flow rate and temperature consistency (Nkele et al., 2023). Hybrid photovoltaic/thermal systems also demonstrate the value of thermo-fluid optimization, since fluid cooling can be used to increase PV efficiency, and simultaneously utilize heat in other applications, requiring highly efficient thermal and hydraulic designs (Tyagi et al., 2012).

With wind energy, kinetic energy conversion to mechanical and electrical is fundamentally aero-thermodynamic. Lift-to-drag optimization and pressure distribution control on turbine blade surfaces determine the aerodynamic design of turbine blades and rotors, with CFD models used to determine turbulence areas, torque distributions and wake influence, which directly influence the efficiency of the system (Outar et al., 2025; Katharikeyan & Hussain, 2025). Recent advances, such as the introduction of more efficient energy grabbing at low wind velocities with the use of improved shapes of air foils such as VIT7510, and the problems of noise reduction and deployment around urban regions have been solved (Outar et al., 2025; Kathikeyan & Hussain, 2025). In addition to aerodynamic performance, thermal management in turbines is also a key factor to extending lifespan and avoiding failure.

The conversion of biomass energy is an example of the combination of thermo-fluid processes and chemical reactions. All other forms of reactor, e.g. gasification, pyrolysis, and combustion, require turbulent mixing, residence time control, temperature control, and fluidized beds possess the finest mass and heat transfer properties. The reactors have been particularly valued due to the possibility to use a huge range of feedstocks, such as agricultural and municipal wastes. Particle behaviour, reaction rates and thermal behaviour FD-based analysis FD-based analysis has been instrumental in the optimal utilization of reactors (avoiding defluidization), biofuel scalability (Wang & Wu, 2023; Khodabandehlou et al., 2025; Xue & Fox, 2014).

Hydrothermal and enhanced geothermal systems (EGS) of geothermal energy systems are also based on thermo-fluid processes. The extraction of heat is controlled by multiphase flow and coupled thermo-hydro-mechanical-chemical processes in reservoirs, and induced seismicity and reservoir depletion are the significant issues of concern to the stability of the system (Kumar & Alam, 2025; Llanos & Blessent, 2025). The fluid mechanics is critical in the design of underground wells and flow circuits, where the stability of flow regimes should be guaranteed in changing geological and working conditions. The use of nanofluids and other engineered working fluids also increases the thermal conductivity and the reliability of the system (Llanos & Blessent, 2025; Assad et al., 2020).

The assessment of thermo-fluid systems in renewable applications depends on a number of essential measures. Thermal efficiency is the standard measure of useful energy conversion compared to the input energy. Exergetic efficiency is a more comprehensive measure as it takes into account irreversibility and lost work potential (Nazir et al., 2025; Gorgulu, 2023; Oyedepo et al., 2025). Direct measurements of the effectiveness of energy exchange processes are heat transfer coefficients, which influences the operation of solar collectors, thermal storage systems, and exchangers (Cabeza et al., 2015). Conversely, the pressure decreases and the fluid friction losses, caused by resistance to flow in conduits and reactor beds, undermine performance and reduce reliability, which are to be offset by using improved materials and system geometry optimization (Yu et al., 2025; Yang & Pan, 2015). In addition to these, parameters such as system stability, durability and reliability are critical factors especially when scaling technologies in the laboratory to commercial systems. These aspects guarantee long-term stability against thermal, mechanical, and environmental stresses, which makes them resistant to integration with energy grids and industrial processes (Tian et al., 2024).

The scientific ground of renewable thermo-fluid systems is provided by the combination of the thermodynamics and fluid mechanics, processes of energy conversion, and the performance measures of systems. They have a significant role to play in the global energy transition because they can be used synergistically to optimize, scale, and integrate technologies in solar, wind, biomass, and geothermal energy, and are necessary (Tian et al., 2024; Gorgulu, 2023).

3. RENEWABLE ENERGY SOURCES AND THERMO-FLUID INTEGRATION

The combination of renewable energy systems and advanced thermo-fluid management has become a key strategy in the quest to maximize energy conversion, storage and delivery in sustainable energy infrastructures. Solar energy is still at the top of the list, and solar thermal collectors are experiencing innovation in component structure and heat transfer improvement. Internal grooved tubes, hybrid and nanofluid-based methods of heat transfer enhancement and selective optical coating have greatly improved absorption efficiency. In particular, hybrid nanofluids where nanoparticles are dispersed in base fluids have a superior thermal conductivity and stability and, therefore, enhance collector functionality in parabolic trough and evacuated tube plants. Although the global solar thermal market is projected to shrink in 2023 partially because of policy changes and market saturation, the use of solar thermal in large-scale purposes in industrial processes and district heating is growing. The focus is still on cost-effective, CO_2-neutral solutions, and thermo-fluid challenges are related to the need to achieve uniform absorber temperature profiles, minimize heat losses, and material durability at industrial scale (Rashid et al., 2024).

In addition to thermal systems, photovoltaic-thermal (PV/T) collectors have developed by combining electricity production with coincident thermal collection. Air-cooled, liquid-cooled, and heat pipe PV/T systems all have their own thermo-fluid issues, such as airflow channel geometry optimization in air-type collectors, the use of advanced heat transfer fluids and exchanger configurations in liquid-type systems, which now achieve thermal efficiencies above 70%. Heat pipe PV/T modules also allow medium-to-high-temperature applications, but higher costs and complexity of maintenance are deterrents. The current optimization is advanced cooling systems, phase-change material (PCM) and modular systems with complementary technologies like heat pumps and ground-source systems. All these approaches reveal the importance of multi-dimensional performance measurement within a variable setting (Ma et al., 2025). Concurrently, concentrated solar power (CdSP) systems are increasingly utilizing more formulations of advanced heat transfer fluids (HTFs) that also incorporate thermal energy storage (TES). Also promising have been quaternary nitrate based molten salts and nanoparticle doped fluids in augmenting thermal conductivity and stability, raising storage efficiency and cycle performance. Environmental and economic considerations point to the need of sustainable, low-cost and long-lasting HTFs, and future research is focused on fluid degradation, material compatibility at high temperatures, and cost-effective storage (Chen & Wang, 2020).

Similar developments may be seen in wind energy where optimization of aerodynamics has been given priority to maximize energy capture and minimize flow losses. Dynamic devices such as plasma actuators and synthetic jets and passive devices such as vortex generators and trailing-edge flaps have enabled improved control in the separation of the flow/lift through the drag ratio. Blade design has also been accelerated with the use of computational fluid dynamics (CFD) and machine learning-based optimization techniques to dynamically optimise to different wind conditions. Moreover, biomimetic designs based on the natural surfaces have the promise of breaking the inefficiencies caused by turbulence, yet there are difficulties in scaling the innovation to large blade designs and in ensuring their reliability under changing operational conditions (Chen et al., 2025). In addition to aerodynamics, hybrid wind-thermal storage systems are finding increased use to overcome intermittency and improve dispatchability. These systems are coupled with thermal storage tanks or batteries to absorb the excess generation and stabilize the output with multi-objective optimization frameworks that often include exergoeconomic analysis. Future trends focus on the combination of heterogeneous storage systems, the dynamic management of multiple-energy flows, and the minimization of exergy costs to make the system more reliable and cost-efficient (Omidi, 2024).

Thermo-fluid innovation is also of vital importance in biomass and bioenergy systems. Thermochemical conversion processes like combustion, gasification, and pyrolysis have been developed with process intensification and catalyst development, and now pyrolysis and hydrothermal liquefaction can produce biochar, bio-oil, and syngas with higher yield and selectivity using a wide range of feedstocks. The combination with biochemical processes has also increased the utilization of resources and compatibility with the circular economy. However, the main issues that still remain are high costs of the process, feedstock heterogeneity, and strict requirements of the emissions control (Begum et al., 2024). Biogas systems, in their turn, are becoming more and more reliant on the accurate fluid flow regulation, reactor design, and optimized storage. New technologies like self-regulated production, real-time control of the biogas flow and improved mixing techniques improve efficiency and environmental impact. Comparisons of wet and dry fermentation processes point to the challenge of fluid management, with wet processes necessitating intensive water consumption and dry processes experiencing uniformity and operating risk problems. Predictive control, based on data, is becoming an interesting option to enhance the performance and stability of reactors (Luo et al., 2024).

Geothermal systems, especially enhanced geothermal systems (EGS), are another thermo-fluid engineering frontier. EGS projects, including FORGE and Fervo, have shown that it is possible to create or augment fracture networks in low-permeability reservoirs through advanced hydraulic and chemical stimulation to improve fluid circulation and heat recovery. The key innovation in the scaling of EGS technologies is the optimization of drilling, risk management at deep conditions, and long-term cycling performance. The thermo-fluid issues are still to control the injectivity of fluids, scaling, corrosion, and thermal breakthrough at depth, with simulation models being increasingly used to optimize multi-well flow regimes and minimize induced seismicity. They are also to reduce water use and come up with adaptive control which would help to make EGS operations more sustainable in the long term (National Renewable Energy Laboratory, 2024).

The ocean and hydrokinetic energy, including wave and tidal systems are progressing with turbine, mooring, and hybrid integration solutions. Improved access to larger renewable resources can be achieved by the deployment of shore-based, near-shore and deep offshore environments, but at a higher cost and complexity. Specific tidal cycle and flow regime adaptive control strategies have become an important component of power optimization, and concerns of capital cost, material longevity in a corrosive marine environment, and performance modeling continue to be a challenge (Kangaji et al., 2025). Complicated fluid-structure interactions also face offshore hydrokinetic systems, where turbulence, wave loading, and biofouling all shorten performance over time. Recent development in hydrodynamic modelling, real-time flow measurement, and sophisticated structural material has contributed to the resilience. The solution here is hybrid wind-tidal platforms, which are considered a future remedy to energy security and grid stability but their cost and operational risks are still a hindrance to their widespread use, and superior simulation, structure optimization, and maintenance solutions are wanted (Kangaji et al., 2025).

Thermo-fluid integration in all these renewable pathways plays a major role in solving bottlenecks in technical and sustainability. The innovative role of engineering in the future of renewable systems can be illustrated by progress in the area of heat transfer fluids, flow distribution, material science and hybrid energy coupling. The opportunities lie in multi-modal hybrid systems, where solar, wind, bioenergy, geothermal and ocean energy is coupled with an advanced storage and active control. The future research avenues should be in the domain of AI-predictive optimization, materials, highly robust techno-economic models, and the comprehensive analysis of life-cycle in order to promote resilient, efficient, and decarbonized renewable infrastructures (Kangaji et al., 2025).

4. HYBRID AND INTEGRATED RENEWABLE THERMO-FLUID SYSTEMS

Hybrid and integrated renewable thermo-fluid systems have become one of the most promising ways to achieve sustainable, resilient, and dispatchable energy infrastructures, as they allow combining several renewable sources with advanced fluidic and thermal management methods. Multi-source coupling plans like solar-biomass, wind-solar-thermal, and other more complex integrated systems will maximize use of resources and improve operational reliability, thus overcoming the intermittency of individual renewable resources. In solar-biomass systems, e.g., solar energy is collected during daylight hours using photovoltaic/thermal (PV/T) collectors, and biomass combustion or gasification provides a reliable source of backup thermal and electrical energy when solar availability is low, and thus enables dispatchable operation and continuous load support. Wind-solar-thermal hybrids also take advantage of the time and space variability of wind and solar resources by adding centralized or distributed thermal storage, enabling flexible energy delivery to end-users in both heating/cooling and electricity systems. These systems have been especially useful in systems like district heating grids or power-to-heat systems, where thermal buffering can be used to smooth out supply variability and enhance the integration of intermittent resources (Li et al., 2024; Yao et al., 2023; Hassan et al., 2023; Rasheed et al., 2021).

The heart of these hybrid systems is thermo-fluid processes that take advantage of synergistic energy conversion processes, using working fluids to provide simultaneous electricity and thermal energy recovery, and to allow efficient storage and downstream use. In PV/T hybrids, the production of electricity is supplemented by the recovery of high quality thermal energy via circulating working fluids water, air, or more recently, hybrid nanofluids with increased thermal conductivity. This two-output strategy does not only increase the total exergy output but also provides heat to industrial uses, space heating, or domestic hot water. To achieve the maximum synergies, however, sophisticated thermo-fluid management is needed, such as optimized flow distribution, multi-phase transport, and heat exchanger design that minimizes energy losses. The ability to control temperature gradients and fluid mixing is core to attaining exergy efficiencies of about 35-55%, and to minimizing renewable intermittency and improving dispatch economics. The empirical evidence shows that wind-solar-thermal hybrids can smooth net load profiles and decrease system costs and emissions by about 5-10% relative to single-source systems, which highlights their techno-economic benefits (Li et al., 2024; Hassan et al., 2023).

Recent developments further indicate the revolutionary role of materials and intelligent controls in hybrid thermo-fluid systems. Hybrid nanofluids have also resulted in up to 22 percent thermal and heat transfer improvements in PV/T modules and predictive dispatch, adaptive load matching, and dynamic resource allocation can be predicted and optimized through AI-based approach to energy management. Multi-source systems like focused solar power (Cd)-biomass gasification to create polygeneration have also increased system flexibility to create electricity, heat, and value-added chemical products, extending the possibility of renewable hybridization (Rasheed et al., 2021).

A key facilitator of this hybridization is the incorporation of thermal energy storage systems, which are needed to decouple supply and demand in time. Sensible heat storage (SHS), where liquids like water or molten salts or solids like rocks are heated, is the most mature and reliable technology, but has a relatively low energy density that can constrain compact applications. The latent heat storage (LHS) using phase change materials (PCMs) like paraffin, salt hydrates, or metallic alloys have higher energy density and almost isothermal behavior during phase transition, which makes it efficient to heat and cool. Nevertheless, supercooling, low thermal conductivity, and degradation during cycling are still obstacles to large-scale application (Pfleger et al., 2015). Thermochemical storage (TCS), on the other hand, has very high energy density and low self-discharge due to the use of reversible endothermic and exothermic reactions, e.g. metal oxide redox or salt hydration/dehydration, which makes it well suited to seasonal storage and long-term balancing of variable renewables. However, there are problems of kinetics of reaction, stability, and integration complexity that limit commercialization. Recent studies have emphasized the necessity of improved thermo-fluid design solutions, such as cascade storage systems, direct connection of molten salt tanks to CSP-biomass facilities, and the use of nanofluid-enhanced PCMs to increase storage density and transfer rates and minimize thermal losses during cycling (Pfleger et al., 2015).

The other application frontier to hybrid renewable thermo-fluid systems is to combine it with hydrogen production and fuel cell applications, a major step towards sector coupling and full decarbonization. With surplus renewable electricity generated by wind and solar, it is increasingly being used in the electrolysis of water to create hydrogen that can be stored and re-illustrated into electricity using proton exchange membrane fuel cells (PEMFCs) or solid oxide fuel cells (SOFCs). Besides electricity, the thermal energy of electrolysis and fuel cell stacks can be converted to district heating or domestic hot water, and this enhances overall exergy efficiency. The typical PV or wind farm systems design is electrolyser-based, coupled with hydrogen storage and FC stacks, usually augmented with batteries to allow short-term stabilization. PEM electrolysers have an efficacy of electric-to-hydrogen of more than about 60-70% round-trip efficacies of less than about 30-40%; co-utilization of waste heat can raise the effective energy of the system significantly (Gulraiz et al., 2025). Recent technological developments are concerned with multi-modal optimization techniques which dynamically optimize power between direct utilization, hydrogen storage, and battery systems depending on market prices, demand profiles and weather. Increasing options in the seasonal energy-shifting boundaries and system resilience, the following generation of hydrogen carriers, such as metal hydrides, liquid organic hydrogen carriers (LOHCs), and geological storage, and AI/ML structures are used more to size optimally, operate optimally, and estimate the economic viability of lifetime operations (Gulraiz et al., 2025).

The large scale demonstrations of hybrid thermo-fluid have already been successfully demonstrated in district heating and cooling (DHS/DCS) networks and desalination plants. Multi-source feeds such as solar thermal, biomass CHP and geothermal are also becoming increasingly popular with district heating systems, with a large-scale SHS or LHS storage tanks, to enable more certain high-renewable penetration with more flexible operation. Solar-driven absorption chillers have been integrated with thermal storage in cooling to generate high coefficients of performance and lower energy demand and emissions. Claims indicate that DHS/DCS systems save primary energy by about 30-40% and reduce CO_2 emissions by up to 50%, and scalable systems can support grids of about 20-100 MW (Kumar et al., 2023; Guelpa & Verda, 2019; Inayat & Raza, 2019). In desalination, solar-thermal and solar-biomass systems have made it possible to use continuous multi-stage flash (MSF), multi-effect distillation (MED) and reverse osmosis (RO) systems, through thermal storage and rejection heat recovery to pre-heat. Photothermal nanomaterials, hybrid membrane distillation, and vapor compression cycle powered by renewable energy are the new trends, and they increase energy efficiency and reduce the cost of freshwater production (Kumar et al., 2023).

In spite of their potential, hybrid renewable thermo-fluid systems remain problematic with regard to operational complexity, cost, and scalability. The variability of resources and the necessity to manage them in real-time and multisource require the advanced forecasting, predictive controls, and resilient architectures (Ukoba et al., 2024; Li et al., 2024). Both storage technologies and hydrogen systems continue to be challenged by costs, durability, and material constraints, as well as round-trip inefficiencies, safety issues, and infrastructure gaps (Saha & Rupam, 2023). Moreover, the expensive capital outlays of advanced materials, control infrastructures, and integration architectures are a barrier to deployment (Ukoba et al., 2024). To overcome these challenges, there is a trend in recent optimization approaches to use multi-objective modeling tools with techno-economic, environmental, and reliability aspects as well as AI/ML-based operational frameworks to predict scheduling and fault detection. Resilience and cost-effective storage have become the focus on hybrid storage systems that combine short-term battery storage, medium-term thermal storage, and long-term hydrogen storage (Saha & Rupam, 2023). Future research priorities will be the creation of functional storage materials with greater energy densities, lower costs, and improved cyclability, as well as increased sector coupling in power, heating, industry, transport, and desalination. Just as important will be the development of smart controls and automation to enable real-time optimization, and policy, market and infrastructure frameworks that can enable the district-scale hybridization of renewable systems and the extension of hydrogen infrastructure to ensure long-term energy security (Ukoba et al., 2024).

5. BIBLIOMETRIC ANALYSIS

To give a detailed description of the scientific progress and research trends in the field of renewable energy applications in thermo-fluid systems, a bibliometric study was performed based on the information about publications obtained in the Lens database. This method does not only measure the productivity of research but also highlights the thematic trends, knowledge networks and the changing interdisciplinarity that is the hallmark of this fast growing field. The analysis combines publication patterns, geographical distribution, publication sources, and author contributors, and a detailed keyword co-occurrence graphical representation conducted using VOSviewer. Collectively, these aspects give a systematic picture of the evolution of the research field in the past 20 years, its present state, and the future directions it is taking.

The temporal distribution of publications as illustrated in Figure 1 shows a clear pattern of development. The number of articles published was low in the early 2000s with less than fifty articles per year until 2006, which was in its initial stages of renewable energy integration into thermo-fluid systems. But with the increased need to find sustainable energy solutions and the development of computational modeling methods, the number of publications started to increase steadily since 2010. A strong increase was observed in the period 2015-2021, with the maximum of about 1,250 publications in 2021, which indicates both the maturity of fundamental ideas and the expansion of application fields. In 2022 and 2023, there was a slight drop, which can be explained by the fact that the global research activity was disrupted, but the trend returned to growth in 2024, which indicates the relevance of this interdisciplinary area and its further growth. This longitudinal growth pattern not only highlights the industrial and environmental necessity behind the research but also the revolutionary nature of computational thermo-fluid methodologies in solving global energy problems.

Figure 1. Year-wise Publication Trend (2000-2025)

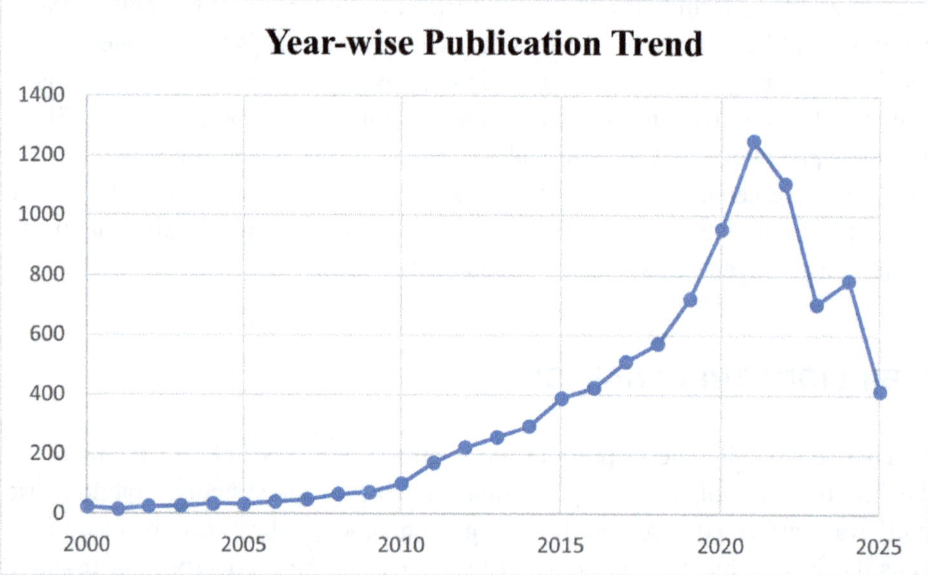

Source: Lens Databsase (Lens, 2025)

The overlay keyword co-occurrence map generated in VOSviewer and data retrieved in Lens.org are more illustrative of the intellectual structure and thematic development of the field as shown in Figure 2. Here, each node represents a keyword found in the scientific corpus, with the size of the node indicating the frequency of the keyword and the thickness of the edge indicating the degree of co-occurrence of the keywords in publications. The geographic location and clustering of the nodes indicate significant thematic clusters and the color gradient shows the temporal development of research fronts. Past research focuses, in darker tones, were on solar energy, thermal energy storage, phase change materials, biomass, and heat transfer enhancement. These basic studies indicate the classical field of thermo-fluid engineering in renewable energy systems. Over time, however, the thematic range has expanded to newer frontiers such as nanoparticles, polymers, tissue engineering, nanomedicine, and sustainability, which are shown in lighter tones, reflecting their relative novelty. This historical turn shows the progressive change of the discipline into a more interdisciplinary discipline that overlaps with materials science, biotechnology and biomedical engineering.

Figure 2. Keyword co-occurrence overlay network on "Thermofluid"

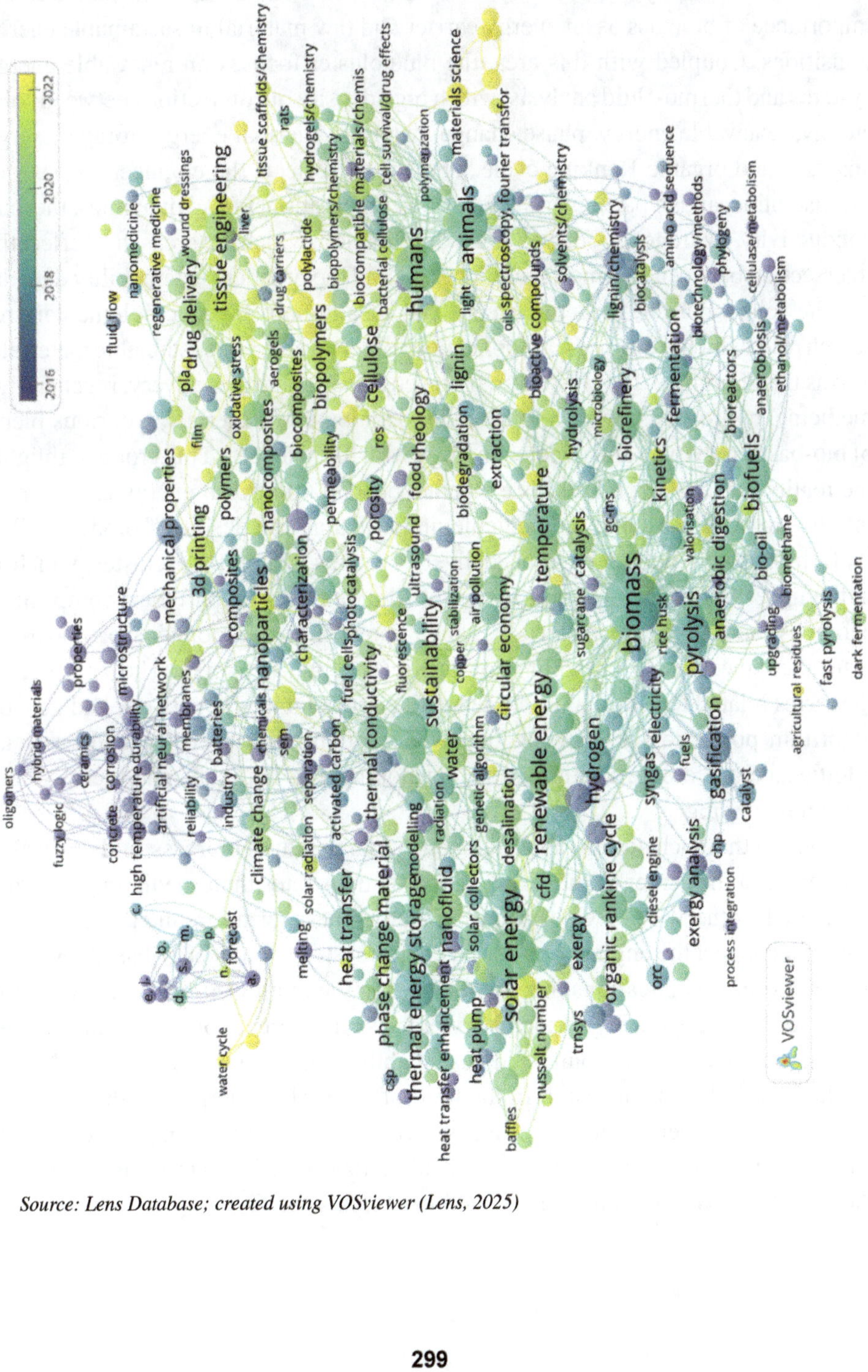

Source: Lens Database; created using VOSviewer (Lens, 2025)

A more detailed examination of the clusters, in Figure 2, shows that there are a number of leading streams of themes. The green cluster in the center of the network includes biomass conversion technologies (pyrolysis, anaerobic digestion, fermentation, and gasification) that are closely related to such keywords as biofuels, lignin, cellulose, biorefinery, and hydrolysis. This unified body of research points to the importance of biomass as an energy carrier and raw material in sustainable energy transitions. Coupled with this area, the blue cluster focuses on renewable energy systems and thermo-fluid analysis, with a high density of connections between solar energy, renewable energy, phase change materials, thermal energy storage, exergy analysis, and organic Rankine cycle. The cluster reflects the engineering basis of the discipline, and in particular the focus on computational fluid dynamics, thermal conductivity, hydrogen production, and desalination, which are areas that reflect the intersection between thermo-fluid processes and the broader sustainability agenda.

The yellow cluster, in its turn, represents an alternative, yet complementary research path, which is biomedical applications of renewable and sustainable materials. Terms that dominate this cluster are tissue engineering, drug delivery, regenerative medicine, hydrogels, nanomedicine and biocompatible materials, a curious blend of bio-based polymers and nanotechnology with life sciences. This group highlights the reality that the discipline has expanded beyond traditional energy engineering into use in healthcare and biomedicine, in particular, the design of biodegradable scaffold and new polymeric drug delivery systems. The purple cluster, which is in the upper left of the network, represents the field of nanoparticles, composites, ceramics, and advanced materials, which spread into the domains of corrosion resistance, mechanical durability and artificial neural networks. The presence of the terms related to artificial intelligence like artificial neural network and genetic algorithm points to the growing importance of machine learning and artificial intelligence-aided optimization in the development of materials and thermo-fluid system modelling.

The fact that such notions as sustainability, temperature, biomass, and renewable energy are at the center of the visualization demonstrates their mediating intention between the otherwise diverse spheres of study. These terms are intellectual moorings that connect the engineering-based energy systems, bioenergy-based research, and biomedical applications thereby reasserting the interdisciplinary nature of the field. Also, the popularity of such keywords as circular economy, climate change, biodegradation, and environmental sustainability indicates a paradigm shift over the last few years: the research in the field of renewable energy and thermo-fluid systems is no longer limited to technical efficiency but is increasingly inclusive of environmental, social, and economic considerations, which makes it consistent with the United Nations Sustainable Development Goals (SDGs).

Further geographical analysis adds to this image by showing the worldwide research activity distribution as shown in Table 1. The European countries predominate in the number of publications, with Switzerland leading with 1,579 publications, followed by the United Kingdom (1,024), the United States (837), Germany (807), and the Netherlands (710). These countries have been enjoying well-developed institutional structure, joint research projects, and stable investments in renewable energy and engineering fields. Asian contributions are on a steep increase, with China (371) and South Korea (206) becoming active contributors, which can be attributed to strategic national investments in clean energy and advanced materials research. Such a geographical distribution demonstrates not only the existing centres of excellence of research in Europe but also the growing research activity in Asia which shows a more globalized research environment.

Table 1. Top contributing countries by publication count (Source: Lens Database) (Lens, 2025).

Country	Count
Switzerland	1579
United Kingdom	1024
United States	837
Germany	807
Netherlands	710
China	371
South Korea	206
Hungary	206
Iran	64
Saudi Arabia	63

The publication of knowledge in this area is also highly centralized to a few large publishers, which indicates the significance of well-established academic platforms in increasing the visibility of research as presented in Table 2. The most publications belong to Springer Science and Business Media LLC (2,826), MDPI AG (1,631), Elsevier BV (1,133), OSTI (559), and the American Chemical Society (177). This distribution indicates that the concentration of influence is in the major international publishing houses that act as the channels of dissemination of the field advancements. At the individual level, the literature shows that there are some prominent authors who have influenced the discourse as presented in Table 3. The most active contributor is Luisa F. Cabeza with 22 publications, followed by Qiuwang Wang (17), Christos N. Markides (16), and Vittorio Verda (15). The distribution of these authors in Europe and Asia emphasizes the collaborative and more transnational research in thermo-fluid systems applied to renewable energy.

Table 2. Top publishers by publication count (Source: Lens Database) (Lens, 2025).

Publisher	Count
Springer Science and Business Media LLC	2826
MDPI AG	1631
Elsevier BV	1133
Office of Scientific and Technical Information (OSTI)	559
American Chemical Society (ACS)	177
Wiley	153
MDPI	146
Royal Society of Chemistry (RSC)	140
Frontiers Media SA	121
InTech	107

Table 3. Top authors by publication count (Source: Lens Database) (Lens, 2025).

Author's Name	Publication Count
Luisa F Cabeza	22
Qiuwang Wang	17
Christos N Markides	16
Vittorio Verda	15
Yulong Ding	14
François Maréchal	13
Francesco Calise	12
Neven Duić	11
Ya-Ling He	10
Elisa Guelpa	9

Collectively, this bibliometric analysis highlights the deep interdisciplinarity and international trend of research in renewable energy applications in thermo-fluid systems. The discipline is both classical in the field of energy system optimization and emerging in the areas of biomaterials, advanced nanotechnologies, and biomedical engineering. The co-occurrence networks and publication trajectories indicate that the future research development will be based on the combination of bioenergy, computational thermo-fluid analysis, material science, nanotechnology, and healthcare applications, all of which are supported by the general imperative of sustainability. In this regard, bibliometric mapping can be used as a descriptive tool, but also as a strategic tool that allows identifying research hotspots, knowledge gaps, and emerging opportunities and directing both academic research and industrial innovation.

6. IONIC LIQUIDS AS ADVANCED THERMOFLUIDS AND THEIR CFD-BASED PERFORMANCE ANALYSIS

The potential promising advanced thermofluids that can be used in renewable energy and heat transfer systems have been attributed to the advantages that can be associated with the physicochemical peculiarities of the ionic liquids (ILs). Their usage also demonstrates the creation of sustainable energy technologies by powering them to transfer the heat with high efficiency, high robustness, and tunability, the demands of the current configurations in solar, storage and cooling (Jiang et al., 2025; Hamdan et al., 2025; Chandra et al., 2025). The ILs are salty liquids comprising of ions only and at room temperature or close to it. Their ionic strength and insignificant vapour pressure make them very thermally stable, to 400 °C, depending on the type of class. Other liquids such as water has boiling point of 100 °C and ethylene glycol has a boiling point of 193 °C boil, whereas ILs do not, but rather undergo degradation at high temperatures thus excluding the chance of vapour formation during operation (Chandra et al., 2025). Their low volatility and flammability are also complemented by nearly zero evil vapor pressure in order to permit the high-temperature and safety operation even in the closed systems (Jiang et al., 2025; Liang et al., 2024). The other distinguishing feature of ILs is that they can be tuned thermophysically because the heat capacity, viscosity and thermal conductivity of ILs can be systematically tuned by using the right cation/anion pairs or by hybridizing nanoparticles to form Ionanofluids or nanoparticle enhanced ILs (NEILs). Among them, it is found that amino acid anion ionic liquids (AAILs) have a higher heat capacity, higher thermal conductivities and lower viscosities than more traditional ILs, e.g., [bmim] [BF_4] and a more significant improvement is observed when they are nanomaterial functionalized e.g., multi-walled carbon nanotubes (Chandra et al., 2025). Additionally, the new ILs, i. e. 1, 2, 4- triazolium alanine derivatives have a heat-storage density which is up to 1.5 higher than that of the traditional heat transfer fluids and a thermal conductivity that is 1.7 higher (Liang et al., 2024). Their chemical stability inherent to them permits them to work in large pH conditions and in bad weather conditions which minimizes the degradation and compatibility issues (Liang et al., 2024).

The ILs possess better safety and performance properties compared to traditional heat transfer fluids, such as water, ethylene glycol and mineral oils (Table 4). One of them would be the fact that water possesses high specific heat and low boiling point and also has high vapor pressure and also ethylene glycol and oils though they increase the temperature range are volatile and are flammable. Rather, ILs are degradation inflammable, thermal-stable and vapor-pressure inert. The maximum achievable heats are approximately 10 J/gK, which is far higher than the required amount of water (4.2 J/gK), ethylene glycol (2.4 J/gK) or even oils (Less than 2 J/gK). Equally, the viscosity of ILs is similar to water and traditional oils, therefore, they can be used more in heat transfer regions where flow is required. They are also versatile in terms of their real life applications due to the customizability of thermal conductivity (0.12 -0.3 W/m.K) (Chandra et al., 2025; Liang et al., 2024).

Table 4. Comparison of ionic liquids with conventional heat transfer fluids (Chandra et al., 2025; Liang et al., 2024).

Property	Water	Ethylene Glycol	Conventional Oil	Ionic Liquids
Boiling Point (°C)	100	193	~ 300	No boiling (decomposes)
Specific Heat (J/g·K)	4.2	2.4	< 2	Up to 10 (AAILs)
Thermal Conductivity (W/m·K)	0.6	0.25	0.1 – 0.15	0.12 – 0.3 (tunable)
Viscosity (mPa·s, at 25°C)	1	16	20–60	8 – 20 (AAILs)
Vapor Pressure at 100°C (kPa)	100	15	up to 10	≈ 0
Flammability	Non-flammable	Flammable	Flammable	Non-flammable

Ills thus guarantee the potential of operating safely with a high temperature and with minimal chances of fire or fluid wastages. They are the new generation thermofluids due to their customizability that can be enhanced by the presence of nano particles (Chandra et al., 2025; Liang et al., 2024).

These favorable characteristics have aided the fast-tracked growth of the IL applications in the renewable energy area and thermal management during the 2020-24 period. As it has been already demonstrated, the nanoparticle improved ILs possess about 23-34% greater heat capacity and about 640% greater thermal conductivity benefits in solar thermal storage over base ILs and conventional fluids with performance advantages in solar thermal power systems (Hamdan et al., 2025;

Chandra et al., 2025). The photothermal system with graphene oxide-IL hybrids was identified to have a thermal efficiency which was around 36.8% greater than the thermal efficiency of water-based system, which implied these hybrid photothermal systems could trap and store solar energy more efficiently (Yusuf et al., 2025). The potential to utilize electronics AAIL-based Ionanofluids to conduct heat at 0-200 °C, where most of the historically used coolants boil or are too viscous to be effectively utilized, is another reason why ILs are the material of choice to be implemented in heat exchangers and cooling the processor (Chandra et al., 2025). Furthermore, ILs are already proving efficient in storage and exchange of both sensible and latent heat of storage in the storage of chemical energy phase-change, as well as other more complex chemical energy storage (Kleoff et al., 2025; Yusuf et al., 2025).

The computational Fluid Dynamics (CFD) is necessarily integrated into the analysis and optimization of the IL-based thermofluids in the renewable energy systems. Flow, temperature, and stress distributions can be delivered in complex geometries, e.g. in solar receivers, storage tanks and heat exchangers, using thermophysical inputs (which may be experimentally measured or calculated) and CFD models such as ANSYS Fluent (Niu, 2021). The comparative CFD experiments have found that ILs, such as potassium bromide solutions, are less effective than ILs, such as [EMIM][DMP]-H_2O mixtures in all conditions when it comes to the heat and mass transfer performance and the overall coefficient of performance in refrigeration when everything remains equal (Niu, 2021). The mechanism of dispersion associated with nanoparticles and its role in contributing to the growing thermal conductivity, viscosity control and heat storage has also been further developed by using advanced multiphase CFD models and thus has dictated the optimal design of the both the Ionanofluids and NEILs (Abd-Elmonem et al., 2025; Srivathsan et al., 2023). In addition, parameters of the operation, i.e. the flow rates, load of nanoparticles, the designs of the heat exchangers, and the external heat fluxes should be optimal to maximize the efficacy and stability of the IL-based thermofluid systems, which can be optimized with the aid of CFD-based optimization (Abd-Elmonem et al., 2025; Srivathsan et al., 2023). The results together bring ILs to the next generation of tunable thermofluids whose tunability and powerful computational-design space, and dominate the research on sustainable energy and heat transfer.

7. CHALLENGES, BARRIERS, AND SUSTAINABILITY CONSIDERATIONS IN RENEWABLE ENERGY APPLICATIONS WITHIN THERMO-FLUID SYSTEMS

The problems, barriers and challenges of sustainability of the renewable energy sources in the thermo-fluid systems can be listed among the most real-life topics of the modern research and policy-making, as the technologies tend to attract more and more attention as the ones that can change the energy infrastructures. However, despite the enormous progress in the field of hybridization strategies, storage integration, and developing sophisticated materials, massive difficulties are yet to be encountered on the technical, economic, policy, and environmental level. Intermittency of the renewable sources such as solar and wind is also one of the technical challenges that lead to the difficulty of incorporating them into the thermo-fluid systems where this constant heat and power is needed. The dynamic quality of the presence of resources may provoke temporal incompatibility between the supply of energy and the requirement of thermal energy, which necessitates the introduction of sophisticated control programs, the schemes of real-time forecast and hybridization to stabilize the work. Even though the addition of solar thermal systems to latent storage or biomass has been shown to mitigate the impact of intermittency, the combination tends to complicate the system and raise the initial cost of installation (Olabi et al., 2023; Alagumalai & Mahian, 2022).

There are also efficiency constraints to large-scale adoption, with many renewable thermo-fluid technologies still limited by fundamental material and process constraints. The thermal storage based on phase change material (PCM) is usually limited by poor thermal conductivity, incomplete phase change, and low heat transfer rates, which impede its scalability in systems like solar thermal power generation, or industrial waste heat recovery. In addition, the variability and distributed character of renewable thermal energy sources makes their effective incorporation into large-scale, centralized thermo-fluid systems problematic, as it introduces concerns of homogeneity, stability, and operational consistency (Alagumalai & Mahian, 2022). Thermal management itself adds further complexity, because controlling temperature accurately and with minimal exergy losses remains a significant challenge with high-temperature renewable processes, such as concentrated solar power (CSP), and with advanced cooling processes. Periodic temperature changes may cause material fatigue and degradation, which reduces the long-term stability of system components, and non-uniform heat distribution in multiphase and dynamic fluid regimes can negatively affect efficiency and reliability. Despite the current progress in nanofluids, composite PCMs, and new heat exchanger geometries, there are still outstanding issues related to long-term stability, compatibility with the existing infrastructure, and the sustainability of the lifecycle that need to be addressed (Alagumalai & Mahian, 2022).

Along with technical bottlenecks, economic and policy barriers also exist to hinder large-scale commercialization. The major impediment is the high capital cost of renewable thermo-fluid systems, especially geothermal, solar thermal, and advanced thermal storage systems, which in many cases make them less competitive than fossil fuel-based systems. To take an example, PCMs and thermochemical storage solutions are more energy-dense and flexible in their functioning, but highly cost-sensitive due to the complexity of the material synthesis and manufacturing requirements (Eze et al., 2025; Alagumalai & Mahian, 2022). Policy and regulatory uncertainties aggravate these financial problems. The lengthy approving processes, environmental impact analysis and regulatory frameworks tend to drag the process of deployment, which are divergent and vary significantly in various jurisdictions. The risks to geothermal energy initiatives are particularly prone to additional risks tied to new demands on induced seismicity, land-use laws, and ecological analyses that lengthen project schedules and financing challenges. The absence of stable, long-term policy support mechanisms (feed-in tariffs, carbon credits, or streamlined integration pathways of the district heating and cooling networks) further deteriorates investor confidence and retards market growth (Eze et al., 2025; Nsude et al., 2025). Nonetheless, new mitigation practices show that policy innovation is one of the keys to busting such barriers. Direct subsidies, feed-in tariffs, carbon trading regimes, and digital permitting systems and standard environmental tests have demonstrated quantifiable benefits in cost competitiveness and are being progressively used to simplify the process and cut the administrative overhead (Nsude et al., 2025).

The environmental considerations add another level of complication to the matter because sustainability analysis would reveal a less apparent trade-off in renewable thermo-fluid systems. Life-cycle assessment (LCA) studies have demonstrated time and again that these technologies have significant advantages as compared to fossil fuel-based systems in the context of greenhouse gas (GHG) emissions, primary energy consumption, and long-term sustainability. Nevertheless, several environmental problems have not been dealt with. The production of the high-technology materials, such as nanofluids, PCMs, and thermochemical compounds, typically consumes much upstream energy and water, and emissions of the process of extracting and processing of the essential raw materials, including rare earths and metallic alloys. Furthermore, the new use of new refrigerants, immersion fluids, and complex coolants is an emerging trend that will require the question of toxicity, persistence and long-term ecological impacts, which cannot be fully presented in the existing LCA frameworks (Cellura et al., 2025; Alissa et al., 2025). The central topic of discussion therefore becomes material sustainability, and high-tech heat transfer fluids and PCMs, which, besides their performance attributes, introduces the question of recyclability, bioaccumulation, and environmental persistence. Using the case of large-scale use of per and polyfluoroalkyl substances (PFAS) in the next-generation cooler technology, the regulatory bodies have been interested due to the bio accumulative and persistent nature in the environment (Alissa et al., 2025). LCA-based strategies are driving more towards early material selection, the use of recycled material and end-of-life recyclability or modular reuse, therefore, rendering technology development compatible with the circular economy. New data center cooling metrics (such as virtual-core measures) are also under discussion so that they could help with the environmental assessment and promote more circular system designs with reduced environmental impact (Cellura et al., 2025; Alissa et al., 2025).

Current research is still providing valuable information on how to overcome these two challenges that are intertwined. Improvements in PCM performance by nanomaterial doping, improved encapsulation methods, and composite formulations are improving the thermal conductivity and cycling stability necessary to scale storage solutions. Parallel techno-economic studies and pilot-scale projects are generating useful data to inform standardized integration strategies, so that systems can be balanced in terms of efficiency, economic feasibility, and policy compatibility in practical implementation. Equally significant, the implementation of the principles of circular economy, i.e., through modular design, recoverable materials, and closed-loop recycling, becomes increasingly popular in research and industry, making thermo-fluid systems not only low-carbon solutions but also the ones that can drive sustainable resource stewardship (Alagumalai & Mahian, 2022; Cellura et al., 2025; Alissa et al., 2025).

Overall, although the potential of using renewable energy sources in thermo-fluid systems is enormous to decarbonize the power and heating industries, the challenges are overwhelming at the technical, economic, policy, and environmental levels. Joint efforts in the materials science, thermal management, system integration and life-cycle analytics will be needed to address these challenges, and the coherent policy frameworks and market incentives will be helpful. The introduction of technology innovation, long-term design principles, and policy support will be of great importance in scaling, widespread, and resilient implementation of the decades to come as the processes of global decarbonization continue to gain momentum (Alagumalai and Mahian, 2022).

8. CONCLUSION

Overall, the use of renewable energy in thermo-fluid systems is a new technology that mitigates the impending problem of climate change and resource sustainability. This chapter has described the concepts of thermodynamics and fluid mechanics underlying energy conversion and management in renewable technologies. The efficiency and reliability of such systems has been significantly enhanced by the development of computational modeling and materials science, and now the thermo-fluid processes can be optimized to a precision. It has been demonstrated in the study of hybrid and integrated renewable systems that these may be used to reduce the intermittency of discrete energy sources thereby offering a more reliable and robust energy supply. These systems may be applied to provide energy on a continuous basis and exploit resources optimally through the synergies of solar, wind, biomass and geothermal resources.

According to the bibliometric research, the active and interdisciplinary character of investigations in this field is also also present and indicates that the interest towards sustainability, developed materials and new control measures is getting bigger. Despite the efforts that have been made, one may still note a series of challenges that include technical constraints, economic constraints and environmental issues. In order to address these obstacles, it will be necessary to closely collaborate in research and development, policy support and industry cooperation. The research of the future should be aimed at the development of superior materials, storage, and effective control systems that could be aligned with the principles of the circular economy. Two concepts that can assist the renewable energy use in thermo-fluid systems to model a decarbonized energy scenario in the future are innovation and sustainability.

REFERENCES

Abd-Elmonem, A., Ahmad, H., Garalleh, H. A., Atta, G., Amjad, M., Suoliman, N. A., Jamshed, W., & Hussain, S. M. (2025). Machine learning and differential transform simulation in analysing the ionic liquid (EtSO4) based ternary hybrid flow over a curved geometry. *Journal of Molecular Liquids, 127109*. Advance online publication. DOI: 10.1016/j.molliq.2025.127109

Ahmadi, M. H., Ramezanizadeh, M., Nazari, M. A., Lorenzini, G., Kumar, R., & Jilte, R. (2018). Applications of nanofluids in geothermal: A review. *Mathematical Modelling and Engineering Problems, 5*(4), 281–285. DOI: 10.18280/mmep.050402

Alagumalai, A., & Mahian, O. (2022). Specialty grand challenge in thermal science and energy systems. *Frontiers in Thermal Engineering, 2*. Advance online publication. DOI: 10.3389/fther.2022.954511

Alissa, H., Nick, T., Raniwala, A., Herranz, A. A., Frost, K., Manousakis, I., Lio, K., Warrier, B., Oruganti, V., DiCaprio, T. J., Oseen-Senda, K., Ramakrishnan, B., Gupta, N., Bianchini, R., Kleewein, J., Belady, C., Fontoura, M., Sinistore, J., Natarajan, M., & Frieze, M. (2025). Using life cycle assessment to drive innovation for sustainable cool clouds. *Nature*. Advance online publication. DOI: 10.1038/s41586-025-08832-3 PMID: 40307558

Assad, M. E. H., Ahmadi, M. H., Sadeghzadeh, M., Yassin, A., & Issakhov, A. (2020). Renewable hybrid energy systems using geothermal energy: Hybrid solar thermal–geothermal power plant. *The International Journal of Low Carbon Technologies, 16*(2), 518–530. DOI: 10.1093/ijlct/ctaa084

Begum, Y. A., Kumari, S., Jain, S. K., & Garg, M. C. (2024). A review on waste biomass-to-energy: Integrated thermochemical and biochemical conversion for resource recovery. *Environmental Science. Advances, 3*(9), 1197–1216. DOI: 10.1039/d4va00109e

Cabeza, L. F., Galindo, E., Prieto, C., Barreneche, C., & Fernández, A. I. (2015). Key performance indicators in thermal energy storage: Survey and assessment. *Renewable Energy, 83*, 820–827. DOI: 10.1016/j.renene.2015.05.019

Cellura, M., Luu, L. Q., Guarino, F., & Longo, S. (2025). A review on life cycle assessment of concentrating solar energy technologies. *Renewable Energy, 249*, 123203. DOI: 10.1016/j.renene.2025.123203

Chandra, A., Sistla, Y. S., Ahmed, M. A., Vasireddy, D. V. S., Jaglan, N., Das, N. K., Banerjee, T., & Sistla, V. S. (2025). Heat transfer fluids: Amino acid anion ionic liquid based IoNanofluids with remarkable thermal conductivity and low viscosity. *RSC Advances, 15*(29), 23146–23164. DOI: 10.1039/d5ra03752b PMID: 40620252

Chen, Q., & Wang, Y. (2020). Research status and development trend of concentrating solar power. *2022 11th International Conference on Renewable Energy Research and Application (ICRERA)*, 390–393. DOI: 10.1109/icrera49962.2020.9242893

Chen, S., Wang, K., Zhao, M., & Gao, Z. (2025). A review of the recent development of innovative wind turbines: Concepts and aerodynamic analyses. *Physics of Fluids, 37*(4). Advance online publication. DOI: 10.1063/5.0256502

Cheng, X., Du, B., He, J., Long, W., Su, G., Liu, J., Fan, Z., & Chen, L. (2025). A Review of thermoplastic composites on wind turbine blades. *Composites. Part B, Engineering, 112411*. Advance online publication. DOI: 10.1016/j.compositesb.2025.112411

Eze, V. H. U., Eze, E. C., Alaneme, G. U., & Bubu, P. E. (2025). Recent progress and emerging technologies in geothermal energy utilization for sustainable building heating and cooling: A focus on smart system integration and enhanced efficiency solutions. *Frontiers in Built Environment, 11*. Advance online publication. DOI: 10.3389/fbuil.2025.1594355

Gorgulu, Y. F. (2023). Evaluating the efficiency of renewable energy technologies through thermodynamic analyzes. *Insight - Mechanics, 6*(1). DOI: 10.18282/m.v6i1.599

Guelpa, E., & Verda, V. (2019). Thermal energy storage in district heating and cooling systems: A review. *Applied Energy, 252*, 113474. DOI: 10.1016/j.apenergy.2019.113474

Gulraiz, A., Bastaki, A. J. A., Magamal, K., Subhi, M., Hammad, A., Allanjawi, A., Zaidi, S. H., Khalid, H. M., Ismail, A., Hussain, G. A., & Said, Z. (2025). Energy advancements and integration strategies in hydrogen and battery storage for renewable energy systems. *iScience, 28*(3), 111945. DOI: 10.1016/j.isci.2025.111945 PMID: 40060896

Hamdan, M., Abujamous, D., Abdelhafez, E., Sukkariyh, M., & Al-Maghalseh, M. M. (2025). Parametric evaluation of sensible heat storage systems using nano-ionic liquids. *Journal of Ecological Engineering, 26*(3), 237–245. DOI: 10.12911/22998993/199322

Hassan, Q., Algburi, S., Sameen, A. Z., Salman, H. M., & Jaszczur, M. (2023). A review of hybrid renewable energy systems: Solar and wind-powered solutions: Challenges, opportunities, and policy implications. *Results in Engineering*, *20*, 101621. DOI: 10.1016/j.rineng.2023.101621

Inayat, A., & Raza, M. (2019). District cooling system via renewable energy sources: A review. *Renewable & Sustainable Energy Reviews*, *107*, 360–373. DOI: 10.1016/j.rser.2019.03.023

Ji, J., Zhang, C., Zhang, X., & Chen, Y. (2024). Recent research advances in wind turbine thermal management technologies. *Renewable & Sustainable Energy Reviews*, *208*, 114983. DOI: 10.1016/j.rser.2024.114983

Jiang, R., Guo, Y., & Guo, X. (2025). Ionic liquids as advanced heat transfer fluids in renewable energy systems. *Solar Energy Materials and Solar Cells*, *294*, 113934. DOI: 10.1016/j.solmat.2025.113934

Kangaji, L. M., Raji, A., & Orumwense, E. (2025). Cutting-edge progress in offshore wind and tidal stream power technology—State-of-the-Art. *AIMS Energy*, *13*(1), 188–230. DOI: 10.3934/energy.2025007

Kariuki, B. W., Hassan, H., Ahmed, M., & Emam, M. (2024). A review on geothermal-solar hybrid systems for power production and multigeneration systems. *Applied Thermal Engineering*, *259*, 124796. DOI: 10.1016/j.applthermaleng.2024.124796

Khan, M. M. K., Chowdhury, A. A., & Hassan, N. M. S. (2017). Application of thermo-fluid processes in energy systems. In *Green energy and technology*. DOI: 10.1007/978-981-10-0697-5

Khodabandehlou, R., Norouzi, H. R., Larimi, A., & Mostoufi, N. (2025). Analysis of biomass gasification in a fluidized bed using CFD-DEM. *Energy*, *320*, 135395. DOI: 10.1016/j.energy.2025.135395

Kleoff, M., Klaucke, F., Voßnacker, P., Weber, R., Neuhoff, K., & Riedel, S. (2025). Utilisation of reactive ionic liquids for energy storage and regulation of the power grid. *Green Chemistry*. Advance online publication. DOI: 10.1039/d5gc02519b

Kumar, A., & Alam, T. (2025). A review on geothermal energy systems and various approaches to enhance the system's performance. *Energy and Building*, *344*, 115962. DOI: 10.1016/j.enbuild.2025.115962

Kumar, L., Soomro, J., Khoharo, H., & Assad, M. E. H. (2023). A comprehensive review of solar thermal desalination technologies for freshwater production. *AIMS Energy*, *11*(2), 293–318. DOI: 10.3934/energy.2023016

Lens, (2025). *Search results for "Renewable Energy Applications In Thermo-Fluid Systems"*. The Lens. Retrieved June 11, 2025, from https://www.lens.org/lens/search/scholar/list?q=Renewable%20Energy%20Applications%20In%20Thermo-Fluid%20Systems&n=10

Li, X., Wang, K., Xu, M., Fu, M., & Miao, S. (2024). Environmental and economic dispatching strategy for power system with the complementary combination of wind-solar-hydro-thermal-storage multiple sources. *Frontiers in Energy Research, 12*. Advance online publication. DOI: 10.3389/fenrg.2024.1338657

Liang, K., Yao, H., Qiao, J., Gao, S., Zong, M., Liu, F., Yang, Q., Liang, L., & Fang, D. (2024). Thermodynamic evaluation of novel 1,2,4-Triazolium alanine ionic liquids as sustainable Heat-Transfer Media. *Molecules (Basel, Switzerland)*, *29*(22), 5227. DOI: 10.3390/molecules29225227 PMID: 39598616

Llanos, E. M., & Blessent, D. (2025). A scoping review of numerical modelling studies of geothermal reservoirs: Trends and opportunities post-COP25. *International Journal of Renewable Energy Development*, *14*(4), 668–693. DOI: 10.61435/ijred.2025.61071

Luo, T., Shen, B., Mei, Z., Hove, A., & Ju, K. (2024). Unlocking the potential of biogas systems for energy production and climate solutions in rural communities. *Nature Communications*, *15*(1). Advance online publication. DOI: 10.1038/s41467-024-50091-9 PMID: 39003261

Ma, Z., Zhao, H., Yan, Z., & Song, J. (2025). Advances and Development Trends in Solar Photovoltaic-Thermal Technology: A review. *Clean Energy*, *9*(4), 98–122. DOI: 10.1093/ce/zkaf020

Nabwey, H. A., Ashraf, M., Nadeem, H., Rashad, A. M., & Chamkha, A. J. (2024). Optimizing renewable energy systems: A comprehensive review of entropy generation minimization. *AIP Advances*, *14*(12). Advance online publication. DOI: 10.1063/5.0245560

National Renewable Energy Laboratory. (2024). *Geothermal – Electricity ATB*. Retrieved August 23, 2025, from https://atb.nrel.gov/electricity/2024/geothermal

Nayak, S., Hassan, M. A., & Paswan, M. (2024). Investigation on thermo-fluid behavior of flat-plate solar water heater employing titanium dioxide-water nanofluid: An experimental and computational approach. *Journal of Mechanical Science and Technology*, *38*(3), 1507–1516. DOI: 10.1007/s12206-024-0242-6

Nazir, M. H., Zaidi, S. Z. J., Hussain, M. M., & Khan, Z. A. (2025). Modelling & Development of Novel Sustainable Thermo-fluids to Maximise the Efficiency of Solar Thermal Applications. *Journal of Engineering Research*. Advance online publication. DOI: 10.1016/j.jer.2025.01.009

Niu, J. (2021). CFD analysis of the influence of ionic liquids on the performances of a refrigeration system. *Fluid Dynamics & Materials Processing*, *17*(5), 1003–1013. DOI: 10.32604/fdmp.2021.015225

Nkele, A., Ikhioya, I., Chime, C., & Ezema, F. (2023). Improving the performance of solar thermal energy storage systems. *Journal of Energy and Power Technology*, *05*(03), 1–25. DOI: 10.21926/jept.2303024

Novozhilov, V. (2001). Computational fluid dynamics modeling of compartment fires. *Progress in Energy and Combustion Science*, *27*(6), 611–666. DOI: 10.1016/s0360-1285(01)00005-3

Nsude, C. C., Connelly, S., Demir, F., Shi, D., & Tsetsura, K. (2025). Drivers, barriers and policy options to renewable energy development in Native American communities: A systematic review. *Energy Research & Social Science*, *125*, 104092. DOI: 10.1016/j.erss.2025.104092

Olabi, A., Elsaid, K., Obaideen, K., Abdelkareem, M. A., Rezk, H., Wilberforce, T., Maghrabie, H. M., & Sayed, E. T. (2023). Renewable energy systems: Comparisons, challenges and barriers, sustainability indicators, and the contribution to UN sustainable development goals. *International Journal of Thermofluids*, *20*, 100498. DOI: 10.1016/j.ijft.2023.100498

Omidi, J. (2024). Advances and opportunities in wind energy harvesting using plasma actuators: A review. *Clean Energy*, *8*(1), 197–225. DOI: 10.1093/ce/zkad085

Outar, H., Ennadafy, H., Jammoukh, M., & Khaili, M. E. (2025). CFD analysis for aerodynamic performance optimization of a vertical axis helical wind turbine using rigid PVC. *E3S Web of Conferences*, *601*, 00101. DOI: 10.1051/e3sconf/202560100101

Oyedepo, S. O., Waheed, M. A., Abam, F. I., Dirisu, J. O., Samuel, O. D., Ajayi, O. O., Somorin, T., Popoola, A. P. I., Kilanko, O., & Babalola, P. O. (2025). A critical review on enhancement and sustainability of energy systems: Perspectives on thermo-economic and thermo-environmental analysis. *Frontiers in Energy Research*, *12*. Advance online publication. DOI: 10.3389/fenrg.2024.1417453

Pfleger, N., Bauer, T., Martin, C., Eck, M., & Wörner, A. (2015). Thermal energy storage – overview and specific insight into nitrate salts for sensible and latent heat storage. *Beilstein Journal of Nanotechnology*, *6*, 1487–1497. DOI: 10.3762/bjnano.6.154 PMID: 26199853

Rasheed, T., Hussain, T., Anwar, M. T., Ali, J., Rizwan, K., Bilal, M., Alshammari, F. H., Alwadai, N., & Almuslem, A. S. (2021). Hybrid nanofluids as renewable and sustainable colloidal suspensions for potential Photovoltaic/Thermal and solar energy applications. *Frontiers in Chemistry*, *9*. Advance online publication. DOI: 10.3389/fchem.2021.737033 PMID: 34646812

Rashid, F. L., Aljibori, H. S., Mohammed, H. I., Ameen, A., Ahmad, S., Hamida, M. B. B., & Al-Rubaye, A. H. (2024). Recent advances and developments of the application of hybrid nanofluids in parabolic solar collector energy systems and guidelines for future prospects. *Journal of Engineering Research*. Advance online publication. DOI: 10.1016/j.jer.2024.04.023

Saha, B. B., & Rupam, T. H. (2023). Specialty grand challenge: Thermal energy storage and conversion. *Frontiers in Thermal Engineering*, *3*. Advance online publication. DOI: 10.3389/fther.2023.1157794

Shahzad, H. M. A., Asim, Z., Khan, S. J., Almomani, F., Mahmoud, K. A., Mustafa, M. R. U., & Rasool, K. (2024). Thermochemical and biochemical conversion of agricultural waste for bioenergy production: An updated review. *Discover Environment*, *2*(1). Advance online publication. DOI: 10.1007/s44274-024-00171-w

Srivathsan, B., G, T., Ram, K. V., & R, H. (2023). Multiphase simulation of sustainable nanoenhanced ionic liquid coolants for improved thermal performance in Ti–6Al–4V alloy drilling. *Heliyon*, *9*(12), e23020. DOI: 10.1016/j.heliyon.2023.e23020 PMID: 38076113

Sutar, S., Rout, S. K., Senapati, J. R., Barik, D., Dennison, M. S., & Praveenkumar, S. (2024b). Numerical investigation on the thermohydraulic performance of a solar air heater duct featuring parabolic rib turbulators on the absorber plate. *Case Studies in Thermal Engineering*, *64*, 105399. DOI: 10.1016/j.csite.2024.105399

Sutar, S., Rout, S. K., Senapati, J. R., & Muduli, K. K. (2024b). Numerical investigation of solar air heater for space heating application. *Applied Solar Energy*, *60*(5), 708–726. DOI: 10.3103/s0003701x24602151

Tian, Z., Zheng, W., Sun, X., Wang, L., Jiang, Y., & Mi, X. (2024). Research on prediction model of two-phase flow and friction pressure drop in spiral pipe based on flow regimes. *Thermal Science and Engineering Progress, 103018*. Advance online publication. DOI: 10.1016/j.tsep.2024.103018

Tyagi, V., Kaushik, S., & Tyagi, S. (2012). Advancement in solar photovoltaic/thermal (PV/T) hybrid collector technology. *Renewable & Sustainable Energy Reviews*, *16*(3), 1383–1398. DOI: 10.1016/j.rser.2011.12.013

Ukoba, K., Olatunji, K. O., Adeoye, E., Jen, T., & Madyira, D. M. (2024). Optimizing renewable energy systems through artificial intelligence: Review and future prospects. *Energy & Environment*, *35*(7), 3833–3879. DOI: 10.1177/0958305x241256293

Wang, Y., & Wu, J. J. (2023). Thermochemical conversion of biomass: Potential future prospects. *Renewable & Sustainable Energy Reviews*, *187*, 113754. DOI: 10.1016/j.rser.2023.113754

Xue, Q., & Fox, R. O. (2014). Computational modeling of biomass thermochemical conversion in fluidized beds: Particle density variation and size distribution. *Industrial & Engineering Chemistry Research*, *54*(16), 4084–4094. DOI: 10.1021/ie503806p

Yang, H., & Pan, M. (2015). Engineering research in fluid power: A review. *Journal of Zhejiang University. Science A*, *16*(6), 427–442. DOI: 10.1631/jzus.a1500042

Yao, S., Wei, Y., Lu, Z., Guo, S., Chen, J., & Yu, Z. (2023). Review of research progress on Concentrated Solar Energy Utilization System. *Renewables*, *1*(4), 397–414. DOI: 10.31635/renewables.023.202200018

Yu, G., Sun, Y., Zhou, Y., & Xu, H. (2025). Review on heat transfer structures for enhancing thermal performance of latent heat thermal energy storage systems. *Journal of Energy Storage*, *132*, 117908. DOI: 10.1016/j.est.2025.117908

Yusuf, A., Bhatti, M., & Ellahi, R. (2025). Study of ionic water/graphene nanofluids in solar panels under the effects of thermal radiation and slip conditions using experimental data. *International Communications in Heat and Mass Transfer*, *164*, 108845. DOI: 10.1016/j.icheatmasstransfer.2025.108845

Zhang, K. (2025). A Review of Biomass Energy: Comparison of utilization methods and future Prospects. *E3S Web of Conferences, 606*, 05007. DOI: 10.1051/e3sconf/202560605007

ADDITIONAL READING

Ellabban, O., Abu-Rub, H., & Blaabjerg, F. (2014). Renewable energy resources: Current status, future prospects and their enabling technology. *Renewable & Sustainable Energy Reviews*, *39*, 748–764. DOI: 10.1016/j.rser.2014.07.113

Khan, M. M. K., Chowdhury, A. A., & Hassan, N. M. S. (2017). Application of thermo-fluid processes in energy systems. In *Green energy and technology*. DOI: 10.1007/978-981-10-0697-5

Montelpare, S., D'Alessandro, V., Zoppi, A., & Costanzo, E. (2017). A Solar Chimney for renewable energy production: Thermo-fluid dynamic optimization by CFD analyses. *Journal of Physics: Conference Series, 923*, 012047. DOI: 10.1088/1742-6596/923/1/012047

Sdringola, P., Proietti, S., Desideri, U., & Giombini, G. (2013). Thermo-fluid dynamic modeling and simulation of a bioclimatic solar greenhouse with self-cleaning and photovoltaic glasses. *Energy and Building, 68*, 183–195. DOI: 10.1016/j.enbuild.2013.08.011

Selimefendigil, F., & Öztop, H. F. (2021). Thermoelectric generation from vented cavities with a rotating conic object and highly conductive CNT nanofluids for renewable energy systems. *International Communications in Heat and Mass Transfer, 122*, 105139. DOI: 10.1016/j.icheatmasstransfer.2021.105139

Walshe, J., Carron, P. M., McLoughlin, C., McCormack, S., Doran, J., & Amarandei, G. (2020). Nanofluid development using silver nanoparticles and Organic-Luminescent molecules for Solar-Thermal and hybrid Photovoltaic-Thermal applications. *Nanomaterials (Basel, Switzerland), 10*(6), 1201. DOI: 10.3390/nano10061201 PMID: 32575601

Zhang, X., & Wang, G. (2017). Thermodynamic analysis of a novel energy storage system based on compressed CO2fluid. *International Journal of Energy Research, 41*(10), 1487–1503. DOI: 10.1002/er.3732

KEY TERMS AND DEFINITIONS

Computational Fluid Dynamics (CFD): A sub-field of fluid mechanics, which employs numerical analysis and algorithms to solve and analyze fluid flows problems.

Energy Conversion: This is the conversion of energy between forms and is important in the better utilization of renewable resources.

Hybrid Systems: Systems that integrate various renewable sources to be more reliable and efficient in response to the intermittency of the source.

Renewable Energy: This energy is natural and is replenished continuously e.g. solar, wind, biomass and geothermal energy.

Sustainability: The capacity to satisfy the current energy demands without deteriorating the capability of future generations to satisfy their own needs, and it focuses on the environmental, economic and social aspects.

Thermal Energy Storage: Technology that stores thermal energy which is used later enhancing the efficiency and reliability of renewable energy systems.

Thermo-fluid Systems: Systems that deal with thermal and fluid interaction dynamics that is essential in the conversion and management of energy in renewable technologies.

Chapter 13

Simulation of Desiccant-Based Air Conditioning System Using Vortex Tube as a Cooling Source

Abhilas Swain
https://orcid.org/0000-0002-2615-0694
School of Mechanical Engineering, KIIT University, India

Aaditya Singh
https://orcid.org/0009-0004-9815-9438
KIIT University, India

Aadhya Vig
https://orcid.org/0009-0009-2984-9896
KIIT University, India

Siba Padarabinda Behera
KIIT University, India

ABSTRACT

The present article focuses on the simulation of a hybrid desiccant wheel based air-conditioning unit coupled with a vortex tube cooler. The desiccant wheel and the vortex tubes are modelled from empirical correlations developed through regression analysis from experimental data. The sensible cooling of the supply air is carried out by the vortex tube cooler after it is dehumidified in the desiccant wheel and releasing the adsorption heat in the energy recovery wheel. The variation of the output parameters with respect to some of the factors are discussed. The variation of temperature and specific humidity of the supply air are illustrated and discussed

DOI: 10.4018/979-8-3373-3621-3.ch013

with respect to the weather conditions of Bhubaneswar city (Odisha, India) during the whole year. The output supply air temperature and humidity levels suggest the feasibility of implementing the vortex tube cooler in conjunction with the desiccant wheel.

INTRODUCTION

Desiccant based air-conditioning systems have been one of the favorites for cooling application in various industries where there is strict requirement of humidity control. This is because of its simple construction, flexibility of arrangements and no harmful impact on environment. Therefore, the research community is persistently working on the possible developments for higher efficiency, better temperature control and utilization of waste heat & renewable energy resources. The advantage of the desiccant-based cooling system are presented in Figure 1.

Figure 1. Features of desiccant based cooling system

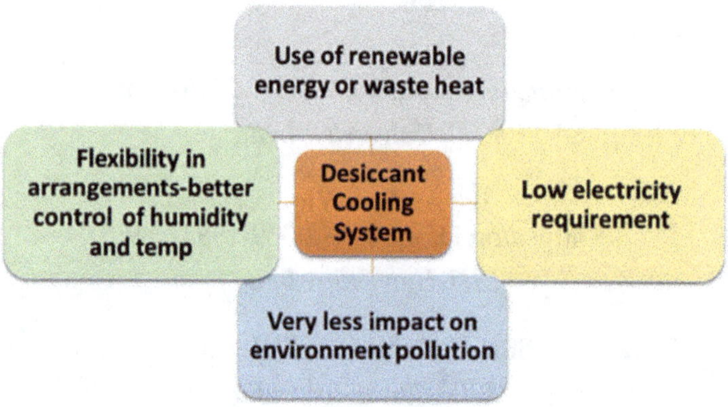

Usually the materials used as solid desiccants are activated carbon, activated alumina, molecular sieve, silica gel, lithium chloride, calcium chloride, etc. In the desiccant based air-conditioning systems, the solid desiccants are used generally in the form of a circular disc like structure, called as desiccant wheel (DW). This is advantageous for the exposure of the desiccant material to the process air flow and the regeneration air stream. Thus, instead of changing the air flow stream, the wheel is rotated to achieve the dehumidification and regeneration. The DW has the porous structure which are coated with the desiccant material for the sorption of water vapors from the air. The desiccant material along with the matrix material are arranged in a honeycomb structure with millimeter level channels which designed for maximum exposure of surface area, lower pressure drop, lower weight and structural stability. Variety of research investigations were carried out on the application of desiccant materials in the refrigeration and air-conditioning systems Major research investigations on the desiccant wheels are discussed in the survey articles by the works (Jani et al., 2016) and (Daou et al., 2006). Few of the research investigations of the last decade are discussed here.

In the works (Ruivo et al, 2013; Ruivo et al, 2015), the authors carried out experimental investigation of a DW based air-conditioning system in which the regeneration heat is received from a micro-generator and if necessary, from a boiler. In another research an empirical correlation is formed from the experimental data for predicting the output air temperature and specific humidity. The work (Chen et al., 2018)] also carried out an experimental investigation in which a precooling facility is provided before the DW to reduce the shoot up of the temperature. Further cooling of the air is achieved by another surface cooler placed after the dehumidification is achieved. In another set of investigations by (Angrisani et al, 2011; Angrisani et al., 2012 & Angrisani et al., 2017), experiments were carried out to determine the effect of various parameters on the output of the desiccant wheel. Then, a semi-empirical mathematical model is developed from the experimental data for predicting the output air temperature and humidity taking different parameters as put for the model such as input air temperature and humidity, ratio of volume flow rates, regeneration temperature, rotations per minutes of the desiccant wheel. The performance of the semi-empirical models are checked with the RMSE and the percentage error values. This semi-empirical model is used in the present investigation for the desiccant wheel. In the work (Parmar & Hindoliya, 2013) the authors have simulated the desiccant wheel based air-conditioning system comprising of the DW, direct evaporative cooler and an energy recovery wheel (ERW). The charts provided by the manufacturer are used for the DW modelling. The modelling is carried out on Engineering Equation Solver (EES) platform. Similarly (Assadi & Mohammadi, 2016) carried out the system simulation on EES and TRNSYS platform for the simulation of a DW based system utilizing solar energy for the regeneration purpose. The work by (Heidari, Roshandel, & Vakiloroaya, 2018) investigated a novel solid desiccant wheel based evaporative cooling system and simultaneously it was used also produce water along with the cooling capability. This system utilized the solar energy for the regeneration purpose. Dynamic simulation was carried out for a building of 60 m^2 area in the hot and humid city of Bandar Abbas, Iran. Furthermore, in addition of achieving desired temperature and humidity level, the system is able to produce 590 liters of water per week out of which 289 L is left for domestic use. The system used 60% less electrical energy as compared to the reference natural gas system. In the TRNSYS programming platform.

The (Saputra et al., 2020) literature carried out an study exploring the better control algorithm for regeneration temperature and the rotational speed of the wheel. The performance are measured with constant regeneration temperature and varying the heat input. In the work (Comino et al, 2020) an experimental study was conducted to examine the effect of rotational speed and the They divided the whole wheel to different sectors and studied their behavior. (Wu et al., 2019) studied the performance of a desiccant wheel made from the wood pulp fiber and they also coated the wheel with silica gel. The moisture removal capacity of the desiccant wheel is improved. (Fong & Lee, 2020) also conducted an experimental study on the clean air heat pump system taking a desiccant wheel as dehumidifier and also tested its performance for removing the volatile organic compounds such as acetone, formaldehyde, toluene and 1, 2-dicholorethance from air. (Caliskan, et al., 2019) conducted a thermodynamic energy and energy analysis of the desiccant-based cooling system. The literatures on the desiccant wheels are thoroughly searched and it is found that the combination of desiccant wheel and vortex tube is not addressed.

The vortex tube is a pipe T-connection structure in which a conical valve is present on one side. A succinct literature review of the articles on vortex tube-based refrigeration system of last five years is also presented here. The principle of vortex tube is well explained in the (Bergmann, 2018) and a schematic is shown to demonstrate the principle of the vortex tube. (Qyyum et al., 2018) carried out a computational investigation on improving the efficiency of Nitrogen expander-based liquefaction process of natural gas (LNG) employing a vortex tube along with the expander. The commercial simulation software Aspen Hysis A(r) is used in the work and is found that the specific energy consumption is reduced by 68.5%. (Kumar et al., 2017) studied the cooling and dehumidification capabilities of the vortex tube by experimentally investigating the variations of (a) drop in cold air temperature, (b) Increase in hot air stream temperature and (c) Percentage change in relative humidity with respect to the cold air mass flow rate and pressure. The different investigations on the vortex tube being engaged in various applications requiring refrigeration are discussed in (Zhang & Guo, 2018). The compressed air is allowed to flow into the vortex tube tangentially and from the circumference and then it divides into two streams out of which one is at a higher temperature and the other is at a lower temperature than the inlet streams.

The literature review suggests that the desiccant wheel based air-conditioning system (DBAS) uses different technologies for reducing the temperature of the process air such as vapour compression system (VCR), direct evaporative cooler (DEC), indirect evaporative cooler (IEC) and surface coolers. Moreover, the literatures related to the vortex tubes reflects that vortex tube might be considered as a source for low temperature so that sensible cooling can be achieved in air-conditioning. As per author ?(tm)s knowledge, there is no literature available on the use of the vortex tube in conjunction with the desiccant wheel for space conditioning system applications. Therefore, the present article focuses on the use of vortex tube for sensible cooling in a desiccant based air-conditioning system. The investigation is primarily consists of two parts. The first part is on the experimental investigation of performance of vortex tube and developing empirical correlation from experimental data. In the second section, simulation of a desiccant based air-conditioning system comprising of desiccant wheel and vortex tube is conducted. For different conditions of outdoor air, the supply air temperature and humidity are predicted. The simulations were carried out on engineering Equation Solver platform (EES).

Some of the latest research and developments are also discussed here related o different aspects of the desiccant wheel. Recent research has extensively expanded the applications, modeling, and synergy of desiccant wheels for air-conditioning, with focus on energy efficiency, renewable integration, and hybrid system configurations.

- (Abdelgaied et al.,2023) demonstrated a novel solid desiccant air conditioning system for thermal comfort with minimized emissions, aligning with sustainability goals in building air conditioning.
- (Khalid et al.,2024) evaluated multi-stage desiccant wheel configurations, showing enhanced thermo-economic performance and increased system COP in industrial dehumidification.
- (Dikshit, 2024) provided a thorough review of solid desiccant systems, outlining their efficacy, economic value, and role in consistent humidity control in air-conditioning environments.
- (Ma et al.,2023) performed experimental effectiveness analysis of a DW dehumidification system, with focus on low-humidity operation, offering theoretical metrics for future DW design and selection.
- (Huang et al., 2024) investigated a hybrid liquid-desiccant air conditioning system, reporting significant energy savings (20.2%) and improved COP (up to 3.1) after optimizing system heat-cold offsets.
- (Iyer et al., 2024) experimentally assessed a compact integrated solid DW/M-cycle cooler, highlighting potential for net-zero emissions and superior humidity reduction in air-conditioning.
- (Singh et al., 2023) introduced a correlation approach using experimental results for predicting DW performance and validating the correlation with robust effectiveness metrics.
- (Zamzamian, 2022) explored theoretical and experimental strategies for heat and moisture transfer in rotary DWs, emphasizing the impact of key operational parameters on system effectiveness.

- (Jing et al., 2025) reviewed refrigerant-based vortex tube (RBVT) separation techniques, underlining the importance of integrated approaches combining DW and vortex tube concepts for energy-efficient cooling.

Hybridization and Advanced Modeling Efforts

- (Awan et al., 2025) established experimentally that hybrid LDAC systems can achieve exceptionally high EER (energy efficiency ratio) in both summer and winter, supporting robust adoption in varied climate zones.
- (Sahlot & Riffat, 2016) advanced DW material science by experimenting on composite coatings (silica gel + MOF-801), achieving increased sorption capacities and enabling lower temperature regeneration.
- (Swami et al., 2022) employed machine learning for predicting DW outlet air properties, reducing calculation error and improving design precision for engineers.

Integration with Vortex Tube Systems

Most desiccant wheel studies focus on standalone or coupled vapor compression and evaporative cooling, but post-2020 publications reveal growing interest in integrating vortex tube cooling for sensible heat extraction.

- (Awan et al., 2025) developed a vortex tube-based cooling system for vaccine containers, showing that parametric optimization of vortex tube inlet pressure and cold flow rates can maintain required cooling loads in energy-constrained environments.
- (Liang et al., 2025) performed numerical analysis of Ranque-Hilsch vortex tube phenomena, demonstrating the sensitivity of fluid choice and turbulence models to device energy-efficiency in refrigeration tasks.
- Experimental investigations by (Jing et al., 2025) and others indicate the wide-ranging potential of DW in combination with advanced cooling strategies, but application in direct hybridization with vortex tubes for space conditioning remains scant.

The objective of the research is to study the limitations of the vortex tube refrigeration system and to explore methods that can improve its performance. The work is aimed at reducing the high energy consumption of the compressor and minimizing the destruction of useful energy inside the system. The research is focused on increasing the cooling capacity of the vortex tube so that the demand in wider applications can be met. The goal is to achieve stable outlet temperatures and better control of cooling. The study also seeks to recover the useful energy lost in the hot stream and to reduce the noise generated during operation. Regular maintenance issues related to filters are considered and solutions are explored to enhance efficiency and reliability.

EXPERIMENTAL INVESTIGATION OF VORTEX TUBE

A vortex tube, also known as a Ranque ?"Hilsch vortex tube, is a simple mechanical device that separates compressed air into hot and cold streams without using any moving parts, refrigerants, or external power apart from the compressed air supply. When high-pressure air is injected tangentially into the vortex chamber through nozzles, it generates a strong swirling motion that creates two distinct airflows: the outer vortex, which spirals toward the hot outlet, gains energy through friction with the tube wall and heats up, while the inner vortex, moving back through the center toward the cold outlet, expands and loses energy, resulting in cooling. By adjusting the control valve at the hot end, the user can regulate the split between hot and cold air, with the cold stream reaching temperatures as low as ?"40 A C and the hot stream exceeding +100 A C depending on conditions. Though the efficiency is relatively low and unsuitable for large-scale refrigeration, the vortex tube is highly reliable because it has no moving parts, and it finds practical use in spot cooling applications such as cooling cutting tools, electronic control panels, worker protective gear, and in industrial settings where conventional refrigeration is impractical.

Figure 2 shows the experimental facility for the determination of the performance of the vortex tube. A compressor is included in the setup for generating compressed air at different pressure values. The compressed air is stored in the surge tank and is allowed to pass through the filter for removal of foreign particles, moisture and oil droplets. A control valve is present to control the flow rate of the air entering into the vortex tube cooler. The pressure and the temperature of the air stream entering to the vortex tube is monitored with a pressure gauge and K-type thermocouple inserted to the hose pipe respectively. The vortex tube converts the single air stream into one hot air stream and one cold air stream. The temperatures of the hot and cold air streams are measured with thermocouples. The flow rates of both the streams are measured through turbine flow meters and also the temperatures are measured.

Figure 2. (a) Schematic of the experimental set up for vortex tube (b) Vortex tube

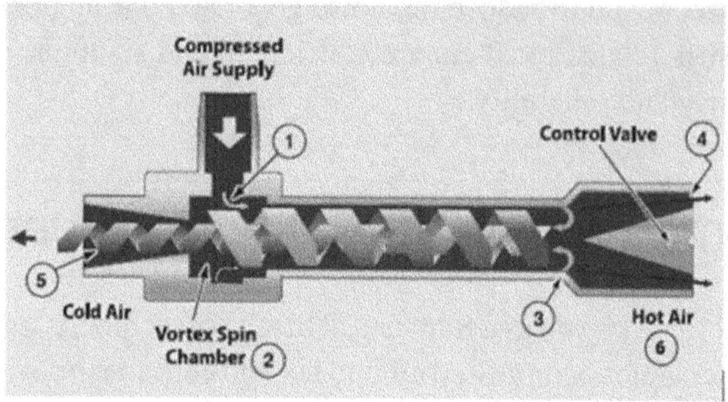

DESICCANT BASED AIR-CONDITIONING SYSTEM

The space conditioning system considered for the present study is comprised of a desiccant wheel, energy recovery wheel, vortex tube cooler and a regenerative heater as shown in Fig 3. The recirculation of the conditioned air is not counted in for the present analysis. As usual the desiccant wheel is divided into the process air section and the regeneration section in equal proportion. The fresh air enters into the process section of desiccant wheel dehumidifier for the removal of moisture (1-2). During this process the adsorption heat released from the desiccant material is absorbed by the process air and its temperature is raised. Therefore, for bringing down the temperature of the process air it is allowed to pass through the energy recovery wheel. As the process air stream passes through the energy recovery wheel, the temperature is reduced without change in the specific humidity value (process 2-3). Further, the air is allowed to pass through the shell side of a heat exchanger in which the cold air from the vortex tube cooler is flowing on the tube side, to get the desired temperature for the comfort condition (3-4). The vortex tube receives the compressed air and divides the stream into two streams out of which one is colder and the other is hotter. In the present analysis the cold air output stream of the vortex tube cooler is considered for the application. The exiting air at the point 4 is supplied to the conditioned space. As the desiccant wheel absorbs the moisture from the process air, it needs to be extracted from the desiccant wheel, so that it again becomes ready for the process air. Therefore, the desiccant material is exposed to the regeneration section. In this section, the air at a higher temperature is allowed to flow through the desiccant wheel leading to the removal of water vapors from the wheel to the air. The moisture content in the desiccant material decreases and it again becomes ready for the process section.

For the purpose of regeneration of the DW and heat extraction of the energy recovery wheel (ERW), the fresh air is used. The fresh air is first allowed to first absorb heat from the energy recovery wheel (5-6). Then, the heated air is further heated by the use of an electric heater for achieving the required regeneration temperature for the DW. The mathematical details of the modelling work are described in the following section.

Figure 3. Schematic of the desiccant wheel based air-conditioning system

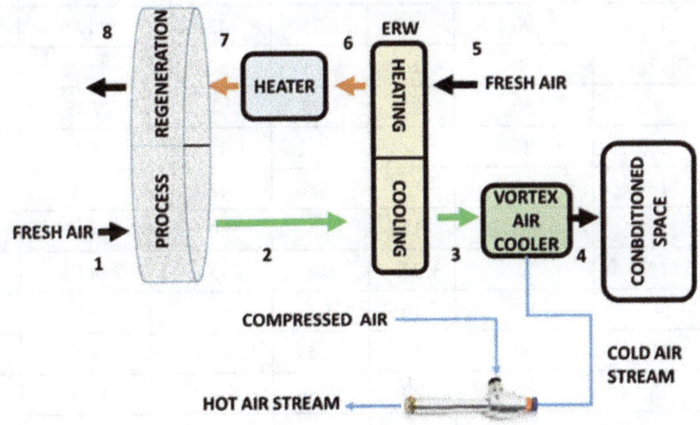

In the present simulation the desiccant wheel is modelled as an empirical correlation. The empirical correlation is developed from the manufacturer data. The manufacturer data are obtained from the Novel Air online data base software. The Table 1 represents the data obtained for a 370 X 220 mm desiccant wheel. Moreover, it is to be noted that the data are obtained in an orthogonal array L25 pattern for five factors and five levels. The process air inlet temperature (Ta1), inlet humidity ratio (w1), regeneration temperature (Treg), rotations per hour of the desiccant wheel and the ratio of volume flow rates are taken as inputs to the software. These data are considered for creating a regression model in the form of a power law correlation by applying least mean square algorithm.

Table 1. Manufacturer data for a 370 x 220 mm desiccant wheel

Ta1 [°C]	w1[g/kg]	Treg [°C]	RPH	V_rat	Ta2[°C]	w2[g/kg]	Wb [°C]
30	9	70	5	0.5	36.7	7.02	19.4
30	11	75	10	0.75	38.1	8.53	21
30	13	80	15	1	51.4	6.79	23.6
30	15	85	20	1.25	54.9	7.69	25.1
30	17	90	25	1.5	58	8.73	26.5
32	9	75	15	1.25	38.3	7.19	20.1
32	11	80	20	1.5	39.6	8.74	21.6
32	13	85	25	0.5	40.9	10.29	23
32	15	90	5	0.75	42.2	11.85	24.4
32	17	70	10	1	50.1	11.28	26
34	9	80	25	0.75	40	7.35	20.7
34	11	85	5	1	41.2	8.93	22.2
34	13	90	10	1.25	42.4	10.52	23.6
34	15	70	15	1.5	50.5	9.96	25.3
34	17	75	20	0.5	43.1	14.03	25.9
36	9	85	10	1.5	41.7	7.5	21.3
36	11	90	15	0.5	43.4	9.12	22.9
36	13	70	20	0.75	42.3	11.04	23.9
36	15	75	25	1	43.5	12.64	25.2
36	17	80	5	1.25	57.1	10.63	27.3
38	9	90	20	1	56	4.68	23.4
38	11	70	25	1.25	43	9.55	23.1
38	13	75	5	1.5	44	11.19	24.4
38	15	80	10	0.5	45.6	12.83	25.8
38	17	85	15	0.75	46.2	14.47	26.9

The empirical correlation formed from the data is as follows in equation (1) and (2) for the outlet temperature and specific humidity of the process air. These two models are able to predict the used data within an error of A15%. The values of the coefficients and powers are presented in table 2. These models are used to predict the performance of the desiccant wheel in the simulation work.

Ta2 = K1*(Ta^a1)*(w1^b1)*(Treg^c1)*(RPH^d1)*(v_rat^e1) (1)

W2 = K2*(Ta^a2)*(w1^b2)*(Treg^c2)*(RPH^d2)*(v_rat^e2) (2)

Table 2. Values of the coefficients for the correlations

Constants for Ta2	**K1**	**a1**	**b1**	**c1**	**d1**	**e1**
	2.4677	0.0752	0.3226	0.3979	0.0336	0.1361
Constants for w2	K2	a2	b2	c2	d2	e2
	0.1612	1.0106	0.8062	-0.3395	-0.0232	-0.1735

Once the process air passes through the desiccant wheel (DW) it gets dehumidified. Thus, the water vapors are absorbed in the desiccant material channels, which releases the heat of absorption. The heat of absorption leads to heating of the process air and its temperature is increased. According to the general comfort conditions necessary for the human beings and also for different food material storage application this process air temperature after exiting from DW is higher. Thus, the temperature needs to be brought down to the required temperature. The decrease in temperature is achieved by passing through energy recover wheel and vortex tube cooler. The effectiveness models are considered for the energy recovery wheel and the vortex tube cooler heat exchanger and the value of the effectiveness is taken as 0.85 and 0.7 respectively. The expressions for effectiveness are shown in equations below (3) and (4).

$$\text{Eff\_erw} = A\frac{T2 - T3}{T2 - T6} \tag{3}$$

$$\text{Eff\_hx\_vc} = A\frac{T3 - T4}{T3 - Tvc} \tag{4}$$

The other temperatures involved in the cycle are determined by the energy balance equation applying to the energy recovery wheel (ERW). The equation is presented in the equation (5). . The energy released by process air will be absorbed by the regeneration air stream. Similarly the energy balance for the DW is also presented in the equation (6).

$$m_a(h_2 - h_3) = m_{reg}(h_5 - h_6) \tag{5}$$

$$m_a(h_1 - h_2) = m_{reg}(h_7 - h_8) \tag{6}$$

Then the air is supplied to the conditioned space. The air will be absorbing the sensible and latent heat load present in the space and will be coming out of the conditioned space at a higher temperature and humidity level.

Assumptions for Simulation

1. Steady-state operation is assumed for all components; transient effects and thermal storage are neglected.
2. Air is modeled as an ideal gas with constant specific heat capacity (c ,s) evaluated at average temperature.
3. Pressure drops across ducts and components (other than the vortex tube) are neglected.
4. Effectiveness of heat exchangers is taken as constant values: 0.85 for the energy recovery wheel and 0.7 for the sensible heat exchanger.
5. Desiccant wheel performance is represented using manufacturer data/correlations with uniform wheel rotation and no axial heat loss.
6. Regeneration heating is supplied by an electric heater with 100% efficiency; heat losses to the environment are ignored.
7. Moisture removal in the desiccant wheel assumes uniform sorption and complete contact between process air and desiccant surface.
8. Vortex tube performance is represented by a semi-empirical correlation derived from experimental data (valid for 2 ?"6 bar inlet pressure and 0.3 ?"0.7 cold mass fraction).
9. Cold and hot outlet streams from the vortex tube are assumed to mix ideally with no further irreversibilities.
10. Ambient reference conditions are fixed at measured outdoor temperature and humidity; no solar gains or building loads are included.

RESULTS AND DISCUSSIONS

1. Performance of Vortex Tube

As shown in the Fig.1 the experiment was conducted for the determination of the cold air temperature available at the exit of the vortex tube which would be used later for the simulation of desiccant based air conditioning. Figure 4 shows the results from the experiment presenting the variation of the change in cold air stream temperature with the mass fraction of cold air. The first observation is that for a certain value of pressure, the reduction in cold air temperature decreases with cold air mass flow fraction. The second observation is that with increase in pressure, the reduction in the temperature is increasing for a particular value of pressure at the inlet of vortex tube. Moreover, the experiments were conducted for both insulated (a) and non-insulated vortex tube (b). From the comparison of Fig. 4(a) and 4(b) it can be understood that the temperature reduction is higher in case of non-insulated vortex tube than the insulated vortex tube. As the compressed air pressure is increased the cold air stream temperature reduces because of more energy transfer between the hot air and the cold air. Thus, there is monotonous decline in the output temperature of the vortex tube output. Therefore, the final supply air temperature is also reduced. However, in practical application the temperature can be maintained by controlling the flow rate.

Figure 4. Variation of the reduction in cold air temperature (^+T_{cold}) with cold mass fraction (A) for vortex tube

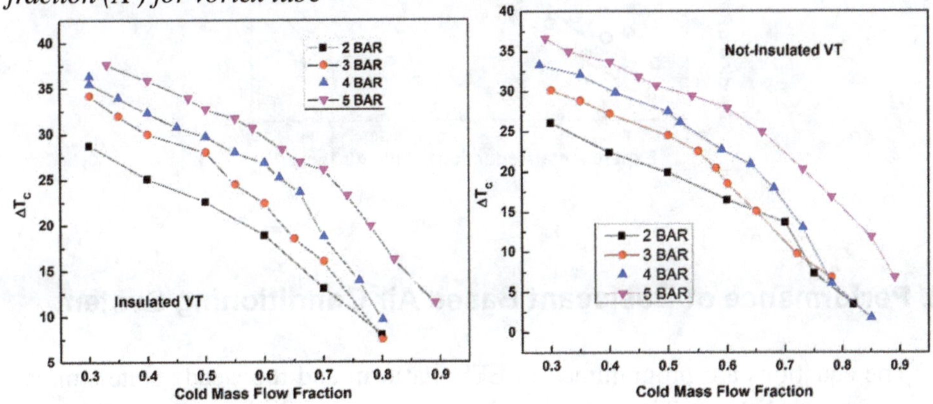

The data obtained from the experiment is converted to a correlation using regression analysis. The semi-empirical correlation obtained is shown in equation (7) which will be considered for modelling the vortex tube in the simulation work. The correlation is able to predict the experimental data within an error of A10% as shown in Fig. 5.

^+*Tcold* = A36.8919A + A3.7536 * A*P*A – A45.2096 * AA A – A0.A2158A

* A*P* * A A (7)

Figure 5. Performance of semi-empirical correlation for prediction of reduction in cold air temperature

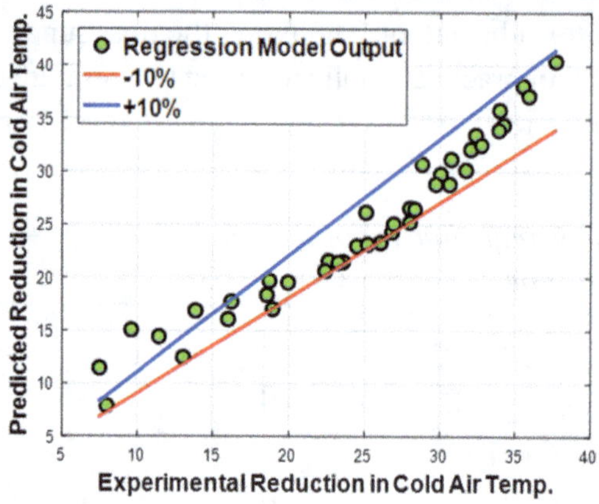

2. Performance of Desiccant Based Air Conditioning System

The equations are programmed in EES platform and the steady state simulations were carried out. The different parameters are varied to study the effect on the output supply air temperature and humidity. Along with this, the performance is also investigated taking the dry bulb temperature and wet bulb temperature as input ambient conditions of Bhubaneswar city for carrying out the simulations. Then, considering the condition during the summer time, the other simulations were carried out. The parameters such as regeneration temperature, ratio of mass flow rate of process and regeneration air streams, and pressure of the compressed air supply to vortex tube are varied.

Figure 6 shows the trend of output supply air temperature and specific humidity with respect to different weather conditions of Bhubaneswar city during a year. The weather conditions are taken from the design conditions of Bhubaneswar, India as per American Society of Heating, Refrigeration and Airconditioning Engineers (2025). It can be observed that the temperature of the output temperature is within 10-20°C which reflects that by controlling the flow rate of the vortex tube air the required temperature can be easily attained. Furthermore, the hybrid air-conditioning unit is also able to reduce the specific humidity of the outside air.

Figure 6. Variation of the temperature and humidity of the supply air during the whole year

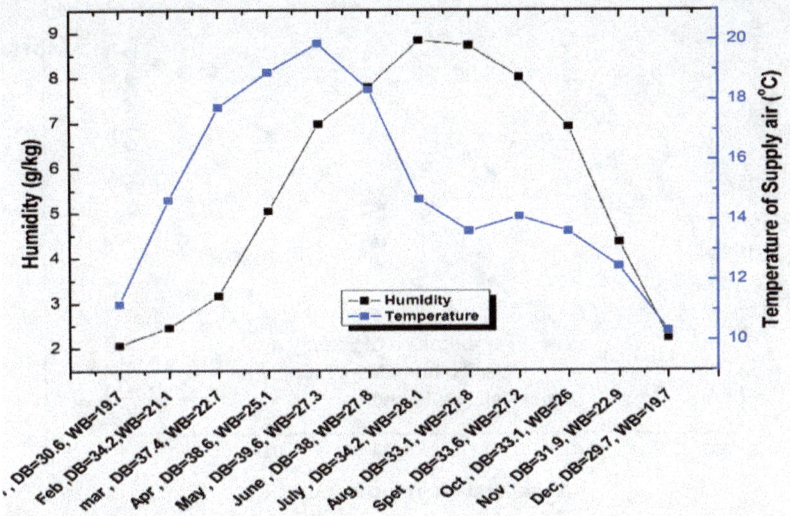

It can be seen in the Figure 6 that the input ambient conditions during the months of April, May and June have higher temperatures. Thus, the focus of the simulation is to test the feasibility of the hybrid air-conditioning system consists of DW and Vortex tube. Therefore, the average of the dry bulb temperature and wet bulb temperature of the summer season is considered for further simulations which is 38.5° C DBT and 27 ° C WBT.

Figure 7. Variation of temperature, specific humidity and relative humidity with respect to the regeneration temperature 38.5° C DBT and 27 ° C WBT.

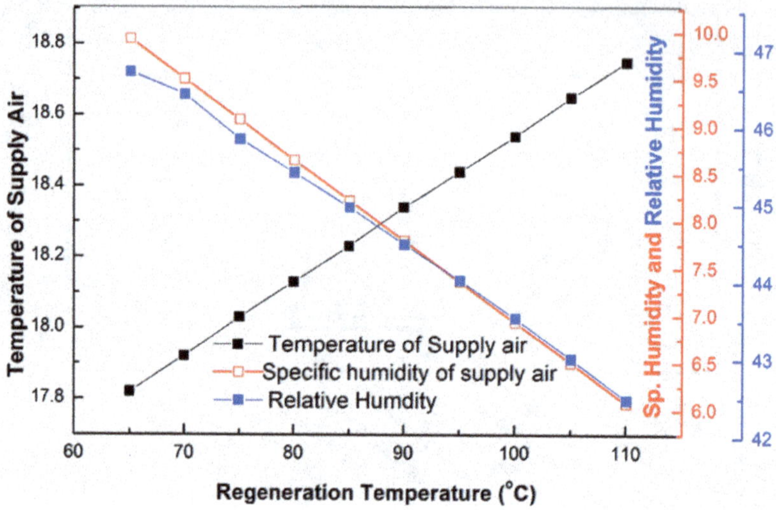

Figure 7 depicts the variation of the supply air temperature, specific humidity and the relative humidity with respect to the regeneration air stream temperature. In all other simulations the regeneration temperature is taken as 90 C. However to study the effect of the regeneration temperature it is varied from 65 C to 105 C, keeping the other parameters fixed such as the air flow rates as 400 m3/ hr for both streams, the RPM of the wheel as 8 and the ambient air at 38.5° C DBT and 27 ° C WBT. It can be observed that the output supply air temperature is increasing with the increase in regeneration temperature. This is because, the present simulation does not include a purge or cooling section for reduction of temperature of desiccant wheel. Further improvements are also need to be investigated for automatic control of the heating of regeneration air or volume flow rate of the air.

The regeneration process consumes energy for heating the air by the use of an electrical heater. In addition to this, the regeneration temperature and air flow should be sufficient for removing the absorbed moisture from the desiccant wheel. Thus, the overall coefficient performance needs optimum regeneration temperature which should be investigated properly and arrangement should be made.

Another important observation is that the desiccant wheel, when regenerated at sufficiently high temperatures (80 ?"90 A C in the present case), shows stable performance throughout the year. The regeneration heater energy demand is higher in humid months but remains within a predictable range, making the system controllable from an operational perspective. It should be noted, however, that the assumed constant effectiveness values for the energy recovery wheel (0.85) and heat exchanger (0.7) influence the magnitude of supply conditions, and sensitivity studies suggest a shift of about 1 ?"2 A C in supply air temperature with variations in these assumptions.

From a practical viewpoint, the system demonstrates strong potential for climates characterized by high humidity, such as coastal or tropical regions. The refrigerant-free nature of the vortex tube and desiccant cycle is aligned with environmental sustainability goals, but the relatively high energy requirement for regeneration remains a limitation. Opportunities for improvement include integrating solar thermal energy or waste-heat streams for regeneration, and optimizing the vortex tube operating pressure and cold fraction to maximize effective cooling while minimizing compressed-air power input.

Figure 8. Variation of the reduction in specific humidity with respect to volume flow rate ratio of process air and the regeneration air

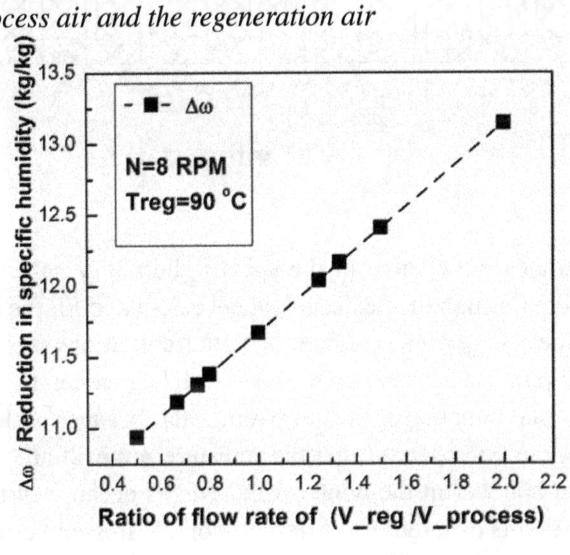

Figure 8 shows the variation of the reduction in the specific humidity for different ratios of the regeneration air and the process air. It can be observed from the figure that the reduction in the specific humidity increases with the increase in the flow ratio. As the flow ratio is increased the amount of the regeneration air flow rate is very much higher as compared to the flow rate of the process air. Thus, the DW is capable of absorbing more moisture from the supply air. However, it is to be kept in mind that the regeneration temperature and RPM of wheel are fixed for the simulation and presently the simulations were carried out for the Hot and humid climate of Bhubaneswar city.

Figure 9. Variation of the moisture removal by DW during the entire year

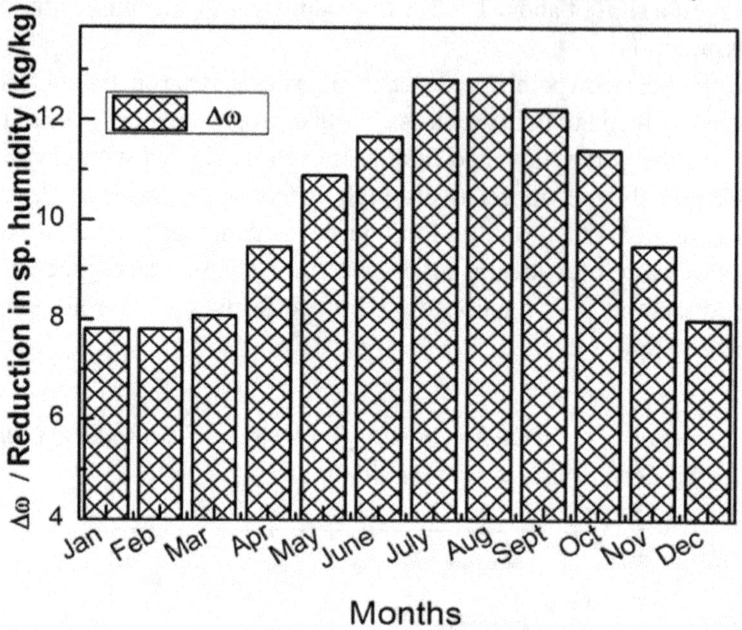

Figure 9 illustrates the change in the specific humidity values of process air stream after passing through the desiccant wheel calculated for the entire year. The parameters of the desiccant wheel are fixed at certain conditions such as volume flow rate ratio as 1, RPM of the desiccant wheel is 8 and the regeneration temperature is 90A C. Thus, this phenomenon of moisture removal is because of the air conditions and properties. It can be observed that the moisture removal amount is higher in the summer and it is lower in the winter time. This is because of the fact that the ambient air condition is different for different months. In summer the temperature and concentration of moisture is high, which drives for better adsorption process. On the other hand, during winter the temperature and moisture concentration is low which in turn slows down the adsorption process occurring in the desiccant material. However, it is also to be noted that the final values of the specific humidity of the output process air is lower in winter and higher in summer due to the initial value. These final values could be controlled in a better way by employing better control algorithms for parameters associated with the desiccant wheel. This is the future scope of the present research investigation. The results indicate that the hybrid system is capable of consistently lowering the humidity ratio of the process air, with values ranging from approximately 7 g/kg in the cooler months to more than 13 g/kg during peak summer. This trend reflects the strong dependence of desiccant wheel performance on the inlet humidity and temperature conditions. Higher outdoor humidity levels during the monsoon and summer months provide greater potential for sorption, which explains the observed peak reduction in July ?"August. Conversely, the relatively lower humidity during winter results in a reduced but still appreciable I"I% of about 7 ?"8 g/kg.

The vortex tube refrigeration system is designed in a simple way and no moving parts are present inside the tube. The system is found to have very low energy efficiency and the coefficient of performance is poor when compared to conventional vapor compression systems. A large portion of compressor work is destroyed through irreversibilities in valves throttling and vortex separation. A continuous supply of high pressure compressed air is required and the use of compressors and surge tanks makes the system costly noisy and bulky. The cooling capacity is limited and the system is mainly used for spot cooling instead of large scale refrigeration or air conditioning. The outlet temperatures are strongly affected by inlet conditions and flow fractions and precise temperature control is difficult. A large portion of usable energy is carried away in the hot stream and it is usually wasted unless recovery is provided. Noise is generated during operation and filters must be maintained regularly to avoid clogging from oil or dust. The system is rarely used in large refrigeration applications and it is mostly applied in small areas where compressed air is already available

CONCLUSION

The article discussed about the modelling and simulation of a hybrid air-conditioning system composed of desiccant wheel and a vortex tube cooler. The trend of output supply air temperature and humidity with respect to the various parameters are studies considering the ambient conditions of summer season in Bhubaneswar city of Odisha, India. In the system no recirculation is considered which will be done in future works. The results obtained shows that the prosed system is capable of maintaining the comfort conditions in Bhubaneswar city. The key results are presented below.

- The temperature and humidity of the supply air obtained are in the range of 10-20° C considering the effectiveness of the cooler as 0.75.
- The temperatures are higher in the summer and it decreases in winter.
- The relative humidity is observed to be varying 42 to 47.5 and the specific humidity ranges from 6 to 10 g/kg.

The article presented the modeling and simulation of a hybrid air-conditioning system composed of a desiccant wheel and a vortex tube cooler. The variation of supply air temperature and humidity with respect to different operating parameters was studied under the ambient summer conditions of Bhubaneswar, Odisha, India. The results indicate that the proposed system is capable of maintaining indoor comfort conditions, with supply air temperature in the range of 10-20 A C (for cooler effectiveness of 0.75), relative humidity between 42 --47.5%, and specific humidity between 6-10 g/kg. Higher supply temperatures were observed during summer months and lower values during winter, which reflects the seasonal influence on system performance.

However, some limitations of the present study must be noted. The energy consumption of the compressor driving the vortex tube and the regeneration heater was not explicitly evaluated, and no overall COP or exergy efficiency analysis was reported. The assumption of constant effectiveness values for the energy recovery wheel and heat exchanger introduces simplifications that may not hold under varying operating conditions. Furthermore, no recirculation of indoor air was considered, and the sustainability impact of electric regeneration heating was not fully quantified.

Future work will address these gaps by conducting a complete energy and exergy analysis of the integrated system, evaluating the practical feasibility in terms of COP and operating cost, and incorporating indoor air recirculation to improve efficiency. Experimental validation of the coupled system under realistic operating conditions will also be pursued. In addition, alternative regeneration strategies such as solar thermal or waste-heat utilization will be explored to improve environmental sustainability. Overall, the study suggests that the integration of a vortex tube with a desiccant wheel offers a promising refrigerant-free solution for space conditioning in hot and humid climates, provided that the identified challenges are systematically addressed.

Box 1. Nomenclature

h	Enthalpy of air
Treg	Regeneration Temperature
Ta1	Temperature of inlet process air
Ta2	Temperature of outlet process air
w1	Specific Humidity of inlet process air
w2	Specific Humidity of outlet process air
m	Mass flow rate
N	Rotations per minute
P	Pressure
v	Volume flow rate

continued on following page

Box 1. Continued

V_rat	Ratio of volume flow rate of process and regeneration air flow
A	Cold air mass flow fraction
Subscripts	
a	Process air stream
reg	Regeneration air stream
Cold	vortex tube cold air
Abbreviations	
DW	Desiccant Wheel
EES	Engineering Equation solver
^+T	Temperature drop
RPM	Rotations Per Minute
RMSE	Root Mean Square Error

REFERENCES

Abdelgaied, M., & Mohamed, S., Bassuoni, M. M., & Khaira, A.M. (2023). Solid desiccant air conditioning system using desiccant dehumidifiers with cooling technique and thermal recovery unit: Experimental investigation and performance analysis. *Journal of Cleaner Production*, 416.

American Society of Heating. (2025). *Refrigeration and Airconditioning Engineers*. ASHRAE Handbook Fundamentals.

Angrisani, G., Capozzoli, A., Minichiello, F., Roselli, C., & Sasso, M. (2011). Desiccant wheel regenerated by thermal energy from a microcogenerator: Experimental assessment of the performances. *Applied Energy*, *88*(4), 1354-1365.

Angrisani, G., Minichiello, F., Roselli, C., & Sasso, M. (2012). Experimental analysis on the dehumidification and thermal performance of a desiccant wheel. *Applied Energy*, *92*, 563-572.

Angrisani, G., Roselli, C., & Sasso, M. (2017). Dehumidification and thermal behaviour of desiccant wheels: Correlations based on experimental and manufacturer data. *Heat Transfer Engineering*, *39*(3), 293-303. DOI: 10.1080/01457632.2017.1295743

Assadi, M. K., & Mohammadi, A. (2016). Simulation of dynamical performance of solar desiccant cooling cycle. *Applied Mechanics and Materials*, *819*, 160-170.

Awan, A., Ge, T. S., Wang, R. Z., & Jing, D. (2025). Flow phenomena inside the Ranque-Hilsch vortex tube: A state-of-the-art review. *Renewable & Sustainable Energy Reviews*, *211*(Feb).

Awan, O. A. A., Ge, T. S., Wang, R. Z., & Jing, D. (2025). Flow phenomena inside the Ranque-Hilsch vortex tube. *Renewable & Sustainable Energy Reviews*, 173.

Bergmann, B. S. (2018, October 31). The theory of the vortex tube. *EXAIR Blog*. https://blog.exair.com/2018/10/31/the-theory-of-the-vortex-tube/

Caliskan, H., Lee, D.-Y., & Hong, H. (2019). Enhanced thermodynamic assessments of the novel desiccant air cooling system for sustainable energy future. *Journal of Cleaner Production*, *211*, 213-221.

Chen, L., Chen, S. H., Liu, L., & Zhang, B. (2018). Experimental investigation of precooling desiccant-wheel air-conditioning system in a high-temperature and high-humidity environment. *International Journal of Refrigeration*, *95*, 83-92.

Comino, Francisco, TA-boas, Francisco, Peci, Federico, & de Adana, Ruiz, Manuel. (2020). Detailed experimental analysis of the energy performance of a desiccant wheel activated at low temperature. *Applied Thermal Engineering*, *178*, 115-580.

Daou, K., Wang, R. Z., & Xia, Z. Z. (2006). Desiccant cooling air conditioning: A review. *Renewable & Sustainable Energy Reviews*, *10*(2), 55-77.

Dikshit, S. V. (2024). A comprehensive review on dehumidification system using solid desiccants. *Proceedings of the Institution of Mechanical Engineers, Part E: Journal of Process Mechanical Engineering*, August.

Fong, K. F., & Lee, C. K. (2020). Solar desiccant cooling system for hot and humid region - A new perspective and investigation. *Solar Energy*, *195*, 677-684.

Heidari, A., Roshandel, R., & Vakiloroaya, V. (2018). Co-production of water and cooling in hot and humid climates. *Energy Conversion and Management*, *185*, 396-409.

. Huang, L., Liu, Y., Li, Y., Han, M., & Zhang, Y. (2024). Performance investigation of a hybrid liquid-desiccant air conditioning system. *Frontiers in Built Environment*, September.

Iyer, P. K., Siva, R., Rose, A., Kumar, S., & Venkatraman, K. (2024). Experimental investigations of a desiccant-coated M-cycle cooler for net zero air-conditioning. *Energy Conversion and Management*, 2024.

Jani, D. B., Mishra, Manish, & Sahoo, P. K. (2016). Solid desiccant air conditioning -A state of the art review. *Renewable & Sustainable Energy Reviews*, *60*, 1451-1469.

Jing, D., Gang, X., Rafiee, M., & Shaji, K. (2025). A review of refrigerant-based vortex tube separation characteristics. *Applied Thermal Engineering*, *173*(February).

Jing, D., Gang, X., Rafiee, M., & Shaji, K. (2025). A review of refrigerant-based vortex tube separation characteristics. *Energy*, 314.

Khalid, A., Mehmood, M. A., Rehman, S., Irfan, M., Farooq, U., Farhan, M., & Riaz, M. (2024). Thermo-economic analysis and comparison of novel multi-stage desiccant wheels. *Energy Reports*, 10.

Kumar, A., Vivekanand, & Subudhi, S. (2017). Cooling and dehumidification using vortex tube. *Applied Thermal Engineering*, *122*, 181-193.

Liang, F., Zhang, L., Wang, S., & Zhou, H. (2025). Experimental investigation on the influence of drainage structure on vortex tube energy separation. *Applied Thermal Engineering*, 2025.

Ma, Z., Zhang, N., Song, S., Zheng, X., Gong, X., & Wang, Y. (2023). Experimental investigation and effectiveness analysis of a desiccant wheel dehumidification system with low air humidity. *International Journal of Refrigeration*, 156.

Parmar, H., & Hindoliya, D. A. (2013). Performance of solid desiccant-based evaporative cooling system under the climatic zones of India. *The International Journal of Low Carbon Technologies*, *8*(1), 52-57.

Qyyum, M. A., Wei, F., Hussain, A., Noon, A. A., & Lee, M. (2018). An innovative vortex-tube turbo-expander refrigeration cycle for performance enhancement of nitrogen-based natural-gas liquefaction process. *Applied Thermal Engineering*, *144*, 117-125.

Ruivo, C. R., Angrisani, G., & Minichiello, F. (2015). Influence of the rotation speed on the effectiveness parameters of a desiccant wheel: An assessment using experimental data and manufacturer software. *Renewable Energy*, *76*, 484-493.

Ruivo, C. R., Carrillo-AndrA(c)s, A., Costa, J. J., & DomA-nguez-MuAoz, F. (2013). Exponential correlations to predict the dependence of effectiveness parameters of a desiccant wheel on the air flow rates and on the rotation speed. *Applied Thermal Engineering*, *51*(2), 442-450.

Sahlot, M., & Riffat, S. B. (2016). Desiccant cooling systems: A review. *The International Journal of Low Carbon Technologies*, *11*(4), 489-505.

Saputra, D. A., Osaka, Y., Tsujiguchi, T., Haruki, M., Kumita, M., & Kodama, A. (2020). Experimental investigation of desiccant wheel dehumidification control method for changes in regeneration heat input. *Energy*, *205*, 118-109.

Singh, R. P., Das, R. K., & Upadhyay, S. N. (2023). Effectiveness approach for predicting desiccant wheel performance: Correlation utilizing experimental results. *International Journal of Refrigeration*, *156*(December).

. Swami, S. M., M.K. Mahesh, D. Kumar, & P. Sahu. (2022). Design and development of a cooling system for a vaccine container by using vortex tube refrigeration. *International Journal of Engineering Research and Technology*, July.

Wu, X. N., Ge, T. S., Dai, Y. J., & Wang, R. Z. (2019). Investigation on novel desiccant wheel using wood pulp fiber paper with high coating ratio as matrix. *Energy*, *176*, 493-504.

Zamzamian, S. A. H. (2022). A new approach for the heat and moisture transfer in rotary desiccant wheel. *Iranian Journal of Chemistry and Chemical Engineering*, *41*(2), 197-208.

Zhang, B., & Guo, X. (2018). Prospective applications of Ranque-Hilsch vortex tubes to sustainable energy utilization and energy efficiency improvement with energy and mass separation. *Renewable & Sustainable Energy Reviews*, *89*, 135-150.

Chapter 14

Studies on Solar-Based Air Conditioning Systems With Latent Thermal Energy Storage

Harshit Gupta
Indian Institute of Technology, Kharagpur, India

Snehangshu Mondal
https://orcid.org/0009-0000-6964-3328
Indian Institute of Technology, Kharagpur, India

Maddali Ramgopal
Indian Institute of Technology, Kharagpur, India

ABSTRACT

Latent Thermal Energy Storage (LTES) in a compact, solar-powered vapour compression air conditioner system with ice as the phase change material is studied numerically. In addition to using propane (R290) as a refrigerant, the system has two VCR systems: a high-capacity unit for charging the LTES and a low-capacity one for chilling during the day. With interior fins to improve heat transfer, a CFD model replicates the storage tank. Findings indicate that the solidification rate does not significantly increase beyond 32 fins. The cooling load of a typical student room is determined, and the necessary power and associated PV panel area are computed using the thermodynamic cycle. The results show that full solar operation is not practical due to the small roof space and that some grid electricity is required. The study backs the development of environmentally friendly cooling solutions for tropical regions.

DOI: 10.4018/979-8-3373-3621-3.ch014

1. INTRODUCTION

Higher temperatures lead to an increased demand for cooling, including the use of air conditioning (AC) units. The demand for AC systems is expected to increase over time as temperatures and populations rise. The demand for AC is limited not only to offices; it is used extensively in individual homes and public transport for thermal comfort. Studies show that the AC accounts for nearly 40% of the total building energy consumption. This energy consumption is higher in tropical climatic zones due to the uncomfortable hot and humid conditions. There are several ways to reduce AC energy consumption. Studies show that the AC energy consumption of office buildings can be reduced by 10 to 15% using Personalized ventilation (PV) systems, as found by Mondal et al. (2024). However, PV systems are currently limited to offices and large buildings that use centralized air conditioning systems. The installation of room air conditioners in residences and small offices is increasing, as there is no scope for centralized systems for such applications. The increasing use of air conditioners can be justified and sustainable if they are run using renewable energy sources, such as solar or wind energy. Among the various forms of renewable energy, solar is the most common and well-developed. However, as solar energy availability is intermittent in nature, suitable storage is needed to ensure air conditioning is available throughout the day. Battery energy storage can store electrical energy. However, according to Fedorová et al. (2022), the initial, operating, and maintenance costs of batteries are very high. Additionally, operating the battery requires suitable environmental conditions. The batteries have to be replaced every few years, which is costly. Hence, solar energy-based air conditioning systems that use electric batteries for energy storage may not be economically viable. Instead of using batteries for storing energy, one can utilize thermal energy storage (TES) devices to store heat or cold. Phase change materials (PCMs) in the form of latent thermal energy storage (LTES) have the capability to store a large quantity of thermal energy in a relatively small mass and volume with a small change in temperature. According to Mohan et al. (2022) and Le et al. (2021), their application is based on transformation, applicability, and project aim. Ravikiran et al. (2023) have classified PCMs based on their phase change temperatures, as depicted in Figure 1.

Figure 1. Classification of PCM based on temperature

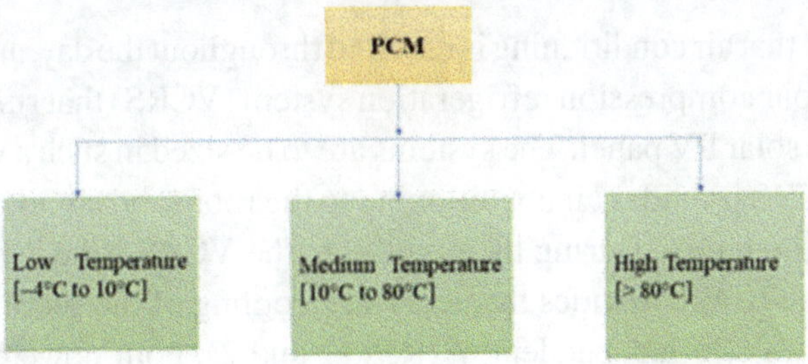

In TES applications, optimal container design is crucial, necessitating comprehensive research on heat transmission in PCM-based systems. Talukdar et al. (2019) conducted a numerical simulation of a cuboidal thermal energy storage system featuring tubes and fins to promote heat transfer, which was experimentally confirmed; however, the study did not account for buoyancy-driven natural circulation. Voller et al. (1987) devised fixed-grid numerical techniques for the solidification and melting of pure substances. Saini et al. (2023) performed 3D models and experiments on a shell-and-tube latent thermal energy storage system utilizing high-temperature phase change materials. Burgess et al. (2022) enhanced portable phase change materials by using a novel energy source term in simulations for cold chain delivery applications. Kumar et al. (2012) introduced a two-dimensional computational fluid dynamics model for lead melting in a cuboid subjected to constant heat flux, which was validated using experimental methods. Dong et al. (2022) investigated spherical phase change material tanks for cold energy storage and determined the ideal diameter for heat transfer efficiency.

According to the literature, a thermal latent energy storage system based on a low-temperature Phase change material (PCM) offers an attractive solution. However, studies on solar-based air conditioning systems with PCM storage are scarce in the literature. Additionally, for the optimal design of the TES system, a detailed CFD simulation has not been conducted. Based on these research gaps, in the present work, the numerical simulation of a tube-in-tube type LTES with water as PCM is carried out. Considering a small room for air conditioning, cooling loads are estimated to size the dimensions of LTES. Based on the cooling load and refrigeration cycle calculations, the required area of the solar PV collectors is calculated. Then, based on the available room area, calculations are performed to determine the solar fraction of the air conditioning system.

2. METHODOLOGY

It is assumed that air conditioning is required throughout the day, and it is provided by a simple vapour compression refrigeration system (VCRS) that runs on the power supply from the solar PV panel. The systems are to be sized in such a way that during daytime, the VCRS provides air conditioning to the room and simultaneously stores cold energy in the LTES. During the nighttime, the VCRS is switched off and the cold stored in the LTES provides the necessary cooling of the room. The proposed system consists of two independent VCRSs (1 and 2), both powered by solar PV and operating during the hours of sunshine. During the daytime, room cooling is provided by VCRS-1, which operates at a relatively high evaporator temperature. In contrast, VCRS-2, which operates at sub-zero temperatures to produce ice, is used to charge the LTES. Propylene Glycol (30% concentration) is used as the secondary fluid for transferring heat from the conditioned space (room) to the evaporator of VCRS-1 (HX-1) during daytime and from the conditioned space to the LTES during nighttime. Glycol is also used during the daytime to transfer heat from LTES to the evaporator of VCRS-2 (HX-2). The flow of glycol is controlled by opening or closing a suitable valve. Figure 2 shows the proposed concept.

Figure 2. Schematic diagram of the system

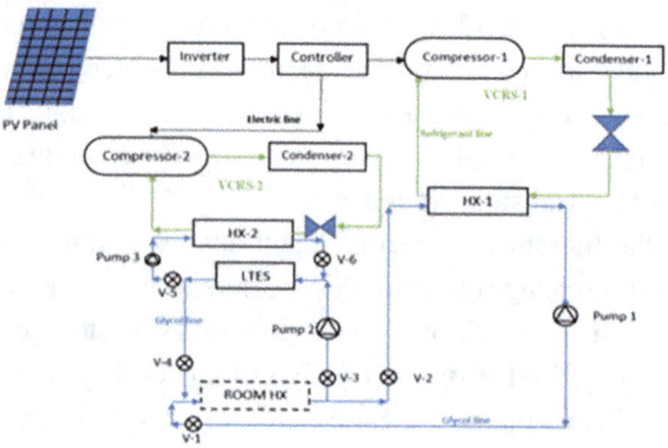

For the sake of modelling, the whole problem is distributed into three parts,

i) Estimation of the required area of the PV panel for the given thermal load profile.
ii) VCRS cycle calculations to estimate the power required for the compressors of VCRS-1 and VCRS-2 based on the room cooling load.
iii) Design and development of internal melt, ice on the coil type LTES.

For the sake of simplicity, it is assumed that both VCRS-1 and VCRS-2 consist of only four basic components: a compressor, condenser, expansion valve, and an evaporator (HX). During daytime, pump-3 is turned on, and valves V-5 & V-6 are kept open while valves V-4 & V-3 are closed, and pump-3 is off. The cold glycol from the evaporator of VCRS-2 flows through the LTES and converts water into ice. During the night time, the valve position is such that the cold glycol flows through the LTES, and the room HX is kept inside the conditioned space, discharging LTES by transferring heat energy from the conditioned space to the LTES. During the daytime, the cooling effect of VCRS-1 is conveyed to the conditioned space through the glycol, which flows through pump-1 when valves V-1 and V-2 are kept open and valves V-3 and V-4 are kept closed. V-1 and V-2 are kept open to allow the flow of glycol through the HX-1 and the room HX.

The operational control strategy of valves and pumps is summarized in Table 1.

Table 1. The operational control strategy of valves and pumps

Day time		Night time	
Pump Status	**Valve Status**	**Pump Status**	**Valve Status**
Pump 1, Pump 3-ON Pump 2-OFF	Valve 1, Valve 2, Valve 5, Valve 6-ON Valve 3, Valve 4-OFF	Pump 1, Pump 3-OFF Pump 2-ON	Valve 1, Valve 2, Valve 5, Valve 6-OFF Valve 3, Valve 4-ON

2.1. Modelling of the room (air-conditioned space) for cooling load calculation

The room shown below in Figure 3 is part of the low-rise apartment, which features multiple rooms on the same level. The given room façade faces south, which is exposed to the outside environment. This 3D room is created by using the SketchUp OpenStudio plugin. All other walls are assumed to be common with other rooms, which are also air-conditioned. To calculate the maximum possible cooling load on the room, it is assumed that heat is transferred only through the exposed roof and south wall, as the other surfaces are surrounded by conditioned spaces. The room's dimensions are such that it is suitable for two occupants. The floor dimensions are 3 m x 3 m, and the ceiling is 3.3 m (11') high. The extended floor space length is in compliance with the National Building Code, which suggests the length of the balcony should lie between 1.21m and 1.82m. In this model, the length of the balcony is 1.5m extending in the south direction.

Figure 3. SketchUp model of the room

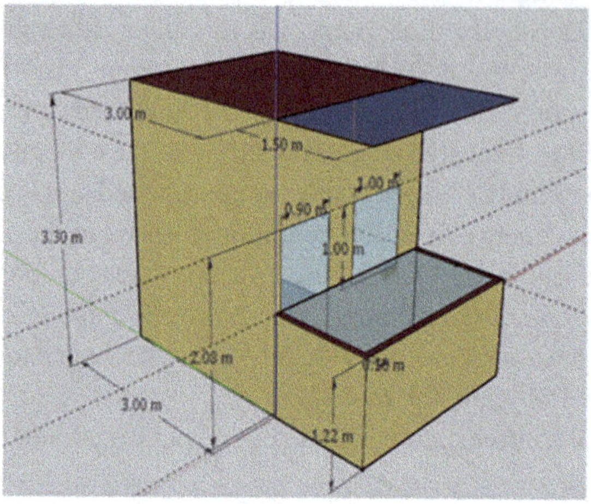

An overhang is also included in the 3D model to provide shade on sunny and monsoon days. The length of the overhang is the same as the length of the balcony.

Table 2. Material properties of the building (ARORA, R. C. (2012))

Sl. No.	Material	Description	Specific heat (kJ/kg-K)	Density Kg/m³	Thermal conductivity (W/m-K)	Thickness (m)
1	Brick	Fire clay brick	0.96	2000	1.06	0.22
2	Plaster	Cement	0.88	1885	8.65	0.025
3	POP	Gypsum	1.088	2320	0.16	0.001
4	Wood	Door	2.39	720	0.158	0.025
5	Concrete	8IN Concrete HW	0.837	2243	1.729	0.2033

Figure 4 shows the details of the walls, and the material specifications are given in Table 2. For measuring the heat transfer through the building elements shown in Figure 2, Thermal Zone 1, which represents the room, has six surfaces, out of which four are walls facing normal to the geographical direction, one surface is the roof, and one is the floor. There are two sub-surfaces on the south wall: one is for the 2.74 m x 0.914 m door, and the other is for the glass window (1m x 1 m). All the other three walls, except the south wall, are at adiabatic boundary conditions, which means that the room next to the thermal zone 1 is air-conditioned at all times. The floor surface of the zone is kept as a thermally grounded boundary condition. Transient heat transfer to the zone happens mainly through the south wall, sub-surfaces, and the roof. Transient heat transfer in buildings occurs due to changes in solar radiation throughout the day, particularly in summer.

For the given room, hourly cooling load calculations are performed using EnergyPlus 23.2.0 software. The building is simulated in EnergyPlus with the above inputs, the Typical Meteorological Year (TMY) weather file of Kolkata (2009), and the thermal set point of the room (23°C).

Figure 4. Cross-section view of building wall

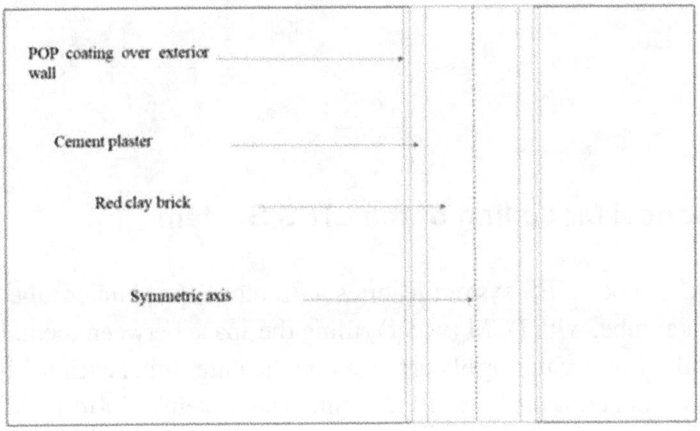

2.2. Modelling of the refrigeration system

For modelling the solar air conditioning system, two refrigeration systems are considered, where VCRS-1 operates at an evaporator temperature of 5°C and VCRS-2 operates at an evaporator temperature of −10°C. The following simplified assumptions are made for calculating the power input to both systems.

1. The systems operate in quasi-steady conditions.
2. The compressors operate at 70% isentropic efficiency.
3. The throttling process is isenthalpic.
4. There is no loss of pressure in the pipes/HX.
5. Degree of superheating at the end of the evaporator is 5°C.
6. Design outside dry bulb temperature is 40°C according to the weather file of Kolkata.

The process flow diagram (PFD) of VCRS at typical design conditions is shown in Figure 5.

Figure 5. PFD of VCRS-1 (left) and VCRS-2 (right)

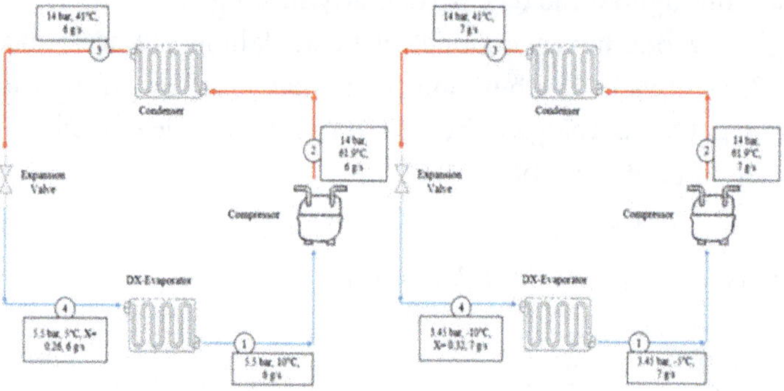

2.3. Numerical Modelling of the LTES System

An annular tube LTES system features a 75 mm copper inner tube and a 150 mm steel outer tube, with PCM (water) filling the space between them. Longitudinal rectangular fins are uniformly arranged on the outer tube, each having a radial length of 25 mm and a thickness of 2.5 mm. The system's fin-to-PCM area ratio is 2.01% for a 4-fin configuration. To assess the impact of the number of fins on the solidification rate, multiple cases are analysed, maintaining the same area ratio while varying the thickness of the fins. The cross-section of the LTES for the 4-fin configuration is shown in Figure 6. The material properties of ice & water, and copper are taken from standard sources.

For the simulation, a laminar flow model is assumed on the ice-water side, as the maximum estimated Rayleigh number (Ra) is less than 10^8. Where,

$$\mathrm{Ra}=\frac{g\gamma_{PCM}(T_{wall}-T_{\infty})L^3}{\alpha_{PCM}\vartheta_{PCM}} \tag{1}$$

Where L is the characteristic length, equal to 230 mm, and Twall is 10°C and 5 °C for solidification and melting simulation, respectively.

Figure 6. Computational domain for the 4-fins geometry

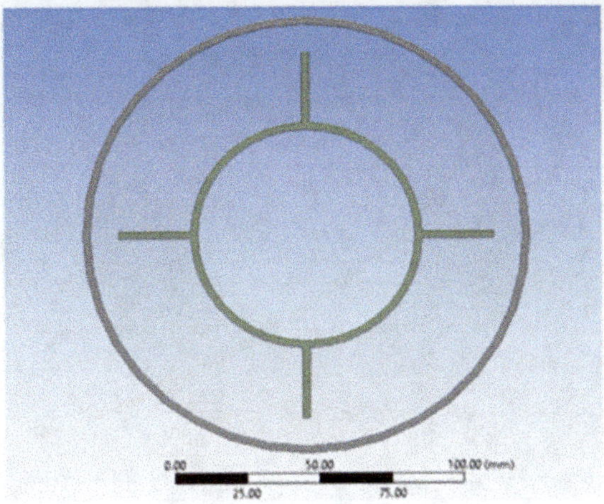

Gravity is included using the Boussinesq approximation for the 2D transient melting problem, while a constant density is assumed for solidification. Continuity, velocity, and energy residuals are 10^{-5}, 10^{-3}, and 10^{-9}, respectively. The SIMPLE algorithm is used for pressure-velocity coupling, with PRESTO for pressure discretization and second-order upwind for momentum and energy. The maximum number of iterations per time step is 50, and the mushy zone constant is set to 10^5, based on previous studies by Yang et al. (2022).

The boundary and initial conditions for all the different cases studied are the same. The initial temperature of the PCM is 1°C in both solidification and melting cases. The inner wall tube temperature is –10°C for ice formation, while for melting, it is 5°C. For the outer wall of the outer tube adiabatic boundary condition is assumed.

A grid independence test was conducted for the 4-fin geometry using orthogonal and tetrahedral elements, with element counts of 2,700, 3,700, and 4,700. Based on the results shown in Figure 7, a 2700-element mesh was selected. The time step size of 0.2s aligns with literature values based on Saini et al. (2023) and Yang et al. (2022).

Figure 7. Grid independence test on 4 fins LTES

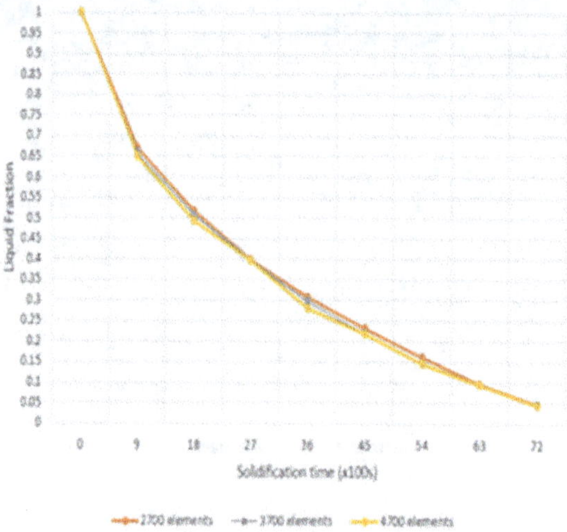

3. RESULT AND DISCUSSION

The CFD simulation uses a SIMPLE-based solver with a default morphological constant and a time step of 0.2s. The continuity equation converges easily, and the melt front is tracked every 100s until 95% solidification. Initially, heat transfer is high due to the lack of ice, but as ice forms, heat transfer decreases. The final 5% of solidification takes significantly longer time, requiring 1,240s compared to the 14,891s for the first 95%. Hence the simulation is stopped at 95% solidification to optimize the number of fins, and later a full-scale simulation is conducted for the optimized configuration.

3.1. Solidification with number of fins and time

The influence of fins on solidification is analysed with constant boundary conditions, initial conditions, and fins-to-PCM area ratio. For no fins, the outer tube's inner diameter is reduced from 150mm to 149mm. The solidification curves of all cases are shown in Figure 8. In the first 50s, solidification rate is similar for all cases except the no-fin case. Solidification time decreases with an increase in fins up to 32, after which there is no significant change. Thus, the 32-fin configuration is selected, taking 29,827 s for full solidification.

Figure 8. Solidification curve of different fin cases (Solidification time is increasing with a decrease in the number of fins)

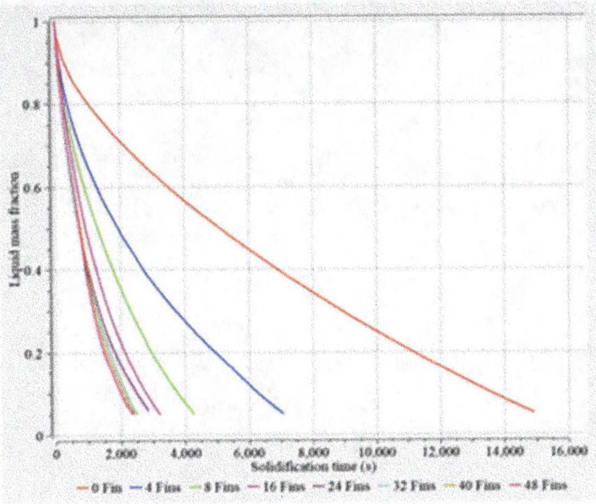

3.2. Rate of heat transfer with number of fins and time

The heat transfer rate increases sharply in the first 50s, then gradually decreases and stabilizes in the final phase. This steady behaviour is observed in the no-fin case. Heat transfer in the PCM is mainly due to conduction, as buoyancy-driven flow is minimal. The buoyancy effect is due to natural convection, a form of heat transfer. Because of natural convection, the density of the fluid layer adjacent to the solid surface changes. Thus, buoyancy comes into action. Even after complete solidification, the heat transfer rate does not reach zero, but continues to reduce the ice temperature until it reaches –10°C. In the 32-fin case, the heat transfer rate remains negative at –573W, indicating ongoing heat transfer despite zero liquid fraction shown in Figure 9.

Figure 9. Heat transfer rate through the inner tube wall to the PCM for different numbers of fins

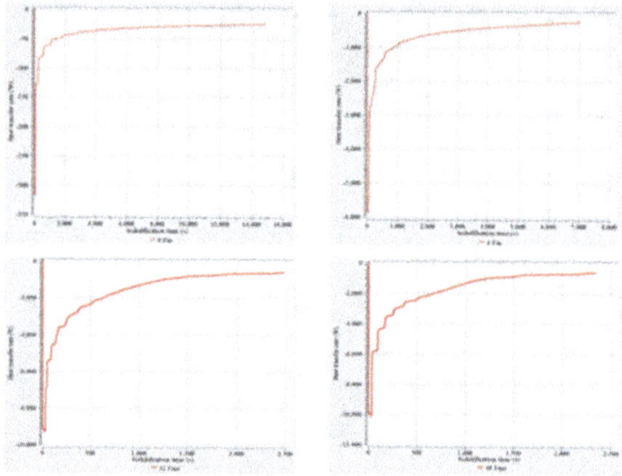

3.3. Liquid fraction contour

The liquid fraction contour illustrates the moving solid front of PCM. For 4 and 8 fins, contours are shown at 300s, 1200s, 2100s, 3000s, and 3900s. For 16, 24, and 32 fins, results are shown at 300s, 1200s, and 2100s, as shown in Figure 10. As shown in these figures, the ice formation begins on the tube and gradually spreads to the fins. After ice forms on the fins, its thickness increases, and the regions between consecutive fins freeze. In the 4 and 8 fin cases, concentric ice rings do not form due to the large space between fins, while in the 24/32/40 fin cases, concentric rings form after 50% solidification. A temperature gradient across the length of the fins creates a tapered ice layer shape. The velocity contours are shown for 32 fins in Figure 11.

Figure 10. Solidification contour of different numbers of fins with a time interval

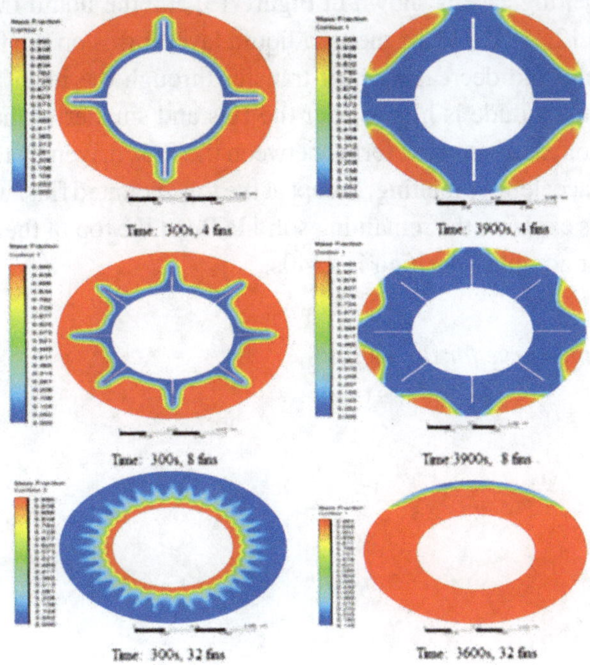

Figure 11. Velocity contour of 32 fins at the 2700s (left) and 1500s (right)

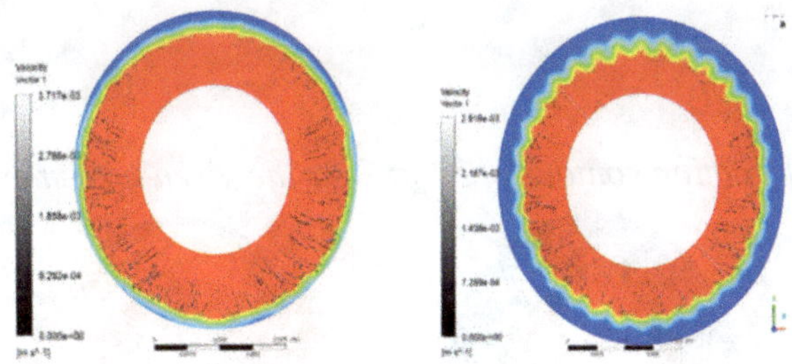

3.4. Melting of PCM

After determining 32 fins as the optimal configuration, PCM melting is studied with an initial temperature of -1°C. The Boussinesq approximation is used to account for heat transfer through natural convection, conduction, and conjugate heat transfer.

Convection accelerates the melting rate near the fins. The liquid fraction contour for the 32-fin melting case is shown in Figure 13, and the liquid fraction vs. time plot is in Figure 12. As the PCM melts, a liquid layer forms on the fins, increasing thermal resistance and decreasing heat transfer through the inner tube wall. The velocity vector magnitude is higher near the fins and smaller farther away due to natural convection. A free vortex forms between the fins, driven by the temperature gradient, which accelerates melting, except at the top-mounted fins, where no vortex is observed. This explains the remaining solid PCM at the top of the tube. The total time required for complete melting is 4340s.

Figure 12. Melting curve for the 32 fins

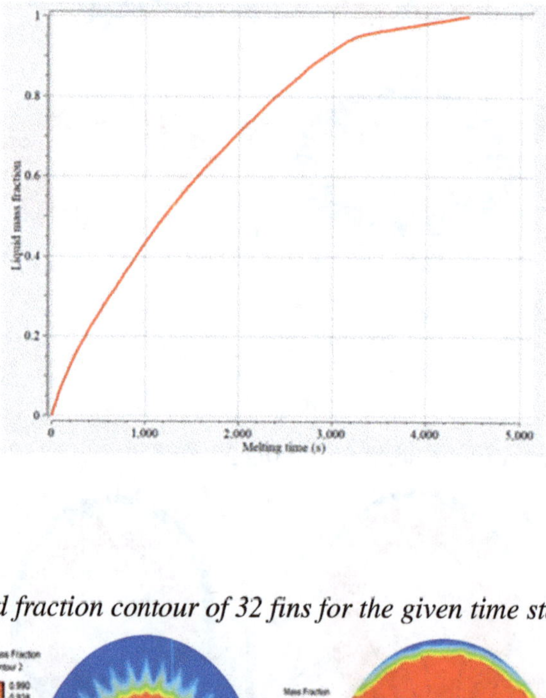

Figure 13. Liquid fraction contour of 32 fins for the given time stamp

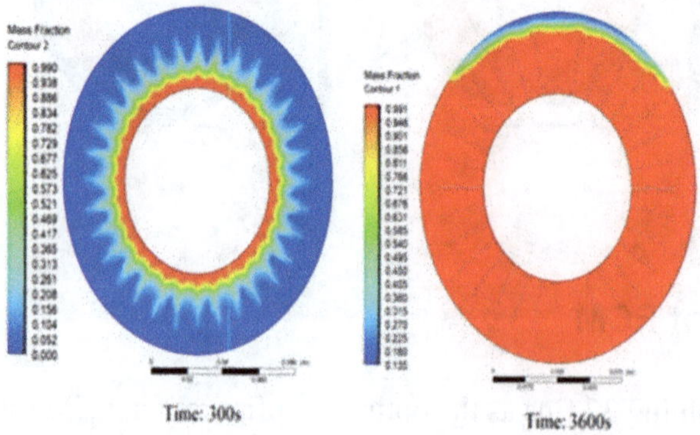

Validation of the CFD mathematical model is performed using experimental investigations conducted by Lokendra Kumar et al. (2012) on the melting of lead in a cuboid with constant heat flux boundary conditions, as observed using thermal neutron radiography. The variation of the liquid fraction with time is reasonable based on the performed experiment. The experimental result is obtained on the 3D geometry of dimension 50mmx50mmx60mm. A constant heat flux of 35.1 kW/m^2 is applied to the vertical wall (50 x 50 mm). The melt front is traveling from right to left, and symmetric results are supposed to get in the direction of the depth of the cuboid. All other faces of the lead cuboid are insulated, with only two exposed to the atmosphere, to calculate the liquid fraction and the movement of the melt front. Properties of Lead are listed in Table 3, with a grid size of 1 mm x 1 mm, as mentioned in Lokendra Kumar et al. (2012). A time step of 0.01 seconds is chosen for this validation to get the proper convergence of the continuity equation during the iteration.

3.5. Building load calculations:

From the simulation of the building in EnergyPlus, the cooling load of the coil for various months is plotted in Figure 14

Figure 14. Cooling coil load throughout the year

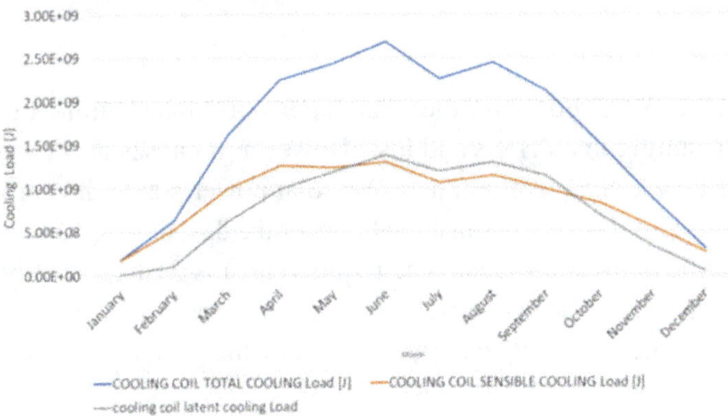

A PV panel supplies electricity to VCRS-1 and VCRS-2, while other appliances run on grid power. The highest electric load occurs on the design day (6th June) due to the maximum thermal load. The PV module area is calculated based on the hourly thermal load profile, and electric power is determined by dividing the cooling rate by the respective COP. Since power production is not continuous during daylight, a charging and discharging strategy ensures the load is met throughout the day. The maximum electric power consumption is 419 W occurs at 19th hours for the conventional AC system with no TES. The load is managed so that both VCRSs operate during peak power production, with only VCRS-1 functioning during the first and last hour. Sufficient power production starts after 7:00 AM and lasts until 4:00 PM. The load versus power production profile for June 6, with zero energy consumption from the grid, is presented in Figure 15.

Figure 15. Power consumption and solar power production profile for 6th June(24Hr) with zero energy from the grid

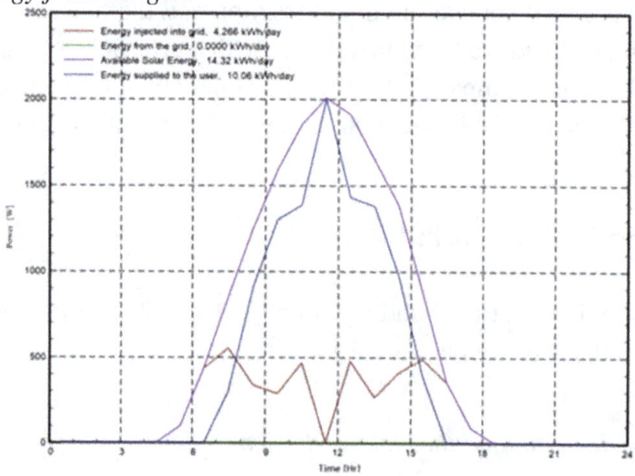

A tilt angle of 8° is chosen for the solar panel, optimizing summer power generation while resulting in a 0.1% yield loss. However, it causes an 11% yield loss in winter and a 2.3% annual loss. For a 4 kWp solar power plant with zero grid power to recharge the TES during the sunshine hours of the day, during this time, VCRS-1 and VCRS-2 are both operational. The required area is 21m^2, but only 13.57m^2 is available on the roof. Alternatively, a 2.1 kWp solar power plant can be installed in 11.6 m^2, with the remaining energy supplied by the grid, as shown in Figure 16.

Figure 16. Power consumption and solar power production profile for the 6th June (24Hr) with energy from grid

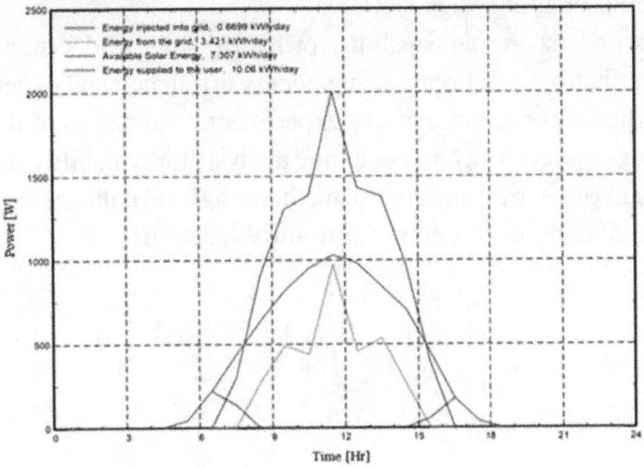

4. CONCLUSION

The study investigates the impact of fins on the performance of a tube-in-tube Latent Thermal Energy Storage (LTES) system. Reducing fin thickness while maintaining the same radial length enhances heat transfer. For the no-fin case, the heat transfer rate at 50% solidification is 35W. With 4 fins, the power increases to 673W, and with 32 fins, it reaches 1855W, demonstrating the advantage of adding fins. However, no significant improvement occurs beyond 32 fins, making this the optimal number.

The proposed charging and discharging strategy is as follows: VCRS-1 operates from 9:00 AM to 5:00 PM, and VCRS-2 charges the LTES for 7 hours (10:00 AM to 4:00 PM) to meet the cooling load from 5:00 PM to 8:00 AM. The total energy to be stored in the LTES for the design day is 71.3 MJ (5.63 Ton-Hr), equivalent to the latent heat of 213.47 kg of PCM (water). The solar plant covers 21.3 m^2, while the available roof area is 13.57 m^2, allowing for a 2.1 kWp solar power system. The remaining energy is supplied by the grid. The grid accounts for 27.4% of the total energy required with the 2.1 kWp solar power plant.

This study quantitatively analyzed a solar-powered air conditioning system including Latent Thermal Energy Storage (LTES) utilizing ice as the phase change material and R290 as the refrigerant. The innovation resides in the dual VCRS arrangement, wherein VCRS-1 provides real-time cooling while VCRS-2 charges the storage, facilitating continuous operation. The CFD simulation of the TES tank with internal fins indicated diminishing returns after 32 fins, providing practical insights for compact TES design.

The results underscore the capability of hybrid solar-grid systems to mitigate photovoltaic limitations while enhancing energy efficiency and dependability. Future research should concentrate on the experimental validation of the dual VCRS + LTES system, conduct a techno-economic analysis for scalability, and investigate other phase change materials and control methodologies to enhance environmentally sustainable air conditioning systems for tropical locations.

REFERENCES

ARORA, R. C. (2012)) Refrigeration and air conditioning. PHI Learning Pvt. Ltd.

Burgess, S., Wang, X., Rahbari, A., & Hangi, M. (2022). Optimisation of a portable phase-change material (PCM) storage system for emerging cold-chain delivery applications. *Journal of Energy Storage*, *52*, 104855.

Dong, X., Gao, G., Zhao, X., Qiu, Z., Li, C., Zhang, J., & Zheng, P. (2022). Investigation on heat transfer and phase transition in phase change material (PCM) balls and cold energy storage tank. *Journal of Energy Storage*, *50*, 104695.

Fedorová, K., & Kvasnica, M. (2022, September). Predictive Thermal Management of an Industrial Battery Energy Storage System. In 2022 Cybernetics & Informatics (K&I) (pp. 1-6). IEEE.

Kumar, L., Manjunath, B. S., Patel, R. J., Markandeya, S. G., Agrawal, R. G., Agrawal, A., & Prabhu, S. V. (2012). Experimental investigations on melting of lead in a cuboid with constant heat flux boundary condition using thermal neutron radiography. *International Journal of Thermal Sciences*, *61*, 15–27.

Kumar, L., Manjunath, B. S., Patel, R. J., Markandeya, S. G., Agrawal, R. G., Agrawal, A., & Prabhu, S. V. (2012). Experimental investigations on melting of lead in a cuboid with constant heat flux boundary condition using thermal neutron radiography. *International Journal of Thermal Sciences*, *61*, 15–27.

Mondal, S., Ramgopal, M., & Mukhopadhyay, S. (2024). Mathematical analysis of personalized air conditioning and ventilation system. In Proceedings of the 27th National and 5th International ISHMT-ASTFE Heat and Mass Transfer Conference December 14-17, 2023, IIT Patna, Patna-801106, Bihar, India. Begel House Inc.

Ravikiran, N., Padya, B., Rao, A., Kali, R., & Jain, P. K. (2023). Classification of phase change materials. In *Multifunctional phase change materials* (pp. 95–112). Woodhead Publishing.

Saini, M. C., & Jakhar, O. P. (2023). CFD simulation and experimental validation of PCM thermal energy storage system for micro trigeneration system application. *International Journal of Refrigeration*, *149*, 119–134.

Talukdar, S., Afroz, H. M. M., Hossain, M. A., Aziz, M. A., & Hossain, M. M. (2019). Heat transfer enhancement of charging and discharging of phase change materials and size optimization of a latent thermal energy storage system for solar cold storage application. *Journal of Energy Storage*, *24*, 100797.

Voller, V. R., & Prakash, C. (1987). A fixed grid numerical modelling methodology for convection-diffusion mushy region phase-change problems. *International Journal of Heat and Mass Transfer*, *30*(8), 1709–1719.

Yang, X., Wang, X., Liu, Z., Luo, X., & Yan, J. (2022). Effect of fin number on the melting phase change in a horizontal finned shell-and-tube thermal energy storage unit. *Solar Energy Materials and Solar Cells*, *236*, 111527.

Chapter 15

The Characteristics of Two-Dimensional Steady Flow Past a Square Cylinder

Satrajit Halder
https://orcid.org/0009-0006-2942-997X
National Institute of Technology, Durgapur, India

Deepak Kumar
National Institute of Technology, Durgapur, India

Sujit Karmakar
National Institute of Technology, Durgapur, India

ABSTRACT

The steady flow characteristics of circular and elliptical cylinders have been extensively examined at Reynolds numbers (Re) well beyond the critical value, where flow transitions from steady to unsteady. However, studies concerning the steady flow over square cylinders have been mainly confined to Reynolds numbers up to 40. While the flow becomes unsteady beyond the critical Re, it remains valuable to numerically investigate the characteristics of steady flow, as this can provide insights into flow stability. In the present work, steady two-dimensional flow past a stationary square cylinder has been computed for Reynolds numbers up to 200. The main characteristics of steady flow, such as bubble length, bubble width, drag coefficient on the cylinder surface, and maximum vorticity on the surface of the cylinder, are investigated. Results indicate that bubble length and bubble width are

DOI: 10.4018/979-8-3373-3621-3.ch015

linear functions of Re up to Re = 80, after which it exhibits nonlinear behaviour well fitted by a second-order polynomial. The drag coefficient shows linear variation with Re^(−0.66).

1. INTRODUCTION

The flow past a bluff body is a classical fluid mechanics problem with several engineering applications such as turbines, chimneys, bridge pillars, heat exchangers, and aerospace structures. A bluff body is an object in fluid flow that causes early boundary layer separation, producing a large wake and high pressure drag. Unlike streamlined bodies, bluff bodies experience unsteady vortex shedding and significant recirculation zones. Most research works on bluff bodies concentrate on circular and elliptic cylinders. However, flow past a non-circular geometry like a square has interesting flow behaviour and offers additional difficulty because of the flow separation at sharp edges. In the past, there have been several attempts to investigate this flow using Numerical [Breuer et al. (2000); Gupta et al. (2003) ; Sharma and Eswaran (2004) ; Sen et al. (2011) ; Jiang and Cheng (2020); Mondal and Alam (2023)] and Experimental [A. Okajima et al. (1982) ; Dutta et al. (2008)] methods. It is well established that different types of flow regimes appear when flow passes a square-shaped cylinder body. Flow remains attached to the body at low Reynolds numbers ($Re \leq 1$), commonly known as creeping flow (mentioned by Breuer et al. (2000)). The Reynolds number (Re) is determined based on edge length of the cylinder and is given by the ratio of the oncoming uniform flow speed (U) multiplied by D, to the kinematic viscosity (ν). As the Re value increases, flow starts to separate from the cylinder body and forms a symmetric wake bubble, which is the characteristic of steady flow [Breuer et al. (2000); Gupta et al. (2003) ; Sharma and Eswaran (2004) ; Sen et al. (2011);]. As we keep increasing the Re value, the steady wake bubble breaks its symmetry and becomes unsteady flow. The famous von Karman vortex shedding appears beyond critical $Re = 54$ founded by K. M. Kelkar et al. (1992). The flow turns unsteady once the Reynolds number passes the critical value because the swirling motion behind the cylinder (shear layers) becomes too strong for viscosity to smooth out. This imbalance causes the wake to lose its symmetry and start flipping from side to side, which produces the repeating pattern of vortices known as vortex shedding. Breuer et al. (2000) numerically investigated the steady ($0.5 \leq Re < 60$) and unsteady ($60 \leq Re \leq 300$) flows past a square cylinder for a blockage ratio 0.125. Here, the blockage ratio is defined as the reciprocal of the domain half width. They presented the characteristics of steady flow, such as velocity profiles, drag coefficient and recirculation length. They reported the linear variation of recirculation bubble length with Re and a strong variation of drag coefficients with *Re*. They found that total drag has an equal order of contribution from pressure and viscous drag. Gupta et al. (2003) computed the steady flow for power-law fluid by varying the power law index (*n*) between 0.5 to 1.4 and *Re* ranging between 5 to 40. They consider the blockage ratio 0.125. They observed that the flow patterns indicated relatively shorter wake regions in shear-thinning fluids and slightly longer recirculation lengths in shear- thickening fluids. Sharma and Eswaran (2004) numerically investigated the steady and unsteady flow past a square cylinder. The Reynolds numbers ranged from 1 to 160. They reported the steady flow characteristics for *Re* up to 40. The bubble length was found to vary as a function of *Re*. Sen et al. (2011) numerically computed the steady and unsteady flow past a square cylinder for Reynolds numbers up to 150. They reported the steady flow characteristics for *Re* up to 40, such as bubble length, bubble width, drag coefficients, and separation angle. They found that drag coefficients vary as a function of $Re^{-0.66}$. They also proposed the empirical equations for bubble parameters up to $Re = 40$. Jiang and Cheng (2020) examined the flow separation phenomena in the flow past a square cylinder for Reynolds numbers ranging from 10 to 400. Recently, Mondal and Alam (2023) provided a review on the confined flow past bluff bodies, including circular, square, and rectangular-shaped cylinders. They investigated the effect of the blockage ratio on steady-to-unsteady wake transition, shear layer transition, boundary layer transition, Strouhal number, aerodynamic forces, vortex formation length, and vortex inversion length. They found that square cylinder experiences a greater blockage effect than a circular cylinder.

Figure 1. Computational domain and boundary conditions: L_d marks the downstream boundary and L_u denotes the upstream boundary, and H signifies the half- width of the cross-stream dimension. The center of the cylinder is used as the reference point for measuring all linear distances.

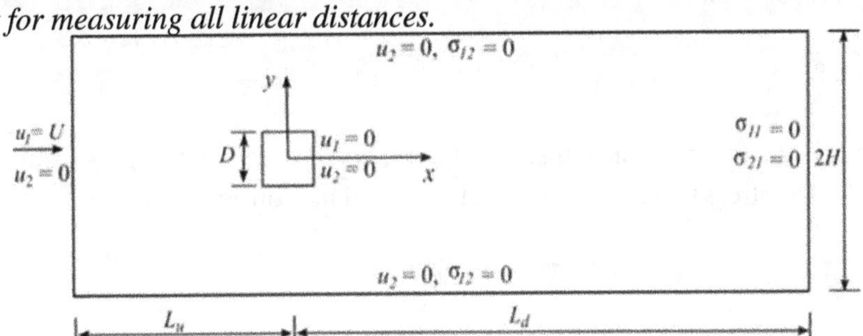

Based on the above literature review, it is evident that the characteristics of steady flow past square cylinders have primarily been studied for low Reynolds numbers up to 40. In this work, we perform a comprehensive analysis of steady flow characteristics, including bubble length, bubble width, cylinder drag coefficient and its components, as well as the vorticity distribution on the cylinder's surface, for Reynolds numbers extending up to 200.Previous studies have reported steady flow past a square cylinder only up to Reynolds numbers of about 40. In this work, the analysis is extended to Reynolds numbers as high as 200, providing new insights into flow behaviour in this unexplored regime. Although the flow becomes unsteady beyond the critical Reynolds number of approximately 54, computing the corresponding steady solutions remains important. These steady states not only enhance the fundamental understanding of the flow but also serve as base flows for linear stability analysis. Similar studies on circular (Kumar and mittal,2006) and elliptic cylinders (Kumar and Kumar,2024) have shown that extending steady solutions beyond the critical Reynolds number yields valuable information on flow dynamics. The present work offers a comparable contribution for the square cylinder, thereby filling a gap in the existing literature and providing benchmark data for future research.

2. METHODOLOGY

A. The Incompressible Flow Equations

Let Ω R^{nsd} be the spatial domain, where $n_{sd} = 2$ is the number of space dimensions. The boundary of Ω is denoted by $\boldsymbol{\Gamma}$ and is assumed to be piecewise smooth. The spatial coordinates are denoted by $\boldsymbol{x}$. The Navier–Stokes equations governing incompressible fluid flow are

$$\rho(u\bullet\nabla u-f)-\nabla\bullet\sigma=0 \text{ on } \Omega \qquad (1)$$

$$\nabla\cdot u=0 \text{ on } \Omega \qquad (2)$$

Here ρ, $\boldsymbol{u}$,$\boldsymbol{f}$, and $\boldsymbol{\sigma}$ are the density, velocity, body force and the stress tensor, respectively. The stress tensor is written as the sum of its isotropic and deviatoric parts:

$$\sigma=-pI+T,T=2\mu\varepsilon(u),\varepsilon(u)=(0.5*(\nabla u+(\nabla u)^{\wedge}T\ ))(3)$$

where p and μ are the pressure and coefficient of dynamic viscosity, respectively. The Dirichlet and Neumann-type boundary conditions are represented as

$$u=g \text{ on } \Gamma g,n.\sigma=h \text{ on } \Gamma h \qquad (4)$$

where Γ_g and Γ_h are complementary subsets of the boundary Γ.

Figure 2. Flow past square cylinder: the finite element mesh consisting 241091 nodes and 239898 quadrilateral elements.

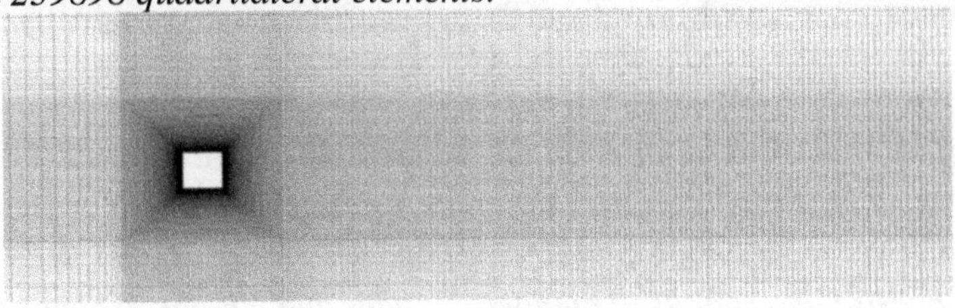

B. Numerical Method

In this work, a stabilized finite element formulation is used to perform the computations of steady flow. This approach, known as the streamline-upwind/Petrov-Galerkin and pressure-stabilizing/Petrov-Galerkin method, has been developed by researchers who are pioneers in the field [(Hughes et al.,1979,1986), Brooks et al. (1982), T. E.Tezduyar et al.(1992), Mittal(2000)]. The formulation utilized for conducting the linear stability analysis was initially introduced by Mittal and Kumar (2003). We use the same formulations in the present work.

C. The Flow Setup, Boundary Conditions and Mesh

Figure 1 illustrates the flow setup and computational domain. A symmetrical setup is used, placing the square cylinder between two lateral boundaries in a rectangular domain. The edge of the cylinder is denoted by *D*, is perpendicular to the flow direction. The Cartesian coordinate system is centered at the origin of the cylinder. The half of the edge length (*D/2*) of the cylinder serves as the reference length scale for presenting the results. In this study, the domain boundaries are placed sufficiently far from the cylinder to ensure minimal influence on the overall flow field. The upstream and downstream boundaries are situated at distances of $L_u = 160$ and $L_d = 200$ units, respectively, from the cylinder's center. The cross-stream dimension is taken as $H = 100$.A typical finite element mesh shown in figure 2 is used to conduct this study. Three different resolutions of mesh, M1 (NN = 192008, NE = 190944), M2 (NN = 210545, NE = 209432) and M3 (NN = 241091, NE = 239898) are used. The NN corresponds to number of nodes and NE corresponds to number of quadrilateral elements. Table1 shows the mesh convergence study and based on these results, it can be safely concluded that the mesh considered here is appropriate for performing the present study. The higher resolution mesh M3 is used to perform all the computations to ensure the accuracy of the results. The mesh is refined near the cylinder surface and in the wake region to accurately capture the separating shear layer, boundary layer, wake bubble of the steady flow, and linear stability modes. As we move away from the cylinder, the mesh is gradually coarsened.

Table 1. Effect of mesh resolution on the steady flow characteristics (LB, WB and Cd) is presented for Re =150. Three different resolutions, M1 (NN = 192008, NE= 190944), M2 (NN = 210545, NE = 209432) and M3 (NN= 241091, NE = 239898) are used. The NN corresponds to number of nodes and NE corresponds to number of quadrilateral elements. The maximum % difference in the numerical data is with respect to M3.

	L_B	W_B	C_d
M1	13.9415	2.2598	1.0758
M2	13.9414	2.2592	1.0756
M3	13.9113	2.2566	1.0754
| max % difference|	0.2171	0.1418	0.0372

Figure 3. Steady flow past square cylinder: schematic representation of the symmetric wake bubble. The bubble length (L_B) is defined as the distance measured along the wake center line between the rear stagnation point and the wake stagnation point. The bubble width (W_B) is the maximum distance between the upper and lower separation streamlines.

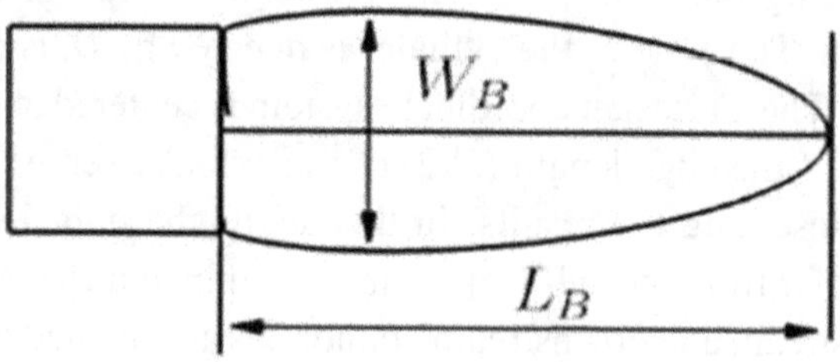

To compute the steady flow, specific boundary conditions are applied. A free-stream value is assigned to the velocity at the upstream boundary. At the downstream boundary, a Neumann-type boundary condition for the velocity is specified that correspond to zero stress vector. On the upper and lower boundaries, the component of velocity normal to and the component of stress vector along these boundaries are prescribed a zero value. The surface of the cylinder is treated as a no-slip boundary, ensuring zero velocity at the surface. The Reynolds number (*Re*) is determined based on edge length of the cylinder and is given by the ratio of the oncoming uniform flow speed (*U*) multiplied by *D*, to the kinematic viscosity (ν). The drag coefficient is calculated by using following equations:

C_d=1/(1/2 ρU^2.(D/2)) ∫_(Γ_cyl)n_x.σdΓ (5)

Here n_x is the x component, of the unit vector n normal to the cylinder boundary, σ and Γ are the stess tensor and cylinder boundary surface.

3. RESULTS AND DISCUSSION

In this section, we present the results obtained by solving the time-independent Navier-Stokes equations for the flow past square cylinders at Reynolds numbers up to 200.

A. Validation

Figure 3 shows the schematic representation of the symmetric wake bubble. The bubble length (L_B) is defined as the distance measured along the wake centreline between the rear stagnation point and the wake stagnation point. The bubble width (W_B) is the maximum distance between the upper and lower separation streamlines. Table 2 compares the L_B, W_B and C_d for flow past square cylinder at $Re = 5$ and 40 with Sen et al. (2011). The close agreement amongst the results is clearly visible. At $Re = 40$, W_B and C_d value matches closely up to two decimal places. Overall the comparison indicates a robust agreement between the findings of this study and prior research.

B. Vorticity Field

Figure 4 shows the vorticity field and the wake bubble computed for flow past square cylinder with increasing *Re*. It is observed that the strength of vorticity in the shear layer increases with increase in *Re*. This causes the wake bubble to grow longer and wider. Therefore, the length and width of the bubble increase with increase in *Re*.

Table 2. Comparison of bubble length, bubble width and drag coefficient for flow past square cylinder with previous studies for different Reynolds number.

	Re = 5			*Re* = 40		
Studies	L_B	W_B	C_d	L_B	W_B	C_d
Sen et al. (2011)	0.2837	–	4.4058	2.8177	1.1717	1.6764
Present study	0.2981	–	4.3900	2.8346	1.1749	1.6661

Figure 4. Vorticity field for flow past square cylinder with increasing Re. Streamlines (in black) show the symmetric wake bubble. The bubble size increases with the increase in Re.

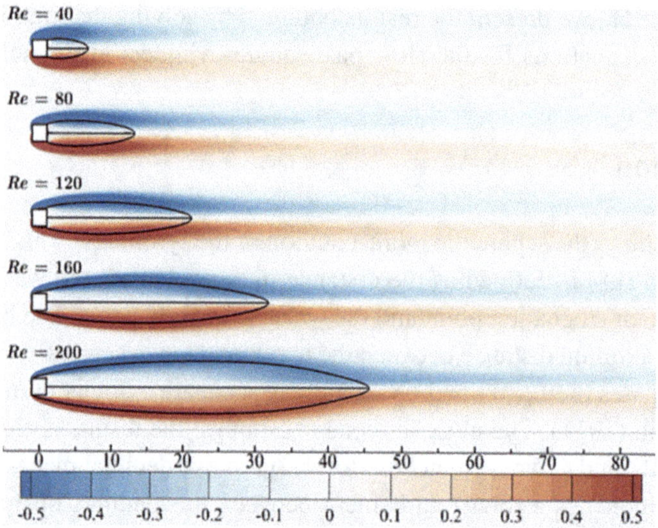

C. Bubble Length

Figure 5 shows the variation of bubble length with *Re*. The bubble length monotonically increases with the increase in *Re*. This is because the vorticity strength in the top and bottom shear layers increases rapidly with increasing *Re*. This is further illustrated by deriving the best-fit equations (using the least squares method) from the data. The equation 5 show that L_B vary linearly with *Re* up to 80. For $Re \geq 90$ second order polynomials better fit the data. It is noticed that, though the coefficients of Re^2 are very small, the contribution of the second order term becomes significant at higher *Re*. The empirical equation used to estimate the value of L_B shows a maximum discrepancy of less than 3.53% compared to the actual value, while the average difference is less than 3.08%. The maximum differences appear for the range $90 \leq Re \leq 200$.

L_B=(0.0676*Re)-0.0486(5≤Re≤20)
L_B=(0.0805*Re)-0.372(25≤Re≤80)
L_B=(0.0005*Re^2)-(0.0009*Re) + 3.2475(90≤Re≤200)

Figure 5. Variation of L_B with Re. The bubble length monotonically increases with the increase in Re.

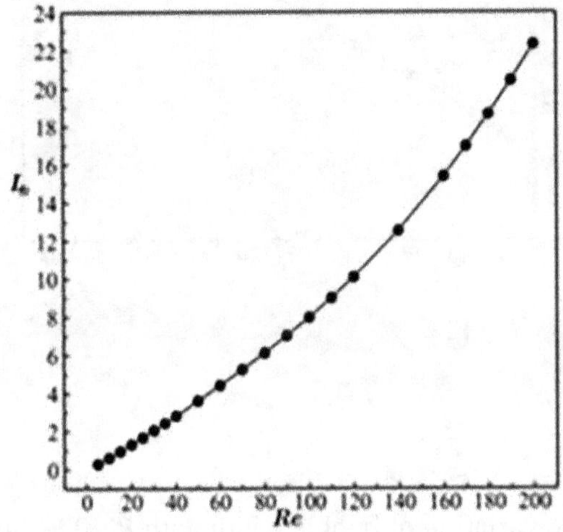

Figure 6. Variation of W_B with Re. The bubble width monotonically increases with the increase in Re.

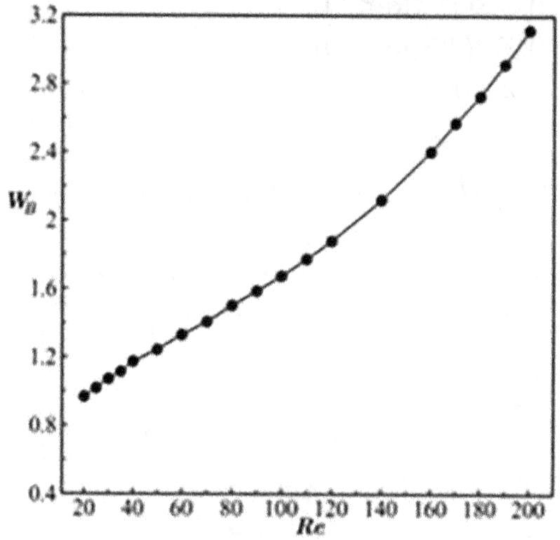

D. Bubble Width

Figure 6 shows the variation of bubble width with *Re*. The bubble width shows non-linear variation with *Re* and it monotonically increases with the increase in *Re*. This happens because of the same reason as the vorticity strength in the top and bottom shear layers increases rapidly with increasing *Re*. The equation 6 show that W_B vary linearly with *Re* up to 80. The empirical equation used to estimate the value of W_B shows a maximum discrepancy of less than 1.86% compared to the actual value, while the average difference is less than 0.63%. For $Re \geq 90$, a non-linear variation with *Re* is observed, which requires a second-order polynomial fit.

$W_B = (0.0087*Re) + 0.805 (20 \leq Re \leq 80)$

E. Drag Coefficients on the Surface of the Cylinder

The contribution of pressure to the total drag is expressed in non-dimensional form as C_{dp}, while the contribution due to shear stress is represented as C_{df}. Consequently, the total drag coefficient is given by $C_d = C_{dp} + C_{df}$. Figure 7 shows that drag coefficients (C_d, C_{dp} and C_{df}) monotonically decreases with increase in *Re*. It is evident that C_{dp} is significantly larger than C_{df}, which is a characteristic feature of bluff body flows. A higher contribution from frictional drag typically indicates streamlined body behaviour, whereas a higher pressure drag is characteristic of bluff body behaviour. Therefore, a higher value of C_{dp} indicates that the square cylinder behaves predominantly as a bluff body. The equation 7 show that C_d vary linearly as a function of $Re^{-0.66}$. The empirical equation used to estimate the value of C_d shows a maximum discrepancy of less than 1.58% compared to the actual value, while the average difference is less than 1.06%.

Figure 7. Variation of total drag coefficient (C_d), pressure drag coefficient (C_{dp}) and skin friction drag coefficient (C_{df}) with Re. The total drag $C_d=C_{dp}+C_{df}$

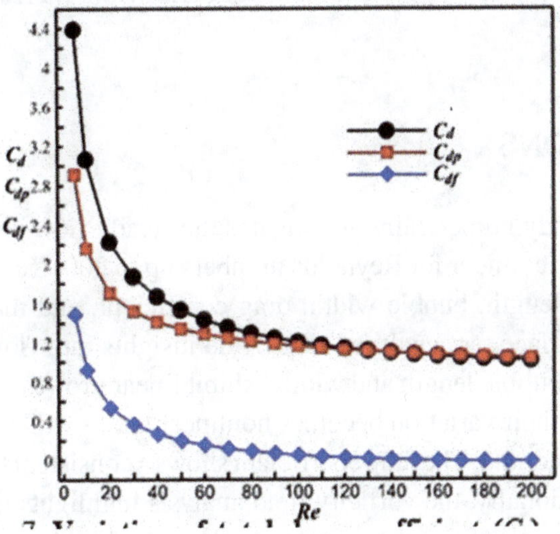

Figure 8. Variation of maximum vorticity on the surface of the cylinder (ω_{max}) with Re.

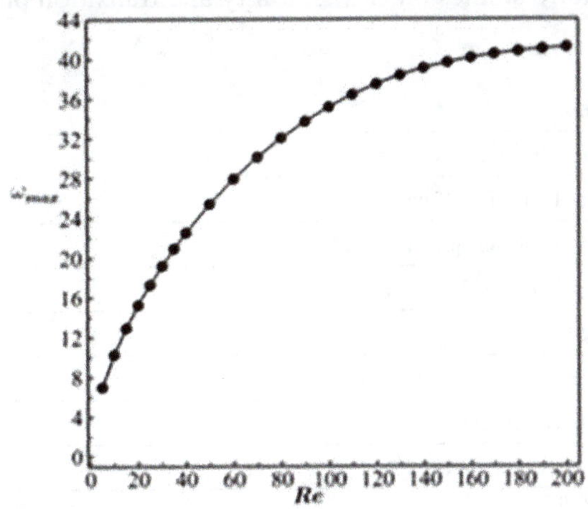

$C_d= (10.8195*(Re^{-0.66}))+0.6960, (5\leq Re\leq 200)$

F. Maximum Vorticity on the Surface of the Cylinder

Figure 8 depicts the variation of the maximum vorticity on the surface of the cylinder (ω_{max}) with increasing *Re*. It is observed that ω_{max} exhibits a non-linear variation with *Re* and increases monotonically as *Re* rises. A high value of ω_{max} on the cylinder surface signifies the presence of a stronger shear layer and the formation of a wider wake. As a result, the bubble length and width increase monotonically with increasing *Re*.

4. CONCLUSIONS

The present study numerically investigates the steady flow characteristics over a stationary square cylinder for Reynolds numbers up to 200. Key flow parameters, including bubble length, bubble width, drag coefficient, and maximum vorticity on the cylinder surface, are analysed to provide insights into flow behaviour. The results reveal that bubble length and width exhibit linear growth with *Re* up to *Re* = 80, beyond which their variation becomes nonlinear and is well approximated by a second-order polynomial. The drag coefficient shows a consistent linear relationship with $Re^{-0.66}$. Additionally, the vorticity field analysis highlights that increasing *Re* intensifies the strength of the shear layers on the cylinder's surface, resulting in a monotonic growth in bubble length and width. This indicates the expansion of the recirculation region behind the cylinder with higher *Re*. These findings contribute to a deeper understanding of steady flow behaviour over square cylinders and provide a foundation for future studies on flow stability and transition phenomena.

Box. Nomenclature

D	edge length of the square cylinder	[m]
H	Computational domain half-width	–
L_u	Computational domain upstream length	–
L_d	Computational domain downstream length	–
L_B	Bubble length	–
W_B	Bubble width	–
C_{dp}	Pressure drag coefficient	–
C_{df}	Skin friction drag coefficient	–

continued on following page

Box. Continued

C_d	Drag coefficient	–
ω_{max}	Maximum vorticity on the surface of cylinder	–
Re	Reynolds number	–

ACKNOWLEDGEMENTS

The authors would like to acknowledge the computational facilities of NIT Durgapur.

REFERENCES

Brooks, A. N., & Hughes, T. J. R. (1982). Streamline upwind/petrov-galerkin formulations for convection dominated flows with particular emphasis on the incompressible navierstokes equations. *Computer Methods in Applied Mechanics and Engineering*, *32*, 199–259.

Breuer, M., Bernsdorf, J., Zeiser, T., & Durst, F. (2000). Accurate computations of the laminar flow past a square cylinder based on two different methods: Lattice-boltzmann and finite-volume. *International Journal of Heat and Fluid Flow*, *21*, 186–196.

Dutta, S., Panigrahi, P., & Muralidhar, K. (2008). Experimental investigation of flow past a square cylinder at an angle of incidence. *Journal of Engineering Mechanics*, *134*, 788–803.

Gupta, A. K., Sharma, A., Chhabra, R. P., & Eswaran, V. (2003). Two-dimensional steady flow of a power-law fluid past a square cylinder in a plane channel: Momentum and heat-transfer characteristics. *Industrial & Engineering Chemistry Research*, *42*, 5674–5686.

Hughes, T. J. R., & Brooks, A. N. (1979). A multidimensional upwind scheme with no crosswind diffusion. In Hughes, T. J. R. (Ed.), *Finite Element Methods for Convection Dominated Flows* (pp. 19–35). ASME.

Hughes, T. J. R., Franca, L. P., & Balestra, M. (1986). A new finite element formulation for computational fluid dynamics: circumventing the babusˇka-brezzi condition: a sta- ble petrov-galerkin formulation of the stokes problem accommodating equal-order interpolations. *Computer Methods in Applied Mechanics and Engineering*, *59*, 85–99.

Jiang, H., & Cheng, L. (2020). Flow separation around a square cylinder at low to moderate reynolds numbers. *Physics of Fluids*, •••, 32.

Kelkar, K. M., & Patankar, S. V. (1992). Numerical prediction of vortex shedding behind a square cylinder. *International Journal for Numerical Methods in Fluids*, *14*, 327–341.

Kumar, B., & Mittal, S. (2006). Prediction of the critical Reynolds number for flow past a circular cylinder. *Computer Methods in Applied Mechanics and Engineering, 195*(44-47), 6046–6058.

Kumar, D., & Kumar, B. (2024). Tertiary wake mode in flows past elliptic cylinders. *European Journal of Mechanics - B/Fluids, 106*, 130–136.

Mittal, S. (2000). On the performance of high aspect ratio elements for incompressible flows. *Computer Methods in Applied Mechanics and Engineering, 188*, 269–287.

Mittal, S., & Kumar, B. (2003). Flow past a rotating cylinder. *Journal of Fluid Mechanics, 476*, 303–334.

Mondal, R., & Alam, M. M. (2023). Blockage effect on wakes of various bluff bodies: A review of confined flow. *Ocean Engineering, 286*, 115592.

Okajima, A. (1982). Strouhal numbers of rectangular cylinders. *Journal of Fluid Mechanics, 123*, 379–398.

Sen, S., Mittal, S., & Biswas, G. (2011). Flow past a square cylinder at low reynolds numbers. *International Journal for Numerical Methods in Fluids, 67*, 1160–1174.

Sharma, A., & Eswaran, V. (2004). Heat and fluid flow across a square cylinder in the two-dimensional laminar flow regime. *Numerical Heat Transfer Part A, 45*, 247–269.

Tezduyar, T. E., Mittal, S., Ray, S. E., & Shih, R. (1992). In- compressible flow computations with stabilized bilinear and linear equal-order-interpolation velocity-pressure elements. *Computer Methods in Applied Mechanics and Engineering, 95*, 221–242.

Chapter 16

Theoretical Investigations on Thermodynamic and Environmental Properties of Zeotropic Mixtures of Carbon Dioxide and Hydrofluoroolefins

Rohit Roy
https://orcid.org/0000-0002-6567-1185
Indian Institute of Technology, Kharagpur, India

Boddepalli Umamahesh
Indian Institute of Technology, Kharagpur, India

Sourav Mitra
Indian Institute of Technology, Kharagpur, India

ABSTRACT

Theoretical investigations have been carried out to evaluate the thermodynamic and environmental properties of different CO_2 (R744) blends with low Global Warming Potential (GWP) Hydrofluoroolefins (HFOs) to assess the suitability of these blends for refrigeration and air conditioning applications. The results show that compared to other available refrigerants, the selected blends offer significantly low GWP values but high indirect emissions due to their poor coefficient of performance (COP).

DOI: 10.4018/979-8-3373-3621-3.ch016

The binary blend yields good thermodynamic characteristics, including large volumetric capacity and low operating pressures and a subcritical cycle of operation. The temperature glide characteristics show the zeotropic behaviour of the blends.

1. INTRODUCTION

Chlorofluorocarbons (CFCs) and Hydrochlorofluorocarbons (HCFCs) were widely used in refrigeration and air-conditioning applications until they were discovered to be harming the stratospheric ozone layer (Montzka et al., 1999; Molina & Rowland, 1974; Engel et al., 2019). As a result, these elements became regulated under the terms of the Montreal Protocol. By 1992, a plan had been developed for gradually eliminating the CFCs and HCFCs. Furthermore, in 1997, the Kyoto Protocol imposed a limit on greenhouse gas concentrations in the surrounding environment to address global warming. Thus, it became imperative to search for refrigerants which have zero Ozone Depletion Potential (ODP) and as low a GWP as possible.

The Global Cooling Pledge, adopted during a general meeting among member parties held in the United Arab Emirates (UAE), aims to reduce cooling-related greenhouse gas emissions by 68% of December 2023 levels by 2050, substantially boost utilisation of feasible cooling by 2030, and improve the average worldwide effectiveness of new air conditioners by 50%. As a consequence, there have been limitations on excessively high GWP compounds like Hydrofluorocarbons (HFCs) (Claesson, 2005). India, as an Article Five nation under the Protocol of Montreal, accepted the agreement in 1992. India has mandated the total phase-out of CFCs by 2010. India declared that CFC manufacture and consumption had been completely phased out in 2008 (Say et al., 2019). Following the phaseout of CFC, the nation was compelled to minimise its HCFC emissions.

During the first stage of the Hydrofluorocarbons Phase Out Management Plan (HPMP), HCFC production and consumption have already been restricted by January 1, 2016, with an entire phase-out by 2040. India is committed to freezing its HCFC usage at the initial level (average for 2009 ?"2010) by 2013, reducing it by 10% (compared to the foundation level) by 2015, and completely phasing it out by 2030 as part of the second stage of the HPMP. The Kigali Amendment, which India ratified in 2016, mandates the country to gradually reduce their HFC manufacturing and usage. With more than a billion people and a rapidly growing economy, India is projected to generate significant amounts of halocarbons. India is predicted to generate 400 tonnes gram carbon dioxide equivalent to HFCs by the year 2050 (Velders et al., 2015).

Researchers pursued two approaches to address the aforementioned problem: finding an alternate single refrigerant (organic or synthetic) or a suitable mixture of low-GWP pure fluids. Many researchers determined that low GWP Hydrofluoroolefins (HFOs), ammonia, carbon dioxide, hydrocarbons, and water are the most promising fluids. Out of these refrigerants, carbon dioxide, with easy availability, non-flammability, non-toxicity, low global warming potential (GWP), zero ozone depletion potential (ODP), and excellent heat transfer properties, makes an ideal refrigerant in many respects. However, despite its merits, CO_2 is currently limited to a few applications only because of its low critical temperature, high operating pressure, and low COP at higher ambient temperatures. HFOs are pure, synthetic refrigerants with low GWP and ODP, providing excellent thermodynamic properties and significant environmental benefits (Ciconkov, 2018). These refrigerants are being explored as possible alternatives in vapour compression systems for vehicle air conditioning, domestic refrigeration units, and other applications (Aprea et al., 2016; Golzari et al., 2017; Belman-Flores et al., 2017; Bobbo et al., 2018). However, these refrigerants lack volumetric refrigeration capacity, and they also have issues related to flammability. To overcome the above-mentioned problems, a simple but effective method is mixing carbon dioxide (CO_2) with these low GWP refrigerants to arrive at a safe, efficient, and environmentally friendly working fluid for various applications (Yelishala et al., 2020). Research in the field of preparing low-GWP refrigerant mixtures as alternatives to high-GWP HFCs and HCFCs is very limited when compared with the set targets for sustainable cooling systems.

Currently, the most popular HFO refrigerant is R1234yf, which is used in mobile air conditioning (MAC) systems as an alternative to R134a, an HFC. The characteristics of this refrigerant are comparable to those of R134a. Compared to R134a, it is distinguished by a reduced latent heat of evaporation, resulting in an increased mass flow rate. Based on the high degree of flame ignition energy and low flame propagation speed, ISO 817 classified R1234yf as belonging to the A2L safety group. However, studies are still being conducted to increase the safety of R1234yf use (Feng et al., 2018). R1234zf is another fluid that may be used in place of R134a in the near future. Two other propane molecules that have been halogenated with four fluorine ?(tm)s are R1234ze(E) and R-1234ze(Z). But because these isomers' characteristics are different from one another, their applications are distinct. A medium-pressure refrigerant R1234ze(E) can be used in medium-temperature refrigeration systems in place of R134a. In addition, its evaporation pressure is roughly 20% less than that of R-134a (Fukuda et al., 2014; Longo et al., 2014; Arat & Arslan, 2017). R1233zd(E) is another refrigerant that may find use in high-temperature units. Its ODP is near 0, and its GWP is less than 7. Compared to earlier generations of refrigerants, this working fluid's harmful influence is minimal, although it is not completely safe for stratospheric ozone. It is thought to be suitable in the organic refrigeration cycle (ORC) because of its high latent heat (196.27 kJ/kg) (MolA(c)s et al., 2014; Guillaume et al., 2017; Welzl et al., 2017; MolA(c) s et al., 2017; Luo et al., 2015). Studies are also conducted on solar-assisted heat pump (Gil & Kasperski, 2018) and centrifugal chiller (Lee et al., 2018) employing R1233zd(E). R1336mzz(Z) is a butane olefin with a GWP value of 2. Similar to the olefins previously discussed, this can be utilized in ORCs (Navarro-EsbrA- et al., 2017) and high-temperature heat pumps because of its high boiling point and critical point. Owing to its low GWP, R-1336mzz(Z) has been utilised in refrigeration as a part of mixture R-514A, which is the refrigerant used in low-pressure centrifugal chiller units to replace R-123. The thermodynamic and thermo-physical properties, environmental and safety properties of various refrigerants are depicted in Table 1.

Table 1. Thermo-dynamic and safety properties of individual refrigerants

Refrigerant name	R1234yf	R1234ze(Z)	R1234ze(E)	R1243zf	R1336mzz(Z)	R744
[1]Molecular weight [g mol ^{A1}]	114.04	114.04	114.04	96.05	164.06	44.01
[1]NBP [A C]	-29.8	9.73	-18.97	-25.42	33.45	-78.5
[1]Critical temperature [A C]	94.70	150.12	109.36	103.78	171.35	31.3
[1]Critical pressure [MPa]	3.62	3.53	3.63	3.52	2.90	7.38
[2]GWP_{100}	<1	<1	<1	<1	2	1
[2]ODP	0	0	0	0	0	0
[2]Atmospheric lifetime (days)	10.5	10.0	16.4	7.0	22.0	-
[3]LFL [kg m^{-3}]	0.289	0.211	0.303	0.144	-	-
[3]Safety Group	A2L	A2L	A2L	A2L	A1	A1

Obtained data from [1]EN 378-1:2016(UNI, 2021); Data from [3]ASHRAE Standard 34:2019 (American Society of Heating, Refrigerating and Air-Conditioning Engineers [ASHRAE], 2019); Data from [2]IPCC Assessment Report V (AR5) (Gebre & Ludwig, 2015);

Carbon dioxide reacts with various refrigerants to produce a refrigerant blend with high critical parameters, lower working pressures, and acceptable GWP values.

These blends exhibit subcritical operation, which reduces the throttling losses. Furthermore, temperature glide can enhance the temperature match, reducing the irreversibility during the heat transfer process (Yelishala, 2019). R744/R1234ze(Z) and R744/R1336mzz(E) have increased volumetric refrigeration capacity (VRC) with an equivalent heating efficiency (HE) to R134a. If the preferred amount of carbon dioxide is greater than 0.6, it eliminates flammability in the mixture. Furthermore, the working pressures of R744/HFO are decreased under typical operating conditions due to HFO accumulation (Luo et al., 2023). R744/R1234yf (44.6/55.4%weight) blend increases COP by 9.7% as compared to pure R744 in the VCR operating cycle (SA nchez et al., 2024). CO_2/R1234yf (94.5/5.5%mass) shows reduced performance in heat pump and water heater applications (Bouteiller et al., 2017). The present investigation projected some interesting binary blends and their important property analysis that includes: operating critical parameters, temperature glide, GWP, TEWI (total equivalent warming impact), volumetric refrigeration capacity and COP.

2. BLEND COMPOSITION ANALYSIS

Pure refrigerants are preferable due to their thermophysical characteristics and azeotropic nature. HFO refrigerant was recognized for its beneficial characteristics, such as exceptionally low GWP, environmental impact, and flammability. However, combining HFCs or natural refrigerants with HFOs improves total system efficiency (Han et al., 2007; Wu et al., 2009; Mota-Babiloni et al., 2015). The mixtures overcome the issue of a limited availability of pure refrigerants. Three categories of working fluid blends exist, these are: azeotropes, near-azeotropes, and zeotropes. Binary blends require careful consideration of component selection, temperature glide, operating parameters, VRC, TEWI, GWP, and COP.

2.1 Selection of Mixture Compositions

The most frequently employed refrigerants in household air conditioning and domestic refrigeration systems are R32 and R134a, that has zero ODP but a high GWP. Finding environmentally acceptable refrigerants is crucial to comply with the Kyoto Protocol.

This study compares the performance of different low-GWP R744 mixtures and their cycle operation. The binary blends include extremely low global warming potential refrigerants R1234yf, R1234ze(Z), R1234ze(E), R1243zf, R1336mzz(Z) and R744. R744 is included as the second element in the binary mixtures to enhance volumetric capacity. The binary blends of R744 and HFOs have been selected for this study based on mass fraction, and all the essential properties are calculated using MATLAB-REFPROP 10.0 interface (Lemmon et al., 2018).

3. PERFORMANCE ANALYSIS OF THE SELECTED BINARY BLENDS

The most commonly used refrigeration cycle is the Vapour Compression Refrigeration Cycle (VCRC). The vapour compression cycle mainly consists of 1). compressor 2). condenser 3). expansion valve 4). evaporator. VCR cycle can be operated as either a subcritical or transcritical cycle. The subcritical cycle indicates phase change process happens during heat extraction as well as heat rejection, but in the transcritical cycle, the phase change process occurs in the evaporator; on the other side, only gas cooling will take place. The simple, subcritical VCR cycle of a zeotropic mixture, along with the P-h diagram, is shown in Figures 1 and 2, respectively.

Figure 1. Schematic diagram of a water-cooled vapour-compressed refrigeration system

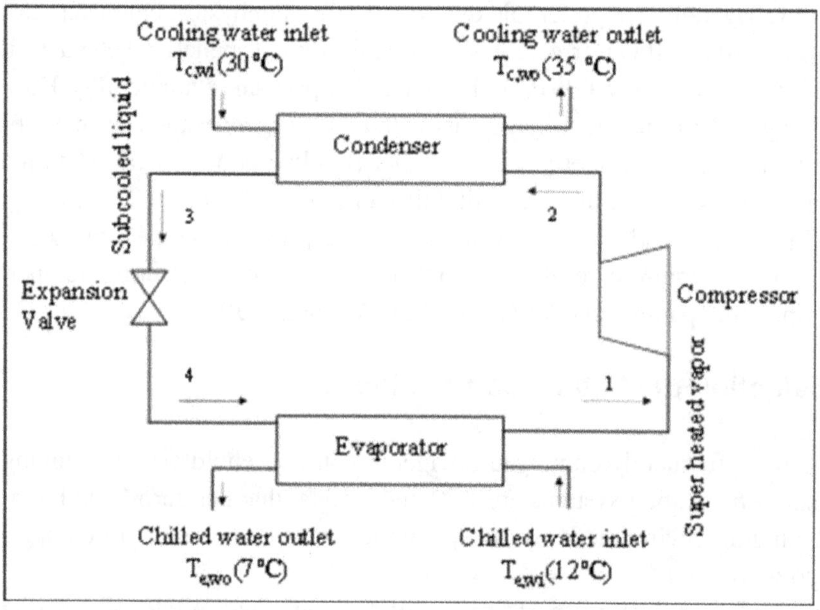

Figure 2. Pressure-entropy diagram of a single-stage refrigeration cycle

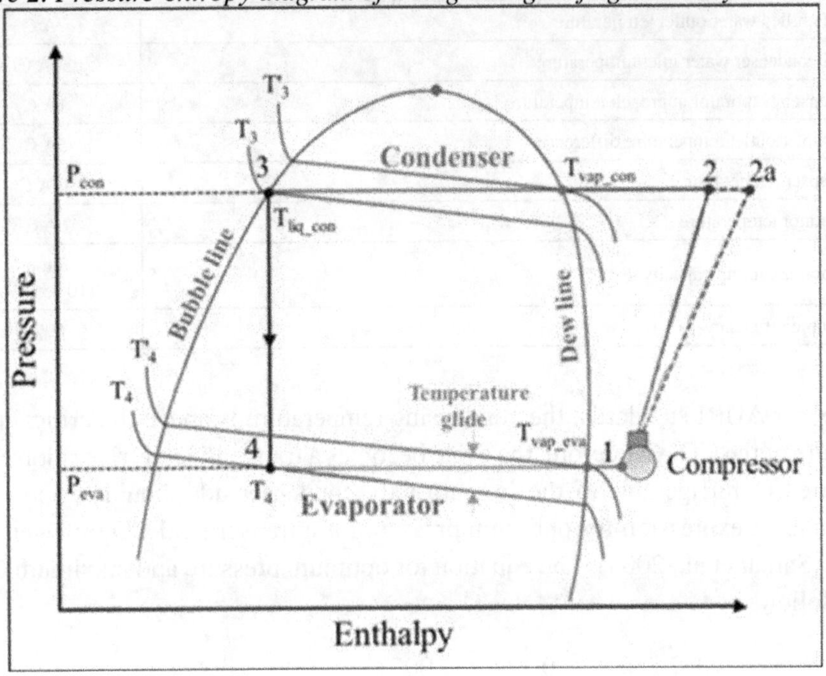

In the above-mentioned P-h diagram, consider the following processes: 1-2a: compression, 2a-3: constant pressure heat rejection, 3-4: isenthalpic expansion process and 4-1: constant pressure heat extraction. The steady temperature line illustrates temperature glide in the evaporator and condenser for blends. The cycle performance can be evaluated in terms of the Coefficient of Performance (COP), defined as the ratio of the refrigeration effect to the energy needed by the compressor, presented in equation 1.

$$COP_A = \frac{RefrigerationAeffect(RE)}{energyArequiredAtoArunAtheAsystem}A \tag{1}$$

The isentropic efficiency of the compressor is defined as the ratio of isentropic compressor work required (1-2) to the actual compressor work required. A water chiller has been considered for this study, and operating cycle parameters as per Air Conditioning Heating and Refrigeration Institute (AHRI) standards were furnished in Table 2 (Poggi et al., 2008; Smith, 2012).

Table 2. Thermo-dynamic and safety properties of individual refrigerants

AHRI chilled water outlet temperature	7A C
AHRI condenser water inlet temperature	30A C
Condenser/evaporator approach temperature	1A C
Water inlet-outlet temperature difference	5A C
Condenser temperature	36A C
Evaporator temperature	6A C
Evaporator cooling capacity	3 ton (10.55 kW)
Isentropic efficiency	69%

As per AHRI standards, the condensing temperature is above the critical temperature of the CO_2. Therefore, the cycle becomes a transcritical refrigeration cycle for pure CO_2 refrigerant. So, the pressure at the condenser side should be optimized to obtain an expression for optimum pressure for a transcritical CO_2 refrigeration cycle (Sarkar et al., 2004). The equation for optimum pressure and maximum COP is as follows:

$$P_{optimal}A = 4.9 + 2.256 t_{cA} - 0.17 t_e + 0.002 t_c^2 \quad (2)$$

$$COP_{max}A = 48.2 + 0.216 t_{eA} + 0.05 t_c (t_c - 50) - 0.0004 t_c^3 \quad (3)$$

$$t_{comp}A = 10.65 + 3.786 t_{cA} - 1.44 t_e - 0.0188 t_c^2 + 0.009 t_e^2 \quad (4)$$

Where $P_{optimal}$ is optimized pressure at the gas cooler, $COP_{maximum}$ is the maximum COP of the cycle, t_c is gas cooler exit temperature, t_e is evaporator temperature, and t_{comp} is discharge temperature at the compressor outlet. These relationships are valid for evaporation temperatures ranging from ^'10 to 10 A C and condensation temperatures between 30 and 50 A C. Based on the above parameters, the Coefficient of performance has been calculated for pure R32, R134a, R744, and different HFO blends using MATLAB-REFPROP 10.0 interface. Several mixtures of R744-HFO have been considered based on their mass fraction, and among them, the best possible COP is depicted in Figures 3 and 4, respectively.

Figure 3. Coefficient of performance of pure refrigerants

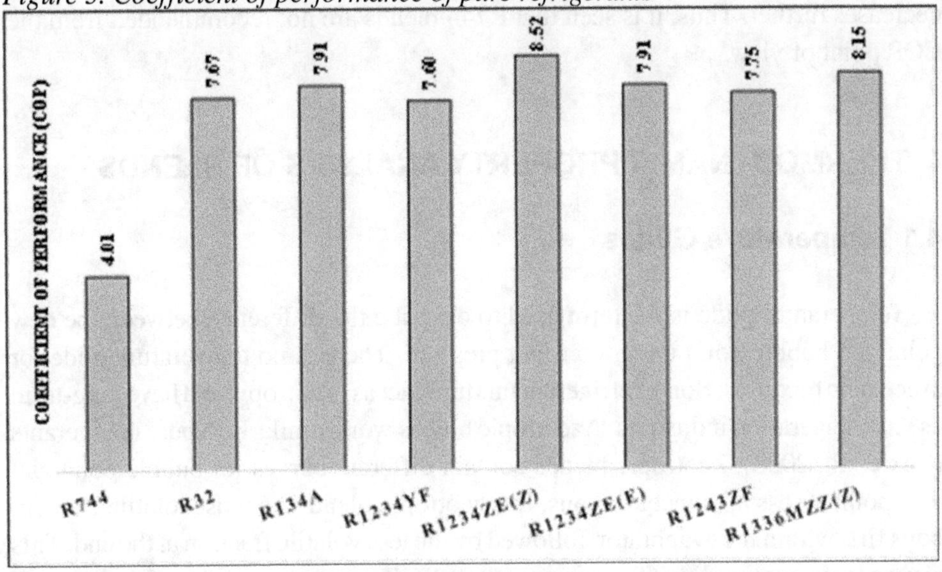

Figure 4. Coefficient of performance of R744-HFO blends

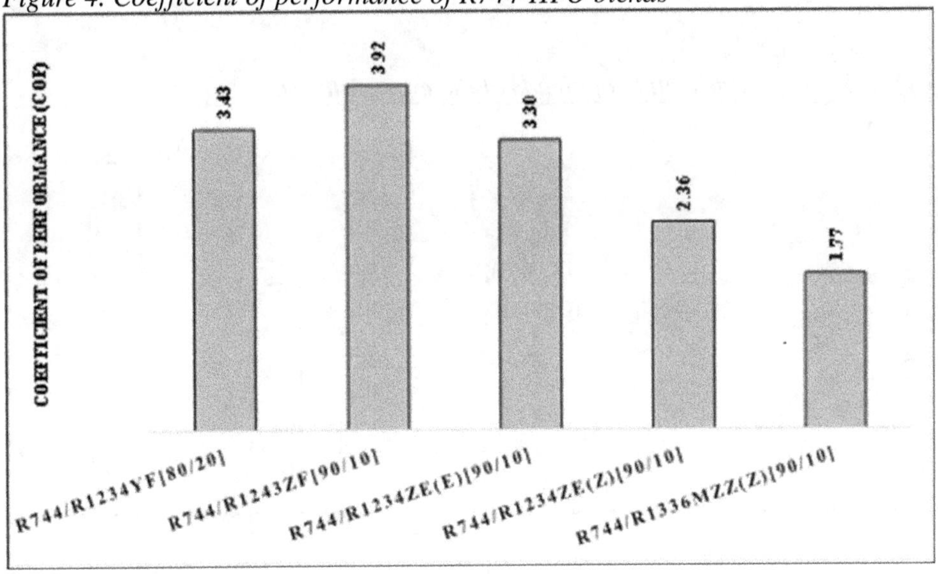

From the above data, it can be concluded that all pure refrigerants operating in subcritical mode show an excellent performance. Due to transcritical operation, pure

CO_2 offers a low COP. When CO_2 is mixed with other pure HFOs, the performance decreases further. Thus, it is seen that R744 blends are not recommended from the COP point of view.

4. THERMODYNAMIC PROPERTY ANALYSIS OF BLENDS

4.1 Temperature Glides

Temperature glide is the term used to describe the difference between the dew point and bubble point under constant pressure. There is no temperature glide for azeotropic mixtures. Some refrigerant mixtures act as azeotropic and have long-term use in refrigeration industries. Azeotropic blends work similarly to pure refrigerants (Cox et al., 2008). Zeotropic blends contain different curves for bubble point and dew point across all concentrations. In a zeotropic blend, the most volatile element boils first within the evaporator, followed by the less volatile fraction at the end. This causes temperature glide (T_{glide}) greater than 2K. Temperature glides of the selected blends in the evaporator and condenser sides are calculated using equation 5. CO_2 and HFO mixture behaviours based on their temperature glides are depicted in Figure 5.

$$T_{glide\_evp} = T_{vap\_evp} - T_4 \tag{5}$$

Figure 5. Temperature glide of blends at the evaporator side

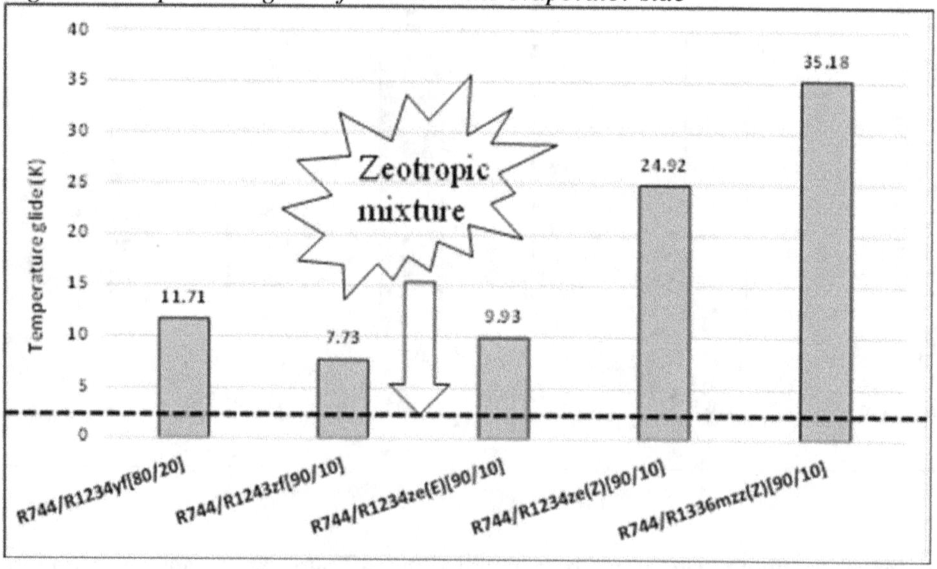

From the above data, it can be inferred that selected CO_2 and HFO blends performed like zeotropic mixtures, as their temperature glides are above 2K.

4.2 Volumetric Refrigeration Capacity

The evaporator's cooling capacity per unit vapour volume is employed for determining volumetric refrigeration capacity at the evaporator side (VRC_{evp}). This is equal to the product of the refrigeration effect ($-3h_{eva}$) and the density of vapour at the evaporator outlet ($I?_v$). A refrigerant with a large volumetric capacity can provide more cooling capability for a given compressor-swept volume (Granryd, 2001). Hydrofluoroolefins have a lower volumetric refrigeration capacity than R744 and other HFCs. As a result, a blend of R744 and HFOs can enhance volumetric capacity subject to the proportion of the mixture. The volumetric refrigeration capacity of pure refrigerants and selected blends has been calculated at an evaporator temperature of 6A C, and the values are depicted in Figure 6.

Figure 6. COP and VRC_{evp} for different refrigerants and selected blends

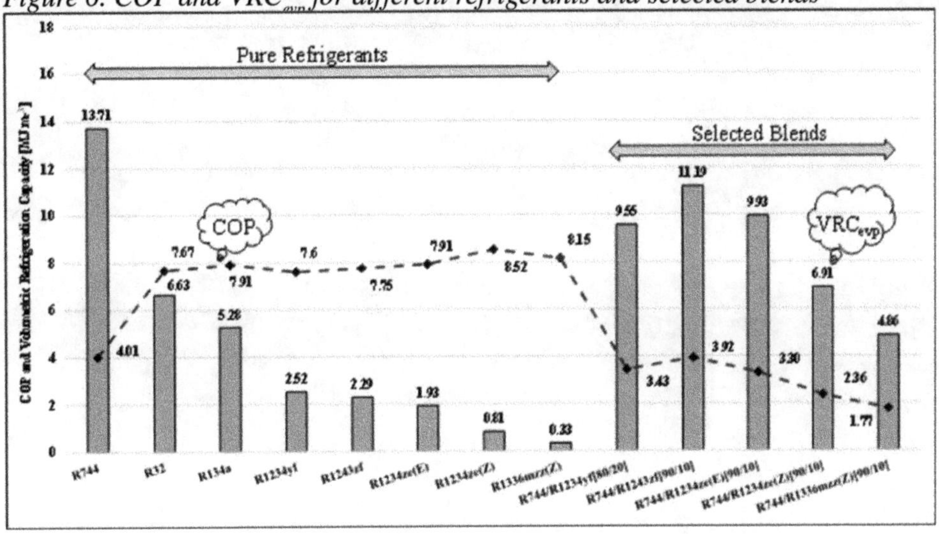

As shown in Fig.6, HFCs and HFOs have very low volumetric capacity as compared to R744. On the other hand, selected blends show comparatively high volumetric capacity as compared to other pure HFCs and HFOs. Thus, in terms of the volumetric capacity, the blends are preferable to pure HFOs.

4.3 Cycle Operating Parameters

Critical temperature and pressure are important thermodynamic parameters of any pure refrigerant or blends. Depending on how their critical point and system operating conditions relate to one another, refrigerants can function in either subcritical or transcritical cycles. In the subcritical refrigeration cycle, the working fluid operates totally at temperatures and pressures below its critical values. The condensation process takes place at a pressure where the refrigerant may transition from vapour to liquid. Transcritical operation takes place when the refrigerant exceeds its critical point throughout a part of the cycle (Pan et al., 2019; Sun et al., 2021; SA nchez et al., 2022). Hence, the critical parameters and the regime of selected blends change with the addition of CO_2 are studied. The critical parameters of the selected blends are obtained from REFPROP 10.0, and the calculated data are depicted in Figure 7.

Operating critical parameters showed that all the selected blends are working under a subcritical refrigeration cycle, although their coefficient of performance is not good because with increasing CO_2 concentration, compressor discharge pressure increases and thereby COP reduces.

Figure 7. Critical and discharge pressures of selected blends

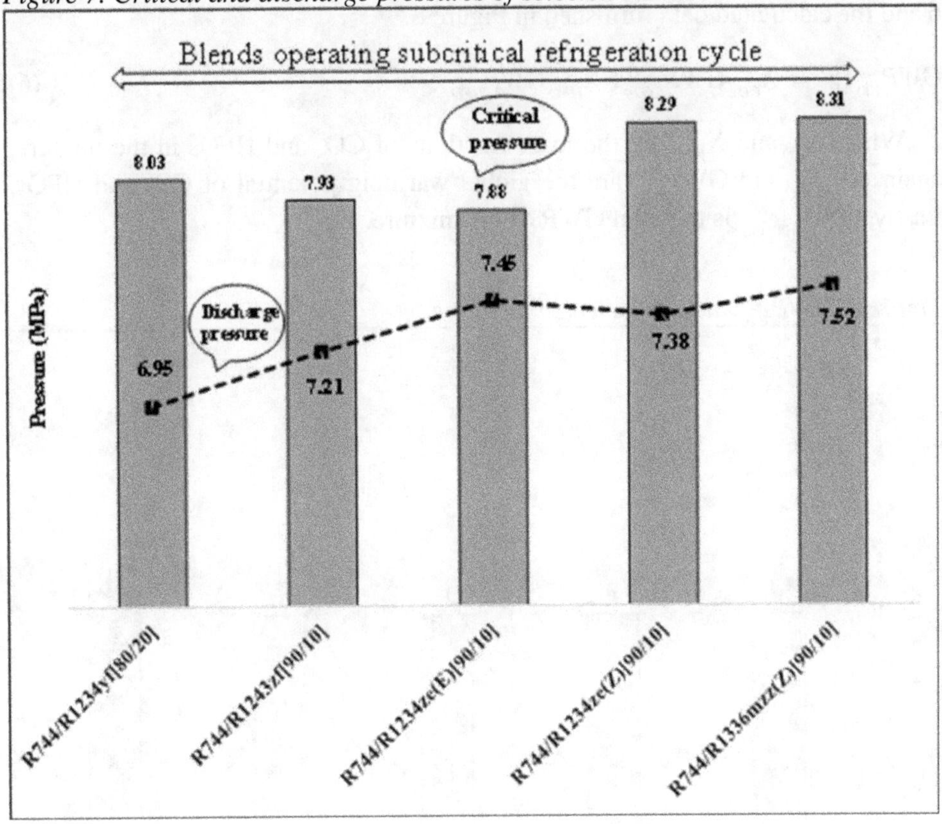

5. ENVIRONMENTAL ASPECTS OF REFRIGERATION BLENDS

The environmental impact of refrigeration blends is evaluated based on two key metrics: Global Warming Potential (GWP) and Total Equivalent Warming Impact (TEWI). These metrics help assess the direct and indirect contributions of refrigerants to climate change.

5.1 Global Warming Potential of Selected Blends

GWP_{100} is a measure of thermal radiation absorbed by greenhouse gases by the identical mass of CO_2 over 100 years. The refrigeration sector is searching for very low global warming potential refrigerants, either pure or mixtures, to comply with international guidelines (United Nations Environment Programme, 1987; SA nchez et al., 2023). The GWP of selected blends is determined in this study using Equation 6 and the calculated data furnished in Figure 8.

$$GWP_{100\_mix} = X_{CO2} GWP_{CO2} + X_{HFO} GWP_{HFO} \tag{6}$$

Where X_{CO_2} and X_{HFO} are the mass fractions of CO_2 and HFOs in the mixture, again GWP_{CO_2} and GWP_{HFO} are the global warming potential of CO_2 and HFOs. Lastly, GWP_{100_mix} is the total GWP of the mixture.

Figure 8. GWP_{100} values of selected mixtures

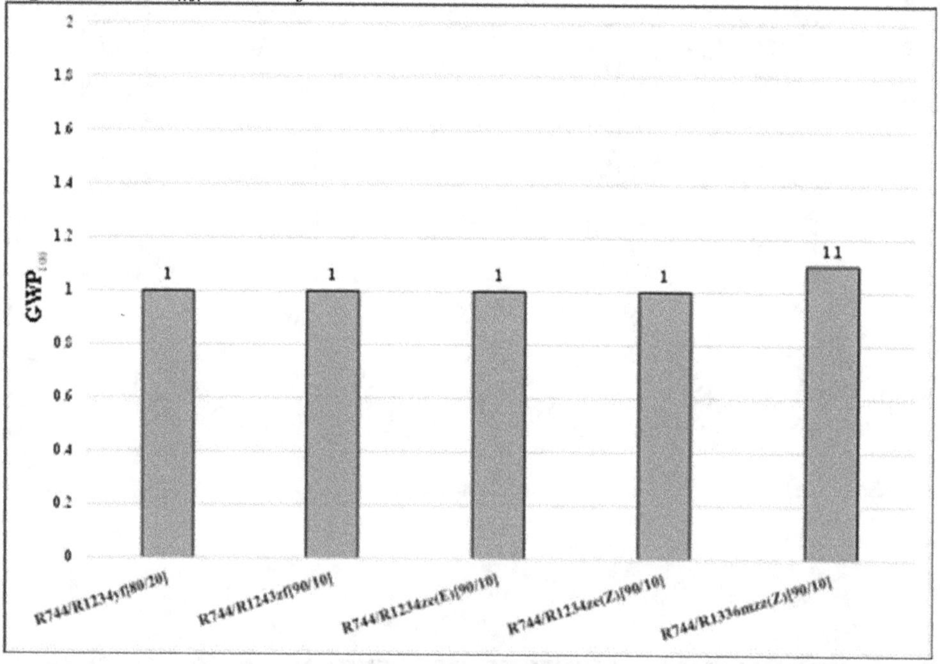

The above results showed that selected blends have much lower GWP than other synthetic blends, making them an environmentally friendly option when it comes to global warming impact over 100 years.

5.2 Total Equivalent Warming Impact (TEWI)

TEWI is a measurement of the total amount of environmental impact of the refrigerant during equipment operation and removal of the fluid that operates at the end of its life. It encompasses both direct and indirect impacts resulting from leakage and disposal, as well as energy consumption during operation. TEWI is different for different countries due to the different emission intensity factors. The emission intensity factor is defined as the mass of CO_2 emitted by the power generator to produce one kWh (kilowatt-hour) of electrical power supplied to the refrigeration system. The emissivity factor is different for different countries. For France, it is very low due to its dependency on nuclear energy. The emission intensity factors of different countries are furnished in Figure 9 (Garg et al., 2017).

Figure 9. Emission intensity factor of different countries per kW power generation

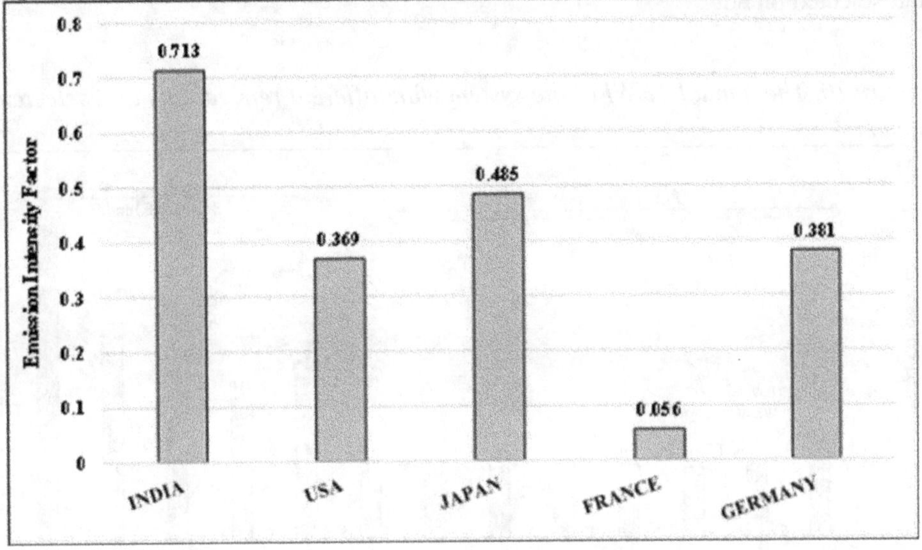

Compared to developed nations, India has a higher emission factor of carbon dioxide per kW of electricity generation. Researchers have elaborated on the Total Equivalent Warming Impact of different refrigerants and performed numerical analysis [45]. According to this study, the TEWI consists of two components: a) Indirect emissions and b) Direct emissions.

$$AnnualAIndirectAWarming = AE_{annual}AA\text{-}AI \quad (7)$$

$$E_{annual} = ApowerAinputAA\text{-}AnumberAofAhoursAworkingAperAdayAA\text{-}A365 \quad (8)$$

$$AnnualADirectAWarming = A\left(GWPAA\text{-}AmA\text{-}L_{annual}\right)$$

$$+A\left(GWPA\text{-}mA\text{-}\left(1A - \mathrm{I}_{recovery}\right)\right) \quad (9)$$

Where I is the emissivity factor, m is the refrigerant charge, L_{annual} is the leakage per year of refrigerant and $I_{recovery}$ = recovery rate (Ceglia et al., 2021).

According to Safety of Carbon Dioxide in Large Refrigeration Systems, Royal Institute of Technology, Sweden (Sawalha, 2008; Yang & Wu, 2013; Devotta et al., 2021; Vaccaro et al., 2024) Carbon dioxide charge in a secondary loop system is 5.25 kg/kW and in cascade systems, it is 1.7 kg/kW. Using the above equations (eqn. 7-9) and data, the TEWI is calculated (Figure 10) for different pure refrigerants and selected blends.

Figure 10. The annual TEWI of the system with different refrigerants and selected blends

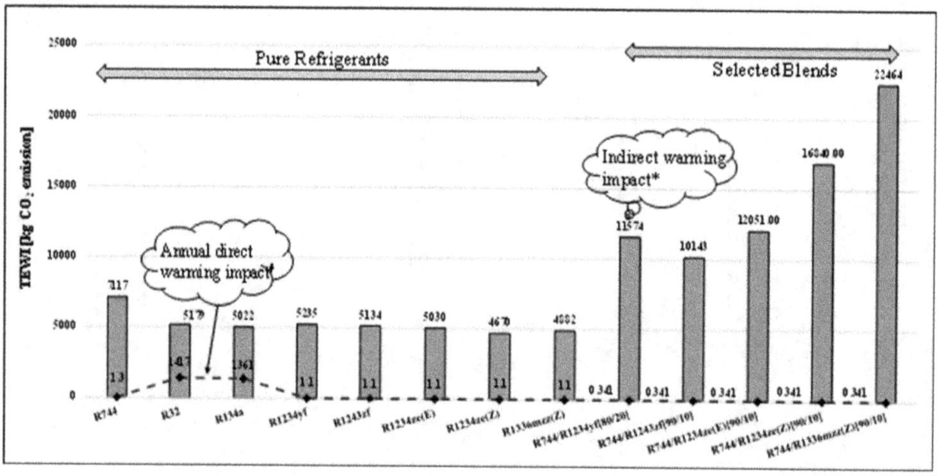

**10-hour system runtime every day for 365 days a year*
#1% leakage rate per annum and 70% refrigerant recovery rate after a 10-year system lifetime

The TEWI data of different refrigerants and blends indicate that the TEWI values of CO_2-based blends are significantly higher compared to other refrigerants due to their low COP. Blends could be a good alternative to synthetic refrigerants when we attain renewable energy-based power generation targets in future. On the other hand, the direct warming impact of these HFCs/HFOs and blends is less than or equal to 1kg of CO_2 emissions per year. Reduction of the indirect warming impact of refrigerant blends can be possible with the improvement of system energy efficiencies, optimised operational parameters, smart technology integration and specific design for different climate conditions.

ACKNOWLEDGEMENT

This theoretical investigation is part of the project ?oDevelopment of low GWP chemicals and their blends as alternatives to HCFCs and HFCs ?? sponsored by the Ministry of Environment (Forest and Climate Change, Govt of India, Ozone Cell, 1st Floor, 9 Institutional Area, Lodhi Road, New Delhi - 100 003). The authors would like to acknowledge the financial support from the Ministry of Environment, New Delhi, India.

REFERENCES

American Society of Heating, Refrigerating and Air-Conditioning Engineers (ASHRAE). (2019). ANSI/ASHRAE Standard 34-2019: Designation and safety classification of refrigerants.

Aprea, C., Greco, A., & Maiorino, A. (2016). An experimental investigation on the substitution of HFC-134a with HFO-1234yf in a domestic refrigerator. *Applied Thermal Engineering*, *106*, 959 ?"967.

Arat, H., & Arslan, O. (2017). Exergoeconomic analysis of a district heating system boosted by a geothermal heat pump. *Energy*, *119*, 1159 ?"1170.

Belman-Flores, J. M., Rangel-HernA ndez, V. H., UsA3n, S., & Rubio-Maya, C. (2017). Energy and exergy analysis of R-1234yf as drop-in replacement for R-134a in a domestic refrigeration system. *Energy*, *132*, 116 ?"125.

Bobbo, S., Di Nicola, G., Zilio, C., Brown, J. S., & Fedele, L. (2018). Low-GWP halocarbon refrigerants: A review of thermophysical properties. *International Journal of Refrigeration*, *90*, 181 ?"201.

Bouteiller, P., Terrier, M.-F., & Tobaly, P. (2017). Experimental study of heat-pump thermodynamic cycles using CO_2-based mixtures ?"Methodology and first results. *AIP Conference Proceedings*, 1814(1).

Ceglia, F., Marrasso, E., Roselli, C., & Sasso, M. (2021). An innovative parameter: Expanded total equivalent warming impact. *International Journal of Refrigeration*, *131*, 980 ?"989.

Ciconkov, R. (2018). Refrigerants: There is still no vision for sustainable solutions. *International Journal of Refrigeration*, *86*, 441 ?"448.

Claesson, J. (2005). Thermal and hydraulic performance of compact brazed plate heat exchangers operating as evaporators in domestic heat pumps.

Cox, N., Mazur, V., & Colbourne, D. (2008). New high-pressure low-GWP azeotropic and near-azeotropic refrigerant blends. Proceedings of the International Refrigeration Conference, Purdue University.

Devotta, S., Chelani, A., & Vonsild, A. (2021). Prediction of flammability classifications of refrigerants by ANN and random forest. *International Journal of Refrigeration, 131*, 947 ?"955.

Engel, A., Rigby, M., Burkholder, J. B., Fernandez, R. P., Froidevaux, L., Hall, B. D., & Yao, B. (2019). Update on ozone-depleting substances (ODSs) and other gases of interest to the Montreal Protocol. In *Scientific Assessment of Ozone Depletion: 2018*. World Meteorological Organization.

Ente Italiano di Normazione (UNI). (2021). *UNI EN 378-1:2021: Refrigerating systems and heat pumps ?"Safety and environmental requirements ?"Part 1: Basic requirements, definitions, classification and selection criteria*. UNI.

Feng, B., Yang, Z., & Zhai, R. (2018). Experimental study on the influence of flame retardants on the flammability of R-1234yf. *Energy, 143*, 212 ?"218.

Fukuda, S., Kondou, C., Takata, N., & Koyama, S. (2014). Low-GWP refrigerants R-1234ze(E) and R-1234ze(Z) for high-temperature heat pumps. *International Journal of Refrigeration, 40*, 161 ?"173.

Garg, A., Shukla, P. R., Kankal, B., & Mahapatra, D. (2017). CO_2 emission in India: Trends and management at sectoral, sub-regional and plant levels. *Carbon Management, 8*(2), 111 ?"123.

Gebre, S. L., & Ludwig, F. (2015). Hydrological response to climate change of the upper Blue Nile River Basin based on IPCC AR5. *Journal of Climatology & Weather Forecasting, 3*(1), 1 ?"15.

Gil, B., & Kasperski, J. (2018). Efficiency evaluation of the ejector cooling cycle using new HFO/HCFO refrigerant as an R-134a replacement. *Energies, 11*(8), 2136.

Golzari, S., Kasaeian, A., Daviran, S., Mahian, O., Wongwises, S., & Sahin, A. Z. (2017). Second-law analysis of an automotive air-conditioning system using HFO-1234yf. *International Journal of Refrigeration, 73*, 134 ?"143.

Granryd, E. (2001). Hydrocarbons as refrigerants ?"An overview. *International Journal of Refrigeration, 24*(1), 15 ?"24.

Guillaume, L., Legros, A., Desideri, A., & Lemort, V. (2017). Performance of a radial-inflow turbine integrated in an ORC system for waste-heat recovery on trucks: Experimental comparison between R-245fa and R-1233zd. *Applied Energy*, *186*, 408 ?"422.

Han, X. H., Wang, Q., Zhu, Z. W., & Chen, G. M. (2007). Cycle performance study on R-32/R-125/R-161 as an alternative to R-407C. *Applied Thermal Engineering*, *27*(14 ?"15), 2559 ?"2565.

Lee, S. J., Shon, B. H., Jung, C. W., & Kang, Y. T. (2018). A novel solar-assisted heat pump using low-GWP R-1233zd(E) with a flexible solar collector. *Energy*, *149*, 386 ?"396.

Lemmon, E. W., Bell, I. H., Huber, M. L., & McLinden, M. O. (2018). *NIST reference fluid thermodynamic and transport properties database (REFPROP), Version 10.0*. National Institute of Standards and Technology.

Longo, G. A., Zilio, C., Righetti, G., & Brown, J. S. (2014). Experimental assessment of the low-GWP refrigerant HFO-1234ze(Z) for high-temperature heat pumps. *Experimental Thermal and Fluid Science*, *57*, 293 ?"300.

Luo, D., Mahmoud, A., & Cogswell, F. (2015). Evaluation of low-GWP fluids for power generation with Organic Rankine Cycle. *Energy*, *85*, 481 ?"488.

Luo, J., Ye, Z., Zhao, Z., Yang, K., Zhang, S., & Wang, Q. (2023). Vapor ?"liquid equilibrium and heating performance of CO_2/R-1234ze(Z) and CO_2/R-1336mzz(E) zeotropic blends. *Applied Thermal Engineering*, *229*, 120576.

MolA(c)s, F., Navarro-EsbrA-, J., Peris, B., Mota-Babiloni, A., BarragA n-Cervera, A?., & Kontomaris, K. K. (2014). Low-GWP alternatives to HFC-245fa in Organic Rankine Cycles: HCFO-1233zd-E and HFO-1336mzz-Z. *Applied Thermal Engineering*, *71*(1), 204 ?"212.

MolA(c)s, F., Navarro-EsbrA-, J., Peris, B., Mota-Babiloni, A., BarragA n-Cervera, A?., & Kontomaris, K. K. (2017). Thermo-economic evaluation of low-GWP alternatives to HFC-245fa in Organic Rankine Cycles. *Energy Procedia*, *142*, 1199 ?"1205.

Molina, M. J., & Rowland, F. S. (1974). Stratospheric sink for chlorofluoromethanes: Chlorine atom-catalysed destruction of ozone. *Nature*, *249*(5460), 810 ?"812.

Montzka, S. A., Butler, J. H., Elkins, J. W., Thompson, T. M., Clarke, A. D., & Lock, L. T. (1999). Present and future trends in the atmospheric burden of ozone-depleting halogens. *Nature*, *398*(6729), 690 ?"694.

Mota-Babiloni, A., Navarro-EsbrA-, J., BarragA n-Cervera, A?., MolA(c)s, F., & Peris, B. (2015). R-450A (R-1234ze(E)/R-134a) as an R-134a replacement: Experimental study. *International Journal of Refrigeration, 51*, 52 ?"58.

Navarro-EsbrA-, J., MolA(c)s, F., Peris, B., Mota-Babiloni, A., & Kontomaris, K. (2017). Experimental study of an Organic Rankine Cycle with HFO-1336mzz-Z for micro-scale low-temperature applications. *Energy, 133*, 79 ?"89.

Pan, L., Ma, Y., Li, T., Li, H., Li, B., & Wei, X. (2019). Cycle performance and combustion characteristics of two CO_2-based binary mixtures for the transcritical power cycle. *Energy, 179*, 454 ?"463.

Poggi, F., Macchi-Tejeda, H., Leducq, D., & Bontemps, A. (2008). Refrigerant charge in systems and strategies for charge reduction. *International Journal of Refrigeration, 31*(3), 353 ?"370.

SA nchez, D., Andreu-NA cher, A., Calleja-Anta, D., Llopis, R., & Cabello, R. (2022). Energy impact of low-GWP alternatives to R-134a in a beverage cooler: Experimental analysis and optimization. *Energy Conversion and Management, 256*, 115388.

SA nchez, D., Larrondo, R., Vidan-Falomir, F., & Cabello, R. (2024). Experimental evaluation of CO_2/R-32, CO_2/R-1234yf and CO_2/R-1270 in a transcritical plant with IHX. *Applied Thermal Engineering, 236*, 121473.

SA nchez, D., Vidan-Falomir, F., Nebot-AndrA(c)s, L., Llopis, R., & Cabello, R. (2023). Alternative blends of CO_2 for transcritical refrigeration systems: Experimental approach and energy analysis. *Energy*.

Sarkar, J., Bhattacharyya, S., & Ram Gopal, M. (2004). Optimization of a transcritical CO_2 heat pump for simultaneous cooling and heating. *International Journal of Refrigeration, 27*(8), 830 ?"838.

Sawalha, S. (2008). Safety of CO_2 in large refrigeration systems. Proceedings of the International Refrigeration Conference, Purdue University, 81 ?"93.

Say, D., Ganesan, A. L., Lunt, M. F., Rigby, M., O ?(tm)Doherty, S., Harth, C., & Bauguitte, S. (2019). Emissions of halocarbons from India inferred through atmospheric measurements. *Atmospheric Chemistry and Physics, 19*(15), 9865 ?"9885.

Smith, S. (2012). AHRI Standard 550 ?"590 ?"Overview of latest updates: Performance rating of water-chilling and heat-pump water-heating packages using the vapor-compression cycle. *ASHRAE Transactions, 118*(1), 317 ?"325.

Sun, X., Shi, L., Tian, H., Wang, X., Zhang, Y., & Shu, G. (2021). A composition-tunable combined cooling and power cycle using CO_2-based binary zeotropic mixture. *Energy Conversion and Management, 244*, 114419.

United Nations Environment Programme. (1987). *Montreal Protocol on Substances that Deplete the Ozone Layer.* U.S. Government Printing Office.

Vaccaro, G., Milazzo, A., & Talluri, L. (2024). A non-flammable, fluorine-free, CO_2-based mixture as a low-TEWI refrigerant. *International Journal of Refrigeration, 158*, 157 ?"163.

Velders, G. J. M., Fahey, D. W., Daniel, J. S., Andersen, S. O., & McFarland, M. (2015). Future atmospheric abundances and climate forcings from scenarios of global and regional HFC emissions. *Atmospheric Environment, 123*, 200 ?"209.

Welzl, M., Heberle, F., & BrA1/4ggemann, D. (2017). Simultaneous experimental investigation of nucleate boiling heat transfer and power output in ORC using R-245fa and R-1233zd(E). *Energy Procedia, 129*, 435 ?"442.

Wu, J., Chu, Y., Hu, J., & Liu, Z. (2009). Performance of R-152a/R-125/R-32 in domestic air-conditioners. *International Journal of Refrigeration, 32*(5), 1049 ?"1057.

Yang, Z., & Wu, X. (2013). Retrofits and options for alternatives to HCFC-22. *Energy, 59*, 1 ?"21.

Yelishala, S. C. (2019). Thermodynamic and combustion study on blends of hydrocarbons and carbon dioxide as alternative refrigerants.

Yelishala, S. C., Kannaiyan, K., Wang, Z., Metghalchi, H., Levendis, Y. A., & Sadr, R. (2020). Thermodynamic study on blends of hydrocarbons and carbon dioxide as zeotropic refrigerants. *Journal of Energy Resources Technology, 142*(8), 082304.

Compilation of References

Abdelgaied, M., & Mohamed, S., Bassuoni, M. M., & Khaira, A.M. (2023). Solid desiccant air conditioning system using desiccant dehumidifiers with cooling technique and thermal recovery unit: Experimental investigation and performance analysis. *Journal of Cleaner Production*, 416.

Abd-Elmonem, A., Ahmad, H., Garalleh, H. A., Atta, G., Amjad, M., Suoliman, N. A., Jamshed, W., & Hussain, S. M. (2025). Machine learning and differential transform simulation in analysing the ionic liquid (EtSO4) based ternary hybrid flow over a curved geometry. *Journal of Molecular Liquids, 127109*. Advance online publication. DOI: 10.1016/j.molliq.2025.127109

Acharya, S., Agrawal, S., & Dash, S. K. (2018). Numerical Analysis of Natural Convection Heat Transfer from a Vertical Hollow Cylinder Suspended in Air. *Journal of Heat Transfer, 140*(5). Advance online publication. DOI: 10.1115/1.4038478

Agarwal, R. S., & Clark, E. (2016). Refrigerant Blends : Calculating Global Warming Potentials. *Ozon Action FACT SHEET, 290*, 2. http://www.unep.fr/ozonaction/information/mmcfiles/7786-e-Calculating_GWPofBlends_post_Kigali.pdf

Ahmadi, M. H., Ramezanizadeh, M., Nazari, M. A., Lorenzini, G., Kumar, R., & Jilte, R. (2018). Applications of nanofluids in geothermal: A review. *Mathematical Modelling and Engineering Problems, 5*(4), 281–285. DOI: 10.18280/mmep.050402

Ahn, S. W. (2001). The effects of roughness types on friction factors and heat transfer in roughened rectangular duct. *International Communications in Heat and Mass Transfer, 28*(7), 933–942. https://doi.org/https://doi.org/10.1016/S0735-1933(01)00297-4

Akafuah, N. K., Salazar, A. J., & Saito, K. (2010). Infrared thermography-based visualization of droplet transport in liquid sprays. *Infrared Physics & Technology, 53*(3), 218–226.

Akula, R., Minnikanti, A., & Balaji, C. (2024). Pin fin–PCM composite heat sink solution for thermal management of cylindrical Li-ion battery. *Applied Thermal Engineering, 248*, 123146. DOI: 10.1016/j.applthermaleng.2024.123146

Alagumalai, A., & Mahian, O. (2022). Specialty grand challenge in thermal science and energy systems. *Frontiers in Thermal Engineering, 2*. Advance online publication. DOI: 10.3389/fther.2022.954511

Alam, T., Saini, R. P., & Saini, J. S. (2014). Experimental investigation on heat transfer enhancement due to V-shaped perforated blocks in a rectangular duct of solar air heater. *Energy Conversion and Management, 81*, 374–383. DOI: 10.1016/j.enconman.2014.02.044

Alawi, A.. (2025). Comprehensive review of thermal management systems for lithium-ion batteries. *Batteries, 11*(3), 221.

Alghassab, M. A.. (2024). Comparative study of aluminum and copper fins in PCM modules. *Journal of Energy Storage, 75*, 109245. DOI: 10.1016/j.est.2024.109245

Alghassab, M. A.. (2024). Comparative study of aluminum and copper fins in PCM-enhanced battery modules. *Journal of Energy Storage, 75*, 109245.

Al-Hallaj, S., & Selman, J. R. (2002). Thermal modeling of secondary lithium batteries for electric vehicle/hybrid electric vehicle applications. *Journal of Power Sources, 110*(2), 341–348. DOI: 10.1016/S0378-7753(02)00196-9

Al-Harrasi, A.. (2019). Extraction and characterization of phytol from algal biomass. *Journal of Applied Phycology, 31*, 1397–1407. DOI: 10.1007/s10811-018-1674-5

Alissa, H., Nick, T., Raniwala, A., Herranz, A. A., Frost, K., Manousakis, I., Lio, K., Warrier, B., Oruganti, V., DiCaprio, T. J., Oseen-Senda, K., Ramakrishnan, B., Gupta, N., Bianchini, R., Kleewein, J., Belady, C., Fontoura, M., Sinistore, J., Natarajan, M., & Frieze, M. (2025). Using life cycle assessment to drive innovation for sustainable cool clouds. *Nature*. Advance online publication. DOI: 10.1038/s41586-025-08832-3 PMID: 40307558

Alma'asfa, S. I., Abdul Aziz, M. S., Khor, C. Y., & Fraige, F. Y. (2025). Thermal management of cylindrical lithium-ion batteries with fin and PCM hybrid systems. *Journal of Thermal Science, 34*(6), 1789–1804.

Altin, R., Çetinkaya, S., & Yücesu, H. S. (2001). Potential of using vegetable oil fuels as fuel for diesel engines. *Energy Conversion and Management, 42*(5–6), 529–538. DOI: 10.1016/S0196-8904(00)00076-1

Altschuh, P., Kunz, W., Bremerich, M., Reiter, A., Selzer, M., & Nestler, B. (2022). Wicking In porous polymeric membranes: Determination of an effective capillary radius to predict the flow behavior in lateral flow assays. *Membranes, 12*(7), 638. PMID: 35877842

Ambirajan, A., Adoni, A. A., Vaidya, J. S., Rajendran, A. A., Kumar, D., & Dutta, P. (2012). Loop heat pipes: A review of fundamentals, operation, and design. *Heat Transfer Engineering*, *33*(4-5), 387–405.

American Society of Heating, Refrigerating and Air-Conditioning Engineers (ASHRAE). (2019). ANSI/ASHRAE Standard 34-2019: Designation and safety classification of refrigerants.

American Society of Heating. (2025). *Refrigeration and Airconditioning Engineers*. ASHRAE Handbook Fundamentals.

Amer, M. M., Shouman, M. A., Salem, M. S., Kannan, A. M., & Hamed, A. M. (2024). Advances in thermal management systems for Li-ion batteries: A review. *Thermal Science and Engineering Progress*, *53*, 102714.

Anderson, J. T., & Saunders, O. A. (1953). Convection from an Isolated Heated Horizontal Cylinder Rotating about Its Axis. *Proceedings of the Royal Society of London. Series A. Mathematical and Physical Sciences*,555–562. https://doi.org/ DOI: 10.1098/rspa.1953.0080

Angrisani, G., Capozzoli, A., Minichiello, F., Roselli, C., & Sasso, M. (2011). Desiccant wheel regenerated by thermal energy from a microcogenerator: Experimental assessment of the performances. *Applied Energy*, *88*(4), 1354 ?"1365.

Angrisani, G., Minichiello, F., Roselli, C., & Sasso, M. (2012). Experimental analysis on the dehumidification and thermal performance of a desiccant wheel. *Applied Energy*, *92*, 563 ?"572.

Angrisani, G., Roselli, C., & Sasso, M. (2017). Dehumidification and thermal behaviour of desiccant wheels: Correlations based on experimental and manufacturer data. *Heat Transfer Engineering*, *39*(3), 293 ?"303. DOI: 10.1080/01457632.2017.1295743

Anon. (1977). Methods of Testing To Determine the Thermal Performance of Solar Collectors. *ANSI Stand B198. 1, 2010*(Ra).

Appavu, V.. (2019). Performance evaluation of biodiesel–alcohol fuel blends in diesel engines. *Applied Energy*, *239*, 1–10. DOI: 10.1016/j.apenergy.2019.01.039

Aprea, C., Greco, A., & Maiorino, A. (2016). An experimental investigation on the substitution of HFC-134a with HFO-1234yf in a domestic refrigerator. *Applied Thermal Engineering*, *106*, 959 ?"967.

Arat, H., & Arslan, O. (2017). Exergoeconomic analysis of a district heating system boosted by a geothermal heat pump. *Energy*, *119*, 1159 ?"1170.

Argade, S., & De, A. (2024). Optimization of Z-type airflow cooling for lithium-ion battery packs. *Applied Thermal Engineering*, *229*, 120834.

ARORA, R. C. (2012)) Refrigeration and air conditioning. PHI Learning Pvt. Ltd.

Arunsandeep, G., Lingayat, A., Chandramohan, V. P., Raju, V. R. K., & Reddy, K. S. (2018). A numerical model for drying of spherical object in an indirect type solar dryer and estimating the drying time at different moisture level and air temperature. *International Journal of Green Energy*, *15*(3), 189–200. DOI: 10.1080/15435075.2018.1433181

Ashok, B.. (2019). Assessment of long-chain alcohol biodiesel blends. *Renewable Energy*, *139*, 1354–1362. DOI: 10.1016/j.renene.2019.03.026

ASHRAE. (2023). Update on new refrigerants designations and safety classifications. *ANSI/ASHRAE Standard 34-2019, August*, 1–5. https://www.ashrae.org/technical-resources/standards-and-guidelines/ashrae-refrigerant-designations

Assadi, M. K., & Mohammadi, A. (2016). Simulation of dynamical performance of solar desiccant cooling cycle. *Applied Mechanics and Materials*, *819*, 160 ?"170.

Assad, M. E. H., Ahmadi, M. H., Sadeghzadeh, M., Yassin, A., & Issakhov, A. (2020). Renewable hybrid energy systems using geothermal energy: Hybrid solar thermal–geothermal power plant. *The International Journal of Low Carbon Technologies*, *16*(2), 518–530. DOI: 10.1093/ijlct/ctaa084

Athar, M. (2003). ISH M. A. Ansari M. & Khan M. A., "Flow characteristics of elliptical orifice plates,". *ISH Journal of Hydraulic Engineering*, *9*(2), 22–35.

Atmanli, A. (2016). Comparative analysis of n-pentanol and diesel–biodiesel–alcohol blends. *Fuel*, *177*, 219–225. DOI: 10.1016/j.fuel.2016.03.029

Atmanli, A. (2016). Influence of higher alcohols on diesel/biodiesel performance. *Fuel*, *182*, 237–244. DOI: 10.1016/j.fuel.2016.04.050

Atmanli, A., & Yilmaz, N. (2018). Phase stability and combustion evaluation of 1-pentanol–diesel blends. *Energy Conversion and Management*, *166*, 332–339. DOI: 10.1016/j.enconman.2018.04.066

Atmanli, A., & Yilmaz, N. (2020). Energy and emission analysis of alcohol–biodiesel fuels. *Energy Reports, 6*, 110–119. DOI: 10.1016/j.egyr.2019.11.010

Awan, A., Ge, T. S., Wang, R. Z., & Jing, D. (2025). Flow phenomena inside the Ranque-Hilsch vortex tube: A state-of-the-art review. *Renewable & Sustainable Energy Reviews, 211*(Feb).

Awan, O. A. A., Ge, T. S., Wang, R. Z., & Jing, D. (2025). Flow phenomena inside the Ranque-Hilsch vortex tube. *Renewable & Sustainable Energy Reviews, ? ? ?*, 173.

Awasthi, A.. (2025). Phase change material-based hybrid BTMS for prismatic lithium-ion batteries: A review. *Journal of Thermal Engineering, 11*(2), 155–172.

Babu, D., & Anand, K. (2017). Experimental study of n-pentanol addition on emission characteristics of CI engines. *Fuel, 196*, 486–495. DOI: 10.1016/j.fuel.2017.01.065

Bagade, V. S., Suryawanshi, P. M., & Nalavade, S. M. (2016). A review of multi-hole orifice plate. *International Journal for Research in Applied Science and Engineering Technology, 7*(4), 3197–3208.

Balakrishnan, J. (2007). "Fuel cell technology." *Proceedings of ICIAFS.*

Balasubramanian, D.. (2025). Hybrid battery thermal management using honeycomb PCM and cooling plates. *Scientific Reports, 15*, 1651.

Bansal, P. (2012). A review - Status of CO 2 as a low temperature refrigerant: Fundamentals and R&D opportunities. *Applied Thermal Engineering, 41*, 18–29. DOI: 10.1016/j.applthermaleng.2011.12.006

Barnwal, B. K., & Sharma, M. P. (2005). Prospects of biodiesel production from vegetable oils in India. *Renewable & Sustainable Energy Reviews, 9*(4), 363–378. DOI: 10.1016/j.rser.2004.04.013

Becker, K. M. (1963). Measurements of Convective Heat Transfer from a Horizontal Cylinder Rotating in a Tank of Water. *International Journal of Heat and Mass Transfer, 6*(12).

Begum, Y. A., Kumari, S., Jain, S. K., & Garg, M. C. (2024). A review on waste biomass-to-energy: Integrated thermochemical and biochemical conversion for resource recovery. *Environmental Science. Advances, 3*(9), 1197–1216. DOI: 10.1039/d4va00109e

Belman-Flores, J. M., Rangel-HernA ndez, V. H., UsA3n, S., & Rubio-Maya, C. (2017). Energy and exergy analysis of R-1234yf as drop-in replacement for R-134a in a domestic refrigeration system. *Energy, 132*, 116 ?"125.

Bergmann, B. S. (2018, October 31). The theory of the vortex tube. *EXAIR Blog*. https://blog.exair.com/2018/10/31/the-theory-of-the-vortex-tube/

Bhushan, B., & Singh, R. (2010). A review on methodology of artificial roughness used in duct of solar air heaters. *Energy*, *35*(1), 202–212. DOI: 10.1016/j.energy.2009.09.010

Bobbo, S., Di Nicola, G., Zilio, C., Brown, J. S., & Fedele, L. (2018). Low-GWP halocarbon refrigerants: A review of thermophysical properties. *International Journal of Refrigeration*, *90*, 181 ?"201.

Bouakkaz, R., Talbi, K., Khelil, Y., Salhi, F., Belghar, N., & Ouazizi, M. (2014). Numerical investigation of incompressible fluid flow and heat transfer around a rotating circular cylinder. *Thermophysics and Aeromechanics*, *21*(1), 87–97. DOI: 10.1134/S0869864314010089

Bouteiller, P., Terrier, M.-F., & Tobaly, P. (2017). Experimental study of heat-pump thermodynamic cycles using CO_2-based mixtures ?"Methodology and first results. *AIP Conference Proceedings*, 1814(1).

Breuer, M., Bernsdorf, J., Zeiser, T., & Durst, F. (2000). Accurate computations of the laminar flow past a square cylinder based on two different methods: Lattice-boltzmann and finite-volume. *International Journal of Heat and Fluid Flow*, *21*, 186–196.

Brooks, A. N., & Hughes, T. J. R. (1982). Streamline upwind/petrov-galerkin formulations for convection dominated flows with particular emphasis on the incompressible navierstokes equations. *Computer Methods in Applied Mechanics and Engineering*, *32*, 199–259.

Bullen, P. R., Cheeseman, D. J., Hussain, L. A., & Ruffel, A. E. (1987). The Determination of Pipe Contraction Coefficients for Incompressible Turbulent Flow. *International Journal of Heat and Fluid Flow*, *8*, 111–118.

Burgess, S., Wang, X., Rahbari, A., & Hangi, M. (2022). Optimisation of a portable phase-change material (PCM) storage system for emerging cold-chain delivery applications. *Journal of Energy Storage*, *52*, 104855.

Cabeza, L. F., Galindo, E., Prieto, C., Barreneche, C., & Fernández, A. I. (2015). Key performance indicators in thermal energy storage: Survey and assessment. *Renewable Energy*, *83*, 820–827. DOI: 10.1016/j.renene.2015.05.019

Cai, Y.. (2024). Multi-objective optimization of PCM–graphite–liquid BTMS. *Energy*, *276*, 127456. DOI: 10.1016/j.energy.2024.127456

Caliskan, H., Lee, D.-Y., & Hong, H. (2019). Enhanced thermodynamic assessments of the novel desiccant air cooling system for sustainable energy future. *Journal of Cleaner Production, 211*, 213 ?"221.

CEA. (2011). CO 2 Baseline Database for the Indian Power Sector User Guide. *Report by Ministry of Power Govt of India, January*, 1–34. http://cea.nic.in/reports/others/thermal/tpece/cdm_co2/user_guide_ver9.pdf

Ceglia, F., Marrasso, E., Roselli, C., & Sasso, M. (2021). An innovative parameter: Expanded total equivalent warming impact. *International Journal of Refrigeration, 131*, 980 ?"989.

Cellura, M., Luu, L. Q., Guarino, F., & Longo, S. (2025). A review on life cycle assessment of concentrating solar energy technologies. *Renewable Energy, 249*, 123203. DOI: 10.1016/j.renene.2025.123203

Çengel, Y. A., & Ghajar, A. J. (2015). Heat and Mass Transfer: Fundamentals & Applications. In *McGraw-Hill Education* (5th ed.). McGraw-Hill Education.

Chandra, A., Sistla, Y. S., Ahmed, M. A., Vasireddy, D. V. S., Jaglan, N., Das, N. K., Banerjee, T., & Sistla, V. S. (2025). Heat transfer fluids: Amino acid anion ionic liquid based IoNanofluids with remarkable thermal conductivity and low viscosity. *RSC Advances, 15*(29), 23146–23164. DOI: 10.1039/d5ra03752b PMID: 40620252

Chandrasekaran, M.. (2025). Experimental study on immersion and boiling-assisted cooling of Li-ion cells. *Journal of Energy Storage, 110*, 112945. DOI: 10.1016/j.est.2025.112945

Chen, Q., & Wang, Y. (2020). Research status and development trend of concentrating solar power. *2022 11th International Conference on Renewable Energy Research and Application (ICRERA)*, 390–393. DOI: 10.1109/icrera49962.2020.9242893

Cheng, X., Du, B., He, J., Long, W., Su, G., Liu, J., Fan, Z., & Chen, L. (2025). A Review of thermoplastic composites on wind turbine blades. *Composites. Part B, Engineering, 112411.* Advance online publication. DOI: 10.1016/j.compositesb.2025.112411

Cheng, Y. J., Liao, Y. H., & Huang, C. K. (2008). Heat transfer on a radially rotating heated cylinder. *International Communications in Heat and Mass Transfer, 35*(10), 1355–1359. DOI: 10.1016/j.icheatmasstransfer.2008.08.002

Chen, L., Chen, S. H., Liu, L., & Zhang, B. (2018). Experimental investigation of precooling desiccant-wheel air-conditioning system in a high-temperature and high-humidity environment. *International Journal of Refrigeration, 95*, 83 ?"92.

Chen, S., Wang, K., Zhao, M., & Gao, Z. (2025). A review of the recent development of innovative wind turbines: Concepts and aerodynamic analyses. *Physics of Fluids*, *37*(4). Advance online publication. DOI: 10.1063/5.0256502

Chen, Z.. (2024). Fin geometry and material optimization in PCM-based battery modules. *Journal of Energy Storage*, *74*, 108921. DOI: 10.1016/j.est.2024.108921

Chisholm, D. (1967). "Flow of incompressible two-phase mixtures through sharped edged orifices," *Mechanical Engine metering. Science*, *9*, 72–78.

Chivu, R. M., Roman, C., & Porumb, R. (2024). Assessment of engine performance and emissions with eucalyptus oil blends. *Energies*, *17*. Advance online publication. DOI: 10.3390/en17010013

Choudhari, V. G., Dhoble, A. S., & Panchal, S. (2020). Numerical analysis of different fin structures in phase change material module for battery thermal management system and its optimization. *International Journal of Heat and Mass Transfer*, *163*, 120434. DOI: 10.1016/j.ijheatmasstransfer.2020.120434

Choudhari, V. G., Dhoble, A. S., & Panchal, S. (2020). Numerical analysis of different fin structures in phase change material modules for battery thermal management and optimization. *International Journal of Heat and Mass Transfer*, *163*, 120434.

Churchill, S. W., & Chu, H. H. S. (1975). Correlating equations for laminar and turbulent free convection from a horizontal cylinder. *International Journal of Heat and Mass Transfer*, *18*(9), 1049–1053. DOI: 10.1016/0017-9310(75)90222-7

Chu, W. X., Lin, Y. C., Chen, C. Y., & Wang, C. C. (2019). Experimental and numerical study on the performance of passive heat sink having alternating layout. *International Journal of Heat and Mass Transfer*, *135*, 822–836. DOI: 10.1016/j.ijheatmasstransfer.2019.02.034

Ciconkov, R. (2018). Refrigerants: There is still no vision for sustainable solutions. *International Journal of Refrigeration*, *86*, 441 ?"448.

Claesson, J. (2005). Thermal and hydraulic performance of compact brazed plate heat exchangers operating as evaporators in domestic heat pumps.

Colbourne, D., Suen, K. O., Li, T. X., Vince, I., & Vonsild, A. (2020). General framework for revising class A3 refrigerant charge limits – a discussion. *International Journal of Refrigeration*, *117*, 209–217. DOI: 10.1016/j.ijrefrig.2020.04.024

Comino, Francisco, TA boas, Francisco, Peci, Federico, & de Adana, Ruiz, Manuel. (2020). Detailed experimental analysis of the energy performance of a desiccant wheel activated at low temperature. *Applied Thermal Engineering*, *178*, 115 ?"580.

Cox, N., Mazur, V., & Colbourne, D. (2008). New high-pressure low-GWP azeotropic and near-azeotropic refrigerant blends. Proceedings of the International Refrigeration Conference, Purdue University.

Cui, M., Zhu, Z., Yang, D., He, Z., & Liu, Y. (2023). Numerical study of battery thermal management with transverse fin–casing composite structures. *Journal of Energy Storage*, *65*, 107462.

Da Silva, F. J. A., & De Matos, J. E. X. (2009). A note on the potential of CNSL in fuel blends for engines in Brazil. *Revista Tecnologia*, *14*(2), 53–58.

Dagdevir, T., & Ding, Y. (2024). Numerical investigation of battery thermal management by using helical fin and composite phase change material. *Journal of Energy Storage*, *75*, 109674. DOI: 10.1016/j.est.2023.109674

Dantas Neto, A. A., Silva, J. P., & Santos, G. A. (2014). Effect of biodiesel/diesel-based microemulsions on the exhaust emissions of a diesel engine. *Brazilian Journal of Petroleum and Gas*, *8*(2), 1–11.

Daou, K., Wang, R. Z., & Xia, Z. Z. (2006). Desiccant cooling air conditioning: A review. *Renewable & Sustainable Energy Reviews*, *10*(2), 55 ?"77.

Das, K. M., Guha, A., & Ramgopal, M. (2024). Studies on CO_2 based two-phase natural circulation loop. In *Proceedings of the 27th National and 5th International ISHMT-ASTFE Heat and Mass Transfer Conference.*

Das, K. M., Guha, A., & Ramgopal, M. (2025). Quantitative assessment of various low-GWP zeotropic mixtures of CO_2 as the working fluid in two-phase natural circulation loops. *Applied Thermal Engineering*, *266*, 125512.

Das, K. M., Ramgopal, M., & Katu, Y. A. (2025). Performance of a refrigeration system with a two-phase CO_2-based natural circulation loop for cold storage applications. *Thermal Science and Engineering Progress*, *58*, 103184.

De Laurier, J. D. (1994). An ornithopter wing design. *Journal of Aircraft*, *31*(6), 1318–1324. DOI: 10.2514/3.46704

Delbeke, J., Runge-Metzger, A., Slingenberg, Y., & Werksman, J. (2019). The paris agreement. *Towards a Climate-Neutral Europe: Curbing the Trend*, 24–45. https://doi.org/DOI: 10.4324/9789276082569-2

Devotta, S., Chelani, A., & Vonsild, A. (2021). Prediction of flammability classifications of refrigerants by ANN and random forest. *International Journal of Refrigeration*, *131*, 947 ?"955.

Dhamudia, K., & Senapati, J. R. (2025). A computational study for enhancement in thermal performance of plus(+)-shaped ribbed solar air heater duct. *International Communications in Heat and Mass Transfer, 161*. Advance online publication. DOI: 10.1016/j.icheatmasstransfer.2024.108411

Dikshit, S. V. (2024). A comprehensive review on dehumidification system using solid desiccants. *Proceedings of the Institution of Mechanical Engineers, Part E: Journal of Process Mechanical Engineering*, August.

Ding, Y., Wei, M., & Liu, R. (2021). Channel parameters for the temperature distribution of a battery thermal management system with liquid cooling. *Applied Thermal Engineering, 186*(September 2020), 116494. DOI: 10.1016/j.applthermaleng.2020.116494

Ding, Y., Ji, H., Wei, M., & Liu, R. (2022). Effect of liquid cooling system structure on lithium-ion battery pack temperature fields. *International Journal of Heat and Mass Transfer, 183*, 122178. DOI: 10.1016/j.ijheatmasstransfer.2021.122178

Dong, X., Gao, G., Zhao, X., Qiu, Z., Li, C., Zhang, J., & Zheng, P. (2022). Investigation on heat transfer and phase transition in phase change material (PCM) balls and cold energy storage tank. *Journal of Energy Storage, 50*, 104695.

Dong, Z., Yang, Y. G., Chen, Q., & Bin, S. (2013). A study of hydraulic characteristics of multi-square-hole orifice plates. *Applied Mechanics and Materials, 256*, 2470–2473.

Dopazo, J. A., & Seara, J. F. (2011). Experimental evaluation of a cascade refrigeration system prototype with CO_2 and NH_3 for freezing process applications. *International Journal of Refrigeration, 34*, 257–267.

Dugdale, D. S. (1997). Viscous flow through a sharp-edged orifice. *International Journal of Engineering Science, 35*(8), 725–729.

Dutta, S., Panigrahi, P., & Muralidhar, K. (2008). Experimental investigation of flow past a square cylinder at an angle of incidence. *Journal of Engineering Mechanics, 134*, 788–803.

Dwivedi, G. C. S. (2020). Hydrotreated vegetable oil and synthetic diesel for CI engines. *Energy Reports, 6*, 185–196. DOI: 10.1016/j.egyr.2020.02.032

Dwivedi, G., & Sharma, M. P. (2014). Potential and limitation of straight vegetable oils as engine fuel – An Indian perspective. *Renewable & Sustainable Energy Reviews, 38*, 1188–1208. DOI: 10.1016/j.rser.2014.05.015

Efficiency, E. (n.d.). *Sustainable Cooling Towards a Cool COP28 " Cool " COP28 : Objectives.*

El-Araby, R., Amin, A., El Morsi, A. K., El-Ibiari, N. N., & El-Diwani, G. I. (2018). Study on the characteristics of palm oil–biodiesel–diesel fuel blend. *Egyptian Journal of Petroleum, 27*(4), 519–528. DOI: 10.1016/j.ejpe.2017.07.001

Elghnam, R. I. (2014). Experimental and numerical investigation of heat transfer from a heated horizontal cylinder rotating in still air around its axis. *Ain Shams Engineering Journal, 5*(1), 177–185. DOI: 10.1016/j.asej.2013.09.008

Elshaer, A. M.. (2024). Comparative performance of copper foam and plate fins for PCM-based battery thermal management. *International Journal of Heat and Mass Transfer, 212*, 124789.

Elshaer, A. M., Abdalrahman, T., & Alghassab, M. A. (2023). Effect of fin and composite PCM integration on thermal regulation in heat sinks. *Scientific Reports, 13*, 11572. DOI: 10.1038/s41598-023-37639-3 PMID: 37414796

Emerson, "*Fundamentals of orifice meter measurement,* White paper," https://www.emerson.com/documents/automation/white-paper-fundamentals-of-orifice-meter-measurement-daniel-en-us-188730.pdf (2017).

Engel, A., Rigby, M., Burkholder, J. B., Fernandez, R. P., Froidevaux, L., Hall, B. D., & Yao, B. (2019). Update on ozone-depleting substances (ODSs) and other gases of interest to the Montreal Protocol. In *Scientific Assessment of Ozone Depletion: 2018*. World Meteorological Organization.

Ente Italiano di Normazione (UNI). (2021). *UNI EN 378-1:2021: Refrigerating systems and heat pumps ?"Safety and environmental requirements ?"Part 1: Basic requirements, definitions, classification and selection criteria.* UNI.

Environemental Protection Agency. (2016). The Montreal Protocol on Substances that Deplete the Ozone Layer: The Kigali Amendment. *The Kigali Amendment (2016): The Amendment to the Montreal Protocol Agreed by the Twenty-Eighth Meeting of the Parties (Kigali, 10-15 October 2016), September 1987.*

Evans, J., Curlin, J., & Clark, E. (2019). IIR UN Environment brief on cold storages and refrigerated warehouse. *UN Environment and IIR* (2019).

Eze, V. H. U., Eze, E. C., Alaneme, G. U., & Bubu, P. E. (2025). Recent progress and emerging technologies in geothermal energy utilization for sustainable building heating and cooling: A focus on smart system integration and enhanced efficiency solutions. *Frontiers in Built Environment, 11*. Advance online publication. DOI: 10.3389/fbuil.2025.1594355

Fan, R., Zheng, N., & Sun, Z. (2021). Evaluation of fin intensified phase change material systems for thermal management of Li-ion battery modules. *International Journal of Heat and Mass Transfer, 166*, 120753. DOI: 10.1016/j.ijheatmasstransfer.2020.120753

Fedorová, K., & Kvasnica, M. (2022, September). Predictive Thermal Management of an Industrial Battery Energy Storage System. In 2022 Cybernetics & Informatics (K&I) (pp. 1-6). IEEE.

Feng, B., Yang, Z., & Zhai, R. (2018). Experimental study on the influence of flame retardants on the flammability of R-1234yf. *Energy, 143*, 212 ?"218.

Fong, K. F., & Lee, C. K. (2020). Solar desiccant cooling system for hot and humid region ?" A new perspective and investigation. *Solar Energy, 195*, 677 ?"684.

Fossa, M., & Guglielmini, G. (2002). Pressure Drop and Void Fraction Profiles during Horizontal Flow through Thin and Thick Orifices. *Experimental Thermal and Fluid Science, 26*, 513–523.

Fossa, M., Guglielmini, G., & Marchitto, A. (2006). Two-Phase Flow Structure Close to Orifice Contractions during Horizontal Intermittent Flows. *International Communications in Heat and Mass Transfer, 33*, 698–708.

Fryer, J., & Lee, K. T. Q. (2012). Methods of calculating Total Equivalent Warming Impact (TEWI). *The Australian Institute of Refrigeration Air Conditioning and Heating*, 21.

Fu, J., Xu, X., & Li, R. (2019). Battery module thermal management based on liquid cold plate with heat transfer enhanced fin. *International Journal of Energy Research, 43*(9), 4312–4321. DOI: 10.1002/er.4556

Fukuda, S., Kondou, C., Takata, N., & Koyama, S. (2014). Low-GWP refrigerants R-1234ze(E) and R-1234ze(Z) for high-temperature heat pumps. *International Journal of Refrigeration, 40*, 161 ?"173.

Gaikwad, A., Yoon, S., & Park, S. (2024). Combustion enhancement and emission reduction in RCCI engine using green-synthesized CuO nanoparticles with Cymbopogon martinii methyl ester and phytol blends. *Industrial Crops and Products, 198*. Advance online publication. DOI: 10.1016/j.indcrop.2023.116627

Gan, G., & Riffat Saffa, B. (1997). Pressure loss characteristics of orifice and perforated plates. *Experimental Thermal and Fluid Science, 14*, 160–165.

Gant, S. E., Pursell, M. R., Lea, C. J., Fletcher, J., Rattigan, W., Thyer, A. M., & Connolly, S. (2011). Flammability of hydrocarbon and carbon dioxide mixtures. *Process Safety and Environmental Protection*, *89*(6), 472–481. DOI: 10.1016/j.psep.2011.06.017

Gao, J., & Wu, F. (2019). Investigation of flow through the two-stage orifice. *Engineering Applications of Computational Fluid Mechanics*, *13*(1), 117–127.

Garg, A., Shukla, P. R., Kankal, B., & Mahapatra, D. (2017). CO_2 emission in India: Trends and management at sectoral, sub-regional and plant levels. *Carbon Management*, *8*(2), 111 ?"123.

Gassan, H., Abdul-Majeed, A., & Maha, R. A. (1991). Correlations developed to predict two phase flow through wellhead chokes. *Journal of Canadian Petroleum Technology*, *30*, 47–55.

Gawande, V. B., Dhoble, A. S., Zodpe, D. B., & Chamoli, S. (2016). Experimental and CFD investigation of convection heat transfer in solar air heater with reverse L-shaped ribs. *Solar Energy*, *131*, 275–295. DOI: 10.1016/j.solener.2016.02.040

Gebre, S. L., & Ludwig, F. (2015). Hydrological response to climate change of the upper Blue Nile River Basin based on IPCC AR5. *Journal of Climatology & Weather Forecasting*, *3*(1), 1 ?"15.

Gedam, P. H., & Sampathkumaran, P. S. (1986). Cashew nut shell liquid: Extraction, chemistry and applications. *Progress in Organic Coatings*, *18*(3–4), 179–205. DOI: 10.1016/0033-0655(86)80019-6

Geng, P.. (2020). Performance analysis of Fischer–Tropsch diesel blends in diesel engines. *Renewable Energy*, *147*, 1721–1730. DOI: 10.1016/j.renene.2019.10.066

Geo, E., Mrad, N., Tazerout, M., & Aloui, F. (2012). Experimental analysis of biofuel as an alternative fuel for diesel engines. *Applied Energy*, *96*, 282–289. DOI: 10.1016/j.apenergy.2012.02.027

Gerald, L. M., Dwayne, T., Carl, B., & Hall, K. R. (2001). Response of a slotted orifice flow meter to an air/water mixture. *Flow Measurement and Instrumentation*, *12*, 175–180. DOI: 10.1016/S0955-5986(01)00018-8

Getu, H. M., & Bansal, P. K. (2008). Thermodynamic analysis of an R744-R717 cascade refrigeration system. *International Journal of Refrigeration*, *31*, 45–54.

Gil, B., & Kasperski, J. (2018). Efficiency evaluation of the ejector cooling cycle using new HFO/HCFO refrigerant as an R-134a replacement. *Energies*, *11*(8), 2136.

Golijanek-Jędrzejczyk, A., Mrowiec, A., Hanus, R., Zych, M., Heronimczak, M., & Świsulski, D. (2021). The assessment of metrological properties of segmental orifice based on simulations and experiments. *Measurement*. Advance online publication. DOI: 10.1016/j.measurement.2021.109601

Golzari, S., Kasaeian, A., Daviran, S., Mahian, O., Wongwises, S., & Sahin, A. Z. (2017). Second-law analysis of an automotive air-conditioning system using HFO-1234yf. *International Journal of Refrigeration*, *73*, 134 ?"143.

Gorgulu, Y. F. (2023). Evaluating the efficiency of renewable energy technologies through thermodynamic analyzes. *Insight - Mechanics, 6*(1). DOI: 10.18282/m.v6i1.599

Grace, H. P., & Lapple, C. E. (1951). Discharge Coefficient of Small Diameter Orifices and Flow Nozzles. *Trans. ASME, 73*, 639–647.

Granryd, E. (2001). Hydrocarbons as refrigerants ?"An overview. *International Journal of Refrigeration*, *24*(1), 15 ?"24.

Gschrey, B., & Zeiger, B. (2015). F-Gas Regulation: Technical Advice to Member States on implementing Article 7(2). *Öko-Recherche, 7*(517). https://ec.europa.eu/clima/policies/f-gas/docs/151023_hfc23_byproduction_en.pdf

Guelpa, E., & Verda, V. (2019). Thermal energy storage in district heating and cooling systems: A review. *Applied Energy*, *252*, 113474. DOI: 10.1016/j.apenergy.2019.113474

Guillaume, L., Legros, A., Desideri, A., & Lemort, V. (2017). Performance of a radial-inflow turbine integrated in an ORC system for waste-heat recovery on trucks: Experimental comparison between R-245fa and R-1233zd. *Applied Energy*, *186*, 408 ?"422.

Gulraiz, A., Bastaki, A. J. A., Magamal, K., Subhi, M., Hammad, A., Allanjawi, A., Zaidi, S. H., Khalid, H. M., Ismail, A., Hussain, G. A., & Said, Z. (2025). Energy advancements and integration strategies in hydrogen and battery storage for renewable energy systems. *iScience*, *28*(3), 111945. DOI: 10.1016/j.isci.2025.111945 PMID: 40060896

Guo, R., & Li, L. (2022). Heat dissipation analysis and optimization of lithium-ion batteries with a novel parallel-spiral serpentine channel liquid cooling plate. *International Journal of Heat and Mass Transfer*, *189*. Advance online publication. DOI: 10.1016/j.ijheatmasstransfer.2022.122706

Gupta, A. K., Sharma, A., Chhabra, R. P., & Eswaran, V. (2003). Two-dimensional steady flow of a power-law fluid past a square cylinder in a plane channel: Momentum and heat-transfer characteristics. *Industrial & Engineering Chemistry Research*, *42*, 5674–5686.

Hamdan, M., Abujamous, D., Abdelhafez, E., Sukkariyh, M., & Al-Maghalseh, M. M. (2025). Parametric evaluation of sensible heat storage systems using nano-ionic liquids. *Journal of Ecological Engineering*, *26*(3), 237–245. DOI: 10.12911/22998993/199322

Hansen, A. C., Zhang, Q., & Lyne, P. W. L. (2005). Ethanol–diesel fuel blends—A review. *Progress in Energy and Combustion Science*, *31*(5), 419–445. DOI: 10.1016/j.pecs.2005.03.001 PMID: 15474927

Hans, V. S., Saini, R. P., & Saini, J. S. (2010). Heat transfer and friction factor correlations for a solar air heater duct roughened artificially with multiple v-ribs. *Solar Energy*, *84*(6), 898–911. DOI: 10.1016/j.solener.2010.02.004

Han, X. H., Wang, Q., Zhu, Z. W., & Chen, G. M. (2007). Cycle performance study on R-32/R-125/R-161 as an alternative to R-407C. *Applied Thermal Engineering*, *27*(14 ?"15), 2559 ?"2565.

Hasan, D. (2021). Effect of conical angle on the hydraulic properties of plate flows: A numerical approach. *Flow Measurement and Instrumentation*, *81*, 102026.

Hassan, Q., Algburi, S., Sameen, A. Z., Salman, H. M., & Jaszczur, M. (2023). A review of hybrid renewable energy systems: Solar and wind-powered solutions: Challenges, opportunities, and policy implications. *Results in Engineering*, *20*, 101621. DOI: 10.1016/j.rineng.2023.101621

He, D. H., Chen, S. L., & Bai, B. F. (2018). New model for measuring stratified gas-liquid flow by Chisholm-model-based V-cone flowmeter. *CIESC Journal*, *69*, 3428–3435. DOI: 10.11949/j.issn.0438-1157.20180136

He, D. H., Chen, S. L., & Bai, B. F. (2019). Numerical simulation of stratified gas-liquid flow inside V cone throttle device. *Zhongguo Shiyou Daxue Xuebao. Ziran Kexue Ban*, *43*, 151–158. DOI: 10.1615/InterJFluidMechRes.2023047456

Hegetschweiler, M. J., Berger, L., Hesse, R., Beeckmann, J., Bariki, C., Pitsch, H., & Linteris, G. T. (2023). Data reduction considerations for the burning velocity of spherical constant volume flames of R32 (CH2F2) with air. *Combustion and Flame*, *254*, 112807. DOI: 10.1016/j.combustflame.2023.112807

Heidari, A., Roshandel, R., & Vakiloroaya, V. (2018). Co-production of water and cooling in hot and humid climates. *Energy Conversion and Management, 185*, 396 ?"409.

Hoang, A. T., & Pham, V. V. (2025). Iso-pentanol–gasoline blends for HCCI engines. *Fuel Communications, 8*, 100083. DOI: 10.1016/j.fuelcom.2025.100083

Hoang, A. T., & Pham, V. V. (2025). *Optimizing combustion efficiency and emission reduction in alcohol-based fuel blends*. Fuel Communications., DOI: 10.1016/j.fuelcom.2025.100094

Ho, Y. S., & Leung, T. P. (1985). Performance of conical entrance orifice plates at low Reynolds numbers. *International Journal of Heat and Fluid Flow, 6*(2), 122–125.

Hughes, T. J. R., & Brooks, A. N. (1979). A multidimensional upwind scheme with no crosswind diffusion. In Hughes, T. J. R. (Ed.), *Finite Element Methods for Convection Dominated Flows* (pp. 19–35). ASME.

Hughes, T. J. R., Franca, L. P., & Balestra, M. (1986). A new finite element formulation for computational fluid dynamics: circumventing the babusˇka-brezzi condition: a sta- ble petrov-galerkin formulation of the stokes problem accommodating equal-order interpolations. *Computer Methods in Applied Mechanics and Engineering, 59*, 85–99.

Huth, M., & Heilos, A. (2013). Fuel flexibility in gas turbine systems: Impact on burner design and performance. In *Modern Gas Turbine Systems: High Efficiency, Low Emission, Fuel Flexible Power Generation*. Woodhead Publishing Limited., DOI: 10.1533/9780857096067.3.635

Hwang, F. S.. (2024). Review of battery thermal management systems in electric vehicles. *Renewable & Sustainable Energy Reviews, 182*, 113298. DOI: 10.1016/j.rser.2023.113298

Idelchik, I. E., Malyavskaya, G. R., Martynenko, O. G., & Fried, E. (1994). *Handbook of Hydraulic Resistances*. CRC Press.

Inayat, A., & Raza, M. (2019). District cooling system via renewable energy sources: A review. *Renewable & Sustainable Energy Reviews, 107*, 360–373. DOI: 10.1016/j.rser.2019.03.023

Indian Society of Heating, Refrigeration and Air Conditioning Engineers. (2015). *ISHRAE position document on refrigerants for Indian refrigeration & air conditioning industries - Challenges & responsibilities.*

Indian Society of Heating. Refrigeration and Air Conditioning Engineers. International Institute of Ammonia Refrigeration. (2019). *The history of ammonia refrigeration* International Institute of Ammonia Refrigeration.

International Energy Institute (IEI). (2023). Statistical Review of World Energy 2023. *BP Energy Outlook, 2023*(70), 8–20.

International Institute of Ammonia Refrigeration. (2008). *Ammonia Data Book: The Profile of a Sustainable Refrigerant*. International Institute of Ammonia Refrigeration.

Islam, M. A., Mitra, S., Thu, K., & Saha, B. B. (2021). Study on thermodynamic and environmental effects of vapor compression refrigeration system employing first to next-generation popular refrigerants. *International Journal of Refrigeration, 131*(August), 568–580. DOI: 10.1016/j.ijrefrig.2021.08.014

Islam, M.. (2018). Microwave-assisted hydrolysis of chlorophyll-rich biomass for phytol production. *Bioresource Technology, 258*, 203–212. DOI: 10.1016/j.biortech.2018.03.051

Iyer, P. K., Siva, R., Rose, A., Kumar, S., & Venkatraman, K. (2024). Experimental investigations of a desiccant-coated M-cycle cooler for net zero air-conditioning. *Energy Conversion and Management*, ? ? ?, 2024.

Jang, D. S., Yun, S., Hong, S. H., Cho, W., & Kim, Y. (2022). Performance characteristics of a novel heat pipe-assisted liquid cooling system for the thermal management of lithium-ion batteries. *Energy Conversion and Management, 251*(November 2021), 115001. DOI: 10.1016/j.enconman.2021.115001

Jani, D. B., Mishra, Manish, & Sahoo, P. K. (2016). Solid desiccant air conditioning ?" A state of the art review. *Renewable & Sustainable Energy Reviews, 60*, 1451 ?"1469.

Jazirian, H., Jafarkazemi, F., & Rabieefar, H. (2023). A numerical model for simulating separated gas-liquid two-phase flow with low GVF in a V-cone flow meter. *Flow Measurement and Instrumentation, 90*, 102329.

Jiang, H., & Cheng, L. (2020). Flow separation around a square cylinder at low to moderate reynolds numbers. *Physics of Fluids*, •••, 32.

Jiang, R., Guo, Y., & Guo, X. (2025). Ionic liquids as advanced heat transfer fluids in renewable energy systems. *Solar Energy Materials and Solar Cells, 294*, 113934. DOI: 10.1016/j.solmat.2025.113934

Ji, J., Zhang, C., Zhang, X., & Chen, Y. (2024). Recent research advances in wind turbine thermal management technologies. *Renewable & Sustainable Energy Reviews, 208*, 114983. DOI: 10.1016/j.rser.2024.114983

Jin, C.. (2019). Low-temperature properties of biodiesel–pentanol blends. *Energy Procedia, 159*, 224–229. DOI: 10.1016/j.egypro.2018.12.021

Jing, D., Gang, X., Rafiee, M., & Shaji, K. (2025). A review of refrigerant-based vortex tube separation characteristics. *Applied Thermal Engineering, 173*(February).

Jing, D., Gang, X., Rafiee, M., & Shaji, K. (2025). A review of refrigerant-based vortex tube separation characteristics. *Energy, ? ? ?*, 314.

Johari, A., Singh, S., & Vidya, S. (2022). Engine performance analysis for diesel engine using hydrogen as an alternative fuel. *Materials Today: Proceedings, 50*, 4478–4484. DOI: 10.1016/j.matpr.2021.12.462

Kadam, A. R., & Chaudhuri, M. B. (2018). Acoustic study and behaviour of flow through orifices. *Procedia Manufacturing, 20*, 154–159.

Kangaji, L. M., Raji, A., & Orumwense, E. (2025). Cutting-edge progress in offshore wind and tidal stream power technology—State-of-the-Art. *AIMS Energy, 13*(1), 188–230. DOI: 10.3934/energy.2025007

Karimi Kisomi, M.. (2025). Comparative analysis of PCM, air, and hybrid cooling systems in lithium-ion batteries. *Journal of Energy Storage, 95*, 112509.

Kariuki, B. W., Hassan, H., Ahmed, M., & Emam, M. (2024). A review on geothermal-solar hybrid systems for power production and multigeneration systems. *Applied Thermal Engineering, 259*, 124796. DOI: 10.1016/j.applthermaleng.2024.124796

Karwa, R., Solanki, S. C., & Saini, J. S. (2001). Thermo-hydraulic performance of solar air heaters having integral chamfered rib roughness on absorber plates. *Energy, 26*(2), 161–176. DOI: 10.1016/S0360-5442(00)00062-1

Kaya, A., Aydin, O., & Dincer, I. (2007). Numerical modeling of forced-convection drying of cylindrical moist objects. *Numerical Heat Transfer Part A, 51*(9), 843–854. DOI: 10.1080/10407780601112753

Kayser, J. C., & Shambaugh, R. L. (1991). Discharge coefficients for compressible flow through small-diameter orifices and convergent nozzles. *Chemical Engineering Science, 46*(7), 1697–1711.

Kays, W. M., & Bjorklund, I. S. (1958). Heat Transfer From a Rotating Cylinder With and Without Crossflow. *Transactions of the American Society of Mechanical Engineers, 80*(1), 70–77.

Kelkar, K. M., & Patankar, S. V. (1992). Numerical prediction of vortex shedding behind a square cylinder. *International Journal for Numerical Methods in Fluids*, *14*, 327–341.

Kennoy, D.H. (2018). ANSI/ASHRAE Addendum g to ANSI/ASHRAE Standard 34-2016. *Ashrae, 8400*, 1–6. www.ashrae.org

Khalid, A., Mehmood, M. A., Rehman, S., Irfan, M., Farooq, U., Farhan, M., & Riaz, M. (2024). Thermo-economic analysis and comparison of novel multi-stage desiccant wheels. *Energy Reports*, ? ? ? , 10.

Khan, M. M. K., Chowdhury, A. A., & Hassan, N. M. S. (2017). Application of thermo-fluid processes in energy systems. In *Green energy and technology*. DOI: 10.1007/978-981-10-0697-5

Khanafer, K., & Aithal, S. M. (2017). Mixed convection heat transfer in a lid-driven cavity with a rotating circular cylinder. *International Communications in Heat and Mass Transfer*, *86*, 131–142. DOI: 10.1016/j.icheatmasstransfer.2017.05.025

Khan, R. A.. (2014). Industrial and biological relevance of phytol and its derivatives. *Industrial Crops and Products*, *53*, 29–37. DOI: 10.1016/j.indcrop.2013.12.024

Khateeb, S. A., Farid, M. M., Selman, J. R., & Al-Hallaj, S. (2004). Design and simulation of a lithium-ion battery with a phase change material thermal management system for an electric scooter. *Journal of Power Sources*, *128*(2), 292–307. DOI: 10.1016/j.jpowsour.2003.09.070

Khodabandehlou, R., Norouzi, H. R., Larimi, A., & Mostoufi, N. (2025). Analysis of biomass gasification in a fluidized bed using CFD-DEM. *Energy*, *320*, 135395. DOI: 10.1016/j.energy.2025.135395

Khuntia, A. K., Misra, S., Alok, P., & Panda, S. K. (2023). Determination of pressure drop through a circular orifice for two-phase flow with Machine learning technique. *Flow Measurement and Instrumentation*, *94*, 102470.

Khuntia, A. K., Panda, S. K., & Misra, S. (2024). Two-phase flow measurement through orifice-meter with regression analysis. *International Journal of Fluid Mechanics Research*, *52*(2), 41–54.

Kim, J., Oh, J., & Lee, H. (2019). Review on battery thermal management system for electric vehicles. *Applied Thermal Engineering*, *149*(September 2018), 192–212. DOI: 10.1016/j.applthermaleng.2018.12.020

Kleoff, M., Klaucke, F., Voßnacker, P., Weber, R., Neuhoff, K., & Riedel, S. (2025). Utilisation of reactive ionic liquids for energy storage and regulation of the power grid. *Green Chemistry*. Advance online publication. DOI: 10.1039/d5gc02519b

Kojasoy, G., Kwame, M. P., & Chang, C. T. (1997). Two-Phase Pressure drop in Multiple Thick and Thin Orifices Plates. *Experimental Thermal and Fluid Science*, *15*, 347–358.

Kong, W., Zhu, K., Lu, X., Jin, J., & Ni, M. (2021). Enhancement of lithium-ion battery thermal management with the divergent-shaped channel cold plate. *Journal of Energy Storage*, *42*(February), 103027. DOI: 10.1016/j.est.2021.103027

Ku, J. (2016). "Loop heat pipe startup behaviors." In *International Conference on Environmental Systems*, no. ICES-2016-24.

Ku, J. (1999). Operating characteristics of loop heat pipes. *SAE Transactions*, •••, 503–519.

Kumar, K. R., Dashora, K., Kumar, S., Dharmaraja, S., Sanyal, S., Aditya, K., & Kumar, R. (2023). A review of drying technology in tea sector of industrial, non-conventional and renewable energy based drying systems. *Applied Thermal Engineering, 224*(December 2022), 120118. https://doi.org/DOI: 10.1016/j.applthermaleng.2023.120118

Kumar, A., Vivekanand, & Subudhi, S. (2017). Cooling and dehumidification using vortex tube. *Applied Thermal Engineering*, *122*, 181 ?"193.

Kumar, A., & Alam, T. (2025). A review on geothermal energy systems and various approaches to enhance the system's performance. *Energy and Building*, *344*, 115962. DOI: 10.1016/j.enbuild.2025.115962

Kumar, B., & Mittal, S. (2006). Prediction of the critical Reynolds number for flow past a circular cylinder. *Computer Methods in Applied Mechanics and Engineering*, *195*(44-47), 6046–6058.

Kumar, D., & Kumar, B. (2024). Tertiary wake mode in flows past elliptic cylinders. *European Journal of Mechanics - B/Fluids*, *106*, 130–136.

Kumari, T., & Pandey, M. (2022). "Study of Physical Characteristics of a Bi-porous Composite Capillary Wick for a Flat Miniature Loop Heat Pipe." In *Conference on Fluid Mechanics and Fluid Power*, pp. 211-221. Singapore: Springer Nature Singapore.

Kumari, V., Chandra, H., Aditya, K., Kumar, K. R., Dashora, K., Krishnan, N., Sanyal, S., Dharmaraja, S., & Kumar, S. (2021). Feasibility study on renewable energy system in tea (Camellia sinensis) estates of North-East India. *Indian Journal of Agricultural Sciences, 91*(11), 1631–1635. DOI: 10.56093/ijas.v91i11.118574

Kumar, K. R., Dashora, K., Krishnan, N., Sanyal, S., Chandra, H., Dharmaraja, S., & Kumari, V. (2021). Feasibility assessment of renewable energy resources for tea plantation and industry in India - A review. *Renewable & Sustainable Energy Reviews, 145*. Advance online publication. DOI: 10.1016/j.rser.2021.111083

Kumar, K., & Ramgopal, M. (2009). Carbon dioxide as a secondary Fluid in natural circulation loops. *Proceedings of the Institution of Mechanical Engineers. Part E, Journal of Process Mechanical Engineering, 223*(3), 189–194.

Kumar, L., Manjunath, B. S., Patel, R. J., Markandeya, S. G., Agrawal, R. G., Agrawal, A., & Prabhu, S. V. (2012). Experimental investigations on melting of lead in a cuboid with constant heat flux boundary condition using thermal neutron radiography. *International Journal of Thermal Sciences, 61*, 15–27.

Kumar, L., Soomro, J., Khoharo, H., & Assad, M. E. H. (2023). A comprehensive review of solar thermal desalination technologies for freshwater production. *AIMS Energy, 11*(2), 293–318. DOI: 10.3934/energy.2023016

Kumar, R., Kumar, A., & Goel, V. (2019). Performance improvement and development of correlation for friction factor and heat transfer using computational fluid dynamics for ribbed triangular duct solar air heater. *Renewable Energy, 131*(February), 788–799. DOI: 10.1016/j.renene.2018.07.078

Kumar, S., Kumar, K. R., & Dashora, K. (2024). Impact study of operating parameters on drying evolution of spherical tea particle under convective influence. *Journal of Food Process Engineering, 47*(8), 1–11. DOI: 10.1111/jfpe.14702

Laguri, V., Kumar, K., & Kumar, P. (2021). Performance analysis of CO_2/natural refrigerants for cascade refrigeration system. In *Sixth International Conference on polygeneration, IISc*.

Le Chatelier, M. H. (2004). Note on mine gas measurement by flammability limits. *Process Safety Progress, 23*(3), 172–174. DOI: 10.1002/prs.10031

Lee, S. J., Shon, B. H., Jung, C. W., & Kang, Y. T. (2018). A novel solar-assisted heat pump using low-GWP R-1233zd(E) with a flexible solar collector. *Energy, 149*, 386 ?"396.

Lee, T. S., Liu, C. H., & Chen, T. W. (2006). Thermodynamic analysis of optimal condensing temperature of cascade condenser in CO_2/NH_3 cascade refrigeration systems. *International Journal of Refrigeration*, *29*, 1100–1108.

Lemmon, E. W., Bell, I. H., Huber, M. L., & McLinden, M. O. (2018). *NIST reference fluid thermodynamic and transport properties database (REFPROP), Version 10.0*. National Institute of Standards and Technology.

Lens, (2025). *Search results for "Renewable Energy Applications In Thermo-Fluid Systems"*. The Lens. Retrieved June 11, 2025, from https://www.lens.org/lens/search/scholar/list?q=Renewable%20Energy%20Applications%20In%20Thermo-Fluid%20Systems&n=10

Liang, F., Zhang, L., Wang, S., & Zhou, H. (2025). Experimental investigation on the influence of drainage structure on vortex tube energy separation. *Applied Thermal Engineering*, ? ? ?, 2025.

Liang, K., Yao, H., Qiao, J., Gao, S., Zong, M., Liu, F., Yang, Q., Liang, L., & Fang, D. (2024). Thermodynamic evaluation of novel 1,2,4-Triazolium alanine ionic liquids as sustainable Heat-Transfer Media. *Molecules (Basel, Switzerland)*, *29*(22), 5227. DOI: 10.3390/molecules29225227 PMID: 39598616

Liang, X.. (2021). Effect of n-pentanol addition on atomization and combustion in biodiesel engines. *Applied Thermal Engineering*, *192*. Advance online publication. DOI: 10.1016/j.applthermaleng.2021.116930

Liger, M., Pornsin-Sirirak, N., Tai, Y.-C., Ho, S., & Ho, C.-M. (2002). "Large-area electrostatic-valved skins for adaptive flow control on ornithopter wings." *AIAA Conference Proceedings*.

Li, H.. (2025). Prismatic battery thermal management using pyrolytic graphite honeycomb fins and liquid plates. *International Journal of Heat and Mass Transfer*, *223*, 125005.

Li, H.. (2025). Prismatic module with pyrolytic graphite honeycomb fins and liquid plates. *International Journal of Heat and Mass Transfer*, *223*, 125005. DOI: 10.1016/j.ijheatmasstransfer.2025.125005

Lin, Z. H. (1982). Two Phase Flow Measurement with Sharp Edge Orifices. *International Journal of Multiphase Flow*, *8*, 683–693.

Liu, F., Zhu, H., Li, C., Wang, Y., Zhang, J., Tang, X., Zhang, T., & Liu, Y. (2025). A combined drying process involving hot air and roasting for improving summer congou black tea quality. *Food Research International, 201*(August 2024), 115584. https://doi.org/DOI: 10.1016/j.foodres.2024.115584

Liu, F., Gan, S., Chen, Y., Qin, W., & Li, J. (2023). Thermal characteristics of Li-ion battery based on phase change material–aluminum plate-fin composite heat dissipation. *Energy Science & Engineering*, *11*(1), 62–78. DOI: 10.1002/ese3.1357

Liu, H.. (2025). A review of lithium-ion battery thermal management strategies. *Energies*, *18*(4), 1654.

Liu, H., Jin, C., Li, H., & Ji, Y. (2023). A numerical study of PCM battery thermal management performance enhancement with fin structures. *Energy Reports*, *9*, 1793–1802. DOI: 10.1016/j.egyr.2023.04.214

Liu, W., Sun, L., Li, Z., Fujii, M., & Dong, L. (2022). Trends and future challenges in hydrogen production and storage research. *Environmental Science and Pollution Research International*, *29*, 33520–33541. DOI: 10.1007/s11356-022-19691-1 PMID: 32529621

Liu, X., Zhang, C. F., Zhou, J. G., Xiong, X., & Wang, Y. P. (2022). Thermal performance of battery thermal management system using fins to enhance the combination of thermoelectric cooler and phase change material. *Applied Energy*, *322*, 119503. DOI: 10.1016/j.apenergy.2022.119503

Liu, X., Zhang, C., Zhou, J., Xiong, X., & Wang, Y. (2022). Thermal performance of hybrid TEC–PCM battery thermal management systems with fin enhancement. *Applied Energy*, *322*, 119503.

Liu, X., Zhang, C., Zhou, J., Xiong, X., & Wang, Y. (2022). Thermal performance of spiral-fin PCM structures for cylindrical lithium-ion batteries. *Energies*, *15*(23), 8847. DOI: 10.3390/en15238847

Li, W., Jishnu, A. K., Garg, A., Xiao, M., Peng, X., & Gao, L. (2020). Heat transfer efficiency enhancement of lithium-ion battery packs by using novel design of herringbone fins. *Journal of Heat Transfer*, *142*(2), 021401. DOI: 10.1115/1.4046160

Li, X., Wang, K., Xu, M., Fu, M., & Miao, S. (2024). Environmental and economic dispatching strategy for power system with the complementary combination of wind-solar-hydro-thermal-storage multiple sources. *Frontiers in Energy Research*, *12*. Advance online publication. DOI: 10.3389/fenrg.2024.1338657

Li, Y., Bai, M., Zhou, Z., Wu, W.-T., & Wei, L.. (2024). Immersion cooling of prismatic lithium-ion batteries using fluorinated liquid. *Applied Thermal Engineering*, *257*, 124453.

Li, Y., Bai, M., Zhou, Z., Wu, W.-T., Wei, L., Hu, C., Liu, X., Gao, S., Li, Y., & Song, Y. (2024). Thermal management for the prismatic lithium-ion battery pack by immersion cooling with fluorinated liquid. *Applied Thermal Engineering*, *257*, 124453. DOI: 10.1016/j.applthermaleng.2024.124453

Li, Y., Wang, X. J., & Geng, Y. F. (2009). Study on wet gas online flow rate measurement based on dual slotted orifice plate. *Flow Measurement and Instrumentation*, *20*, 168–173. DOI: 10.1016/j.flowmeasinst.2009.04.002

Llanos, E. M., & Blessent, D. (2025). A scoping review of numerical modelling studies of geothermal reservoirs: Trends and opportunities post-COP25. *International Journal of Renewable Energy Development*, *14*(4), 668–693. DOI: 10.61435/ijred.2025.61071

Longo, G. A., Zilio, C., Righetti, G., & Brown, J. S. (2014). Experimental assessment of the low-GWP refrigerant HFO-1234ze(Z) for high-temperature heat pumps. *Experimental Thermal and Fluid Science*, *57*, 293 ?"300.

Lu, M., Jiang, Y., Zhao, W., Zhang, J., Chen, Z., & Ning, J. (2025). Effects of different drying methods on color difference, taste and chemical components of Yunnan Congou black tea. *Lebensmittel-Wissenschaft + Technologie*, *223*(March), 117696. DOI: 10.1016/j.lwt.2025.117696

Luo, D., Mahmoud, A., & Cogswell, F. (2015). Evaluation of low-GWP fluids for power generation with Organic Rankine Cycle. *Energy*, *85*, 481 ?"488.

Luo, J., Ye, Z., Zhao, Z., Yang, K., Zhang, S., & Wang, Q. (2023). Vapor ?"liquid equilibrium and heating performance of CO_2/R-1234ze(Z) and CO_2/R-1336mzz(E) zeotropic blends. *Applied Thermal Engineering*, *229*, 120576.

Luo, T., Shen, B., Mei, Z., Hove, A., & Ju, K. (2024). Unlocking the potential of biogas systems for energy production and climate solutions in rural communities. *Nature Communications*, *15*(1). Advance online publication. DOI: 10.1038/s41467-024-50091-9 PMID: 39003261

Lyu, Z., Su, J., Li, Z., Yan, H., & Chen, L. (2024). A compact hybrid battery thermal management system using nanofluid cooling, PCM, and pulsed flow strategies. *Energy Conversion and Management*, *301*, 118213. DOI: 10.1016/j.enconman.2024.118213

Lyu, Z., Su, J., Li, Z., Yan, H., & Chen, L. (2024). Compact hybrid BTMS with nanofluid, PCM, and pulsed flow strategies. *Energy Conversion and Management*, *301*, 118213.

Ma, H., Ding, Z., Cao, Y., Lv, X., Lu, W., Shen, X., & Yin, L. (2015). Characteristics of the heat transfer from a horizontal rotating cylinder surface. *Experimental Thermal and Fluid Science*, *66*, 235–242. DOI: 10.1016/j.expthermflusci.2015.03.031

Ma, H., Xia, W., Zhou, W., Liu, W., Guo, Z., & Zhang, C. (2014). Experimental investigation on the steady, external laminar mixed convection heat transfer characteristics around a large diameter horizontal rotating cylinder. *International Communications in Heat and Mass Transfer*, *57*, 239–246. DOI: 10.1016/j.icheatmasstransfer.2014.08.006

Mahanand, Y., & Senapati, J. R. (2020). Thermal enhancement study of a transverse inverted-T shaped ribbed solar air heater. *International Communications in Heat and Mass Transfer*, *119*(October), 104922. DOI: 10.1016/j.icheatmasstransfer.2020.104922

Makareviciene, V.. (2013). Biodiesel from algae: Promising sustainable alternative fuel. *Environmental Science and Biotechnology Reviews*, *12*, 55–78.

Makareviciene, V., Skorupskaite, V., & Andruleviciute, V. (2023). Biodiesel fuel from microalgae—Promising alternative for the future. *Reviews in Environmental Science and Biotechnology*, *22*, 215–232. DOI: 10.1007/s11157-022-09634-3

Mane, N. S.. (2024). Thermal control of electronics using composite PCM with long fins. *Energy Storage Technology*, *12*(6), 105014.

Mane, N. S., Kodancha, P., Hemadri, V., & Tripathi, S. (2024). Investigation on cooling performance of composite PCM and graphite fin for battery thermal management system of electric vehicles. *Energy Storage*, *6*(6), e70024. DOI: 10.1002/est2.70024

Ma, Z., Zhang, N., Song, S., Zheng, X., Gong, X., & Wang, Y. (2023). Experimental investigation and effectiveness analysis of a desiccant wheel dehumidification system with low air humidity. *International Journal of Refrigeration*, *? ? ?*, 156.

Ma, Z., Zhao, H., Yan, Z., & Song, J. (2025). Advances and Development Trends in Solar Photovoltaic-Thermal Technology: A review. *Clean Energy*, *9*(4), 98–122. DOI: 10.1093/ce/zkaf020

McAdams, W. H. (1954). *Heat Transmission*. McGraw-Hill.

McLinden, M. O., Seeton, C. J., & Pearson, A. (2020). New refrigerants and system configurations for vapor-compression refrigeration. *Science*, *370*(6518), 791–796. DOI: 10.1126/science.abe3692 PMID: 33184206

Mehdi, A., Najaf, B., & Rinaldi, A. F. (2014). Exergetic, economic and environmental (3E) analyses, and multi-objective optimization of a CO_2/NH_3 cascade refrigeration system. *Applied Thermal Engineering*, *65*, 42–50.

Mittal, S. (2000). On the performance of high aspect ratio elements for incompressible flows. *Computer Methods in Applied Mechanics and Engineering*, *188*, 269–287.

Mittal, S., & Kumar, B. (2003). Flow past a rotating cylinder. *Journal of Fluid Mechanics*, *476*, 303–334.

Mohammed, A. N., Juhany, K. A., Kumar, S., Kishore, V. R., & Mohammad, A. (2017). Effects of CO2/N2 dilution on laminar burning velocity of stoichiometric DME-air mixture at elevated temperatures. *Journal of Hazardous Materials*, *333*, 215–221. DOI: 10.1016/j.jhazmat.2017.03.042 PMID: 28359037

MolA(c)s, F., Navarro-EsbrA-, J., Peris, B., Mota-Babiloni, A., BarragA n-Cervera, A?., & Kontomaris, K. K. (2014). Low-GWP alternatives to HFC-245fa in Organic Rankine Cycles: HCFO-1233zd-E and HFO-1336mzz-Z. *Applied Thermal Engineering*, *71*(1), 204 ?"212.

MolA(c)s, F., Navarro-EsbrA-, J., Peris, B., Mota-Babiloni, A., BarragA n-Cervera, A?., & Kontomaris, K. K. (2017). Thermo-economic evaluation of low-GWP alternatives to HFC-245fa in Organic Rankine Cycles. *Energy Procedia*, *142*, 1199 ?"1205.

Molina, M. J., & Rowland, F. S. (1974). Stratospheric sink for chlorofluoromethanes: Chlorine atom-catalysed destruction of ozone. *Nature*, *249*(5460), 810 ?"812.

Mondal, S., Ramgopal, M., & Mukhopadhyay, S. (2024). Mathematical analysis of personalized air conditioning and ventilation system. In Proceedings of the 27th National and 5th International ISHMT-ASTFE Heat and Mass Transfer Conference December 14-17, 2023, IIT Patna, Patna-801106, Bihar, India. Begel House Inc.

Mondal, R., & Alam, M. M. (2023). Blockage effect on wakes of various bluff bodies: A review of confined flow. *Ocean Engineering*, *286*, 115592.

Montreal protocol on substances that deplete the Ozone layer final act 1987. (1989). *Journal of Environmental Law, 1*(1), 128–136. https://doi.org/DOI: 10.1093/jel/1.1.128

Montzka, S. A., Butler, J. H., Elkins, J. W., Thompson, T. M., Clarke, A. D., & Lock, L. T. (1999). Present and future trends in the atmospheric burden of ozone-depleting halogens. *Nature*, *398*(6729), 690 ?"694.

Morrison, G. L., Hall, K. R., Holste, J. C., Macek, M. L., Ihfe, L. M., & De Otte, R. E.Jr. (1994). Comparison of orifice and slotted plate flow meters. *Flow Measurement and Instrumentation*, *5*, 71–77. DOI: 10.1016/0955-5986(94)90039-6

Morris, S. D. (1996). Liquid flow through safety valves: Diameter ratio effects on discharge coefficients, sizing and stability. *Journal of Loss Prevention in the Process Industries*, *9*(3), 217–224.

Mota-Babiloni, A., Navarro-EsbrA-, J., BarragA n-Cervera, A?., MolA(c)s, F., & Peris, B. (2015). R-450A (R-1234ze(E)/R-134a) as an R-134a replacement: Experimental study. *International Journal of Refrigeration*, *51*, 52 ?"58.

Mueller, T. J., & De Laurier, J. D. (2001). "An overview of micro air vehicle aerodynamics." In *AIAA Fixed and Flapping Wing Aerodynamics*, pp. 15–52.

Mund, C., Rathore, S. K., & Sahoo, R. K. (2021). A review of solar air collectors about various modifications for performance enhancement. *Solar Energy*, *228*(August), 140–167. DOI: 10.1016/j.solener.2021.08.040

Murugan, M.. (2025). Recent developments in battery thermal management: Challenges and opportunities. *Energy Science & Engineering*, *13*(1), 105–132.

Nabwey, H. A., Ashraf, M., Nadeem, H., Rashad, A. M., & Chamkha, A. J. (2024). Optimizing renewable energy systems: A comprehensive review of entropy generation minimization. *AIP Advances*, *14*(12). Advance online publication. DOI: 10.1063/5.0245560

Narkhede, S.. (2023). Design and thermal analysis of fin–PCM integrated battery thermal management systems. *International Journal of Applied Mechanics and Engineering*, *28*(4), 755–770.

Nashine, C., Pandey, M., & Baraya, K. K. (2025). Experimental studies on the transient characteristics and start-up behaviour of a miniature loop heat pipe. *Applied Thermal Engineering*, *259*, 124814.

Nasiri, M.. (2025). Phase change material systems for lithium-ion battery thermal management: Advances and perspectives. *Journal of Energy Storage*, *91*, 112345.

National Renewable Energy Laboratory. (2024). *Geothermal – Electricity ATB*. Retrieved August 23, 2025, from https://atb.nrel.gov/electricity/2024/geothermal

Nations, U., & Programme, E. (2016). *Sources, fates, toxicity, and risks of trifluoroacetic acid and its salts : relevance to substances regulated under the Montreal and Kyoto protocols A report prepared by the UNEP Environmental Effects*. 1–26.

Navarro-EsbrA-, J., MolA(c)s, F., Peris, B., Mota-Babiloni, A., & Kontomaris, K. (2017). Experimental study of an Organic Rankine Cycle with HFO-1336mzz-Z for micro-scale low-temperature applications. *Energy, 133*, 79 ?"89.

Nayak, S., Hassan, M. A., & Paswan, M. (2024). Investigation on thermo-fluid behavior of flat-plate solar water heater employing titanium dioxide-water nanofluid: An experimental and computational approach. *Journal of Mechanical Science and Technology, 38*(3), 1507–1516. DOI: 10.1007/s12206-024-0242-6

Nazir, M. H., Zaidi, S. Z. J., Hussain, M. M., & Khan, Z. A. (2025). Modelling & Development of Novel Sustainable Thermo-fluids to Maximise the Efficiency of Solar Thermal Applications. *Journal of Engineering Research*. Advance online publication. DOI: 10.1016/j.jer.2025.01.009

Niu, J. (2021). CFD analysis of the influence of ionic liquids on the performances of a refrigeration system. *Fluid Dynamics & Materials Processing, 17*(5), 1003–1013. DOI: 10.32604/fdmp.2021.015225

Nkele, A., Ikhioya, I., Chime, C., & Ezema, F. (2023). Improving the performance of solar thermal energy storage systems. *Journal of Energy and Power Technology, 05*(03), 1–25. DOI: 10.21926/jept.2303024

No, S. Y. (2016). Application of alcohols as alternative fuels for internal combustion engines. *Renewable & Sustainable Energy Reviews, 54*, 1331–1345. DOI: 10.1016/j.rser.2015.10.044

Novozhilov, V. (2001). Computational fluid dynamics modeling of compartment fires. *Progress in Energy and Combustion Science, 27*(6), 611–666. DOI: 10.1016/s0360-1285(01)00005-3

Nsude, C. C., Connelly, S., Demir, F., Shi, D., & Tsetsura, K. (2025). Drivers, barriers and policy options to renewable energy development in Native American communities: A systematic review. *Energy Research & Social Science, 125*, 104092. DOI: 10.1016/j.erss.2025.104092

Okajima, A. (1982). Strouhal numbers of rectangular cylinders. *Journal of Fluid Mechanics, 123*, 379–398.

Olabi, A., Elsaid, K., Obaideen, K., Abdelkareem, M. A., Rezk, H., Wilberforce, T., Maghrabie, H. M., & Sayed, E. T. (2023). Renewable energy systems: Comparisons, challenges and barriers, sustainability indicators, and the contribution to UN sustainable development goals. *International Journal of Thermofluids, 20*, 100498. DOI: 10.1016/j.ijft.2023.100498

Omidi, J. (2024). Advances and opportunities in wind energy harvesting using plasma actuators: A review. *Clean Energy*, *8*(1), 197–225. DOI: 10.1093/ce/zkad085

Outar, H., Ennadafy, H., Jammoukh, M., & Khaili, M. E. (2025). CFD analysis for aerodynamic performance optimization of a vertical axis helical wind turbine using rigid PVC. *E3S Web of Conferences, 601*, 00101. DOI: 10.1051/e3sconf/202560100101

Oyedepo, S. O., Waheed, M. A., Abam, F. I., Dirisu, J. O., Samuel, O. D., Ajayi, O. O., Somorin, T., Popoola, A. P. I., Kilanko, O., & Babalola, P. O. (2025). A critical review on enhancement and sustainability of energy systems: Perspectives on thermo-economic and thermo-environmental analysis. *Frontiers in Energy Research*, *12*. Advance online publication. DOI: 10.3389/fenrg.2024.1417453

Panchal, S., Dincer, I., Agelin-Chaab, M., Fraser, R., & Fowler, M. (2017). Transient electrochemical heat transfer modeling and experimental validation of a large sized $LiFePO_4$/graphite battery. *International Journal of Heat and Mass Transfer*, *109*, 1239–1251. DOI: 10.1016/j.ijheatmasstransfer.2017.03.005

Panda, S. K. (2023). An orifice flow analysis on the basis of density and viscosity effects of fluids. *WSEAS Transactions on Heat and Mass Transfer*, *18*, 140–146.

Panda, S. K. (2024, February). Sector orifice analysis with a correlation to determine flow coefficient. *Journal of Engineering and Thermal Sciences*, *4*(1), 1–9. DOI: 10.21595/jets.2024.23641

Panda, S. K., Choudhury, B. K., & Rath, K. C. (2022). A Literature Review on Orifice as a Flow Measuring Device. *ECS Transactions*, *107*(1), 815. DOI: 10.1149/10701.0815ecst

Panda, S. K., & Patra, A. (2021). Determination of Coefficient of Contraction of Orifice with Variation of Geometrical Parameter. In Palanisamy, M., Ramalingam, V., & Sivalingam, M. (Eds.), *Theoretical, Computational, and Experimental Solutions to Thermo-Fluid Systems. Lecture Notes in Mechanical Engineering*. Springer., DOI: 10.1007/978-981-33-4165-4_38

Panda, S. K., Rath, K. C., & Choudhury, B. K. (2023). A two phase analysis of orifice based on a non-dimensional number. *Flow Measurement and Instrumentation*, *94*(102470).

Panda, S. K., Rath, K. C., & Choudhury, B. K. (2023). Determining the flow correlation for an orifice with a non-dimensional number. *Flow Measurement and Instrumentation*, *90*, 102338. DOI: 10.1016/j.flowmeasinst.2023.102338

Pan, L., Ma, Y., Li, T., Li, H., Li, B., & Wei, X. (2019). Cycle performance and combustion characteristics of two CO_2-based binary mixtures for the transcritical power cycle. *Energy, 179*, 454 ?"463.

Pan, M.. (2019). Combustion analysis of diesel–alcohol blends. *Energy, 172*, 154–163. DOI: 10.1016/j.energy.2019.01.055

Papagiannakis, R. G., Dinos, N., & Rakopoulos, C. D. (2017). Combustion characteristics of natural gas and alternative fuels in CI engines. *Fuel, 206*, 18–28. DOI: 10.1016/j.fuel.2017.05.094

Parfitt, P., Barthel, M., & Macnaughton, S. (2010). Food waste within food supply chains: quantification and potential for change to 2050. *Philosophical Transactions of Royal Society*, 3065-3081. Arvindraj, K., Chinna, A. R., & Paul, J. (2020). A review: Present scenario of cold chain storage facilities in India. *AIP Conference Proceedings*, •••, 2207–2215.

Parmar, H., & Hindoliya, D. A. (2013). Performance of solid desiccant-based evaporative cooling system under the climatic zones of India. *The International Journal of Low Carbon Technologies*, *8*(1), 52 ?"57.

Patel, V., Panchal, D., Prajapati, A., Mudgal, A., & Davies, P. (2019). An efficient optimization and comparative analysis of cascade refrigeration system using NH_3/CO_2 and C_3H_8/CO_2 refrigerant pairs. *International Journal of Refrigeration, 102*, 62–76.

Pearson, A. (2008). Refrigeration with ammonia. *International Journal of Refrigeration*, *31*(4), 545–551. DOI: 10.1016/j.ijrefrig.2007.11.011

Perumal, K., Michael, W., & Ming, B. (2011) A CFD study of low pressure wet gas metering using slotted orifice meters, *Flow Measurement and Instrumentation*, 22, 33-42, ISSN 0955-5986, DOI: 10.1016/j.flowmeasinst.2010.12.002

Pesaran, A. A., Burch, S., & Keyser, M. (1999). An Approach for Designing Thermal Management Systems for Electric and Hybrid Vehicle Battery Packs Preprint. *The Fourth Vehicle Thermal Management Systems Conference and Exhibition 24-27, January*, 1–18.

Pfleger, N., Bauer, T., Martin, C., Eck, M., & Wörner, A. (2015). Thermal energy storage – overview and specific insight into nitrate salts for sensible and latent heat storage. *Beilstein Journal of Nanotechnology*, *6*, 1487–1497. DOI: 10.3762/bjnano.6.154 PMID: 26199853

Ping, P., Peng, R., Kong, D., Chen, G., & Wen, J. (2018). Investigation on thermal management performance of PCM-fin structure for Li-ion battery module in high-temperature environment. *Energy Conversion and Management*, *176*, 131–146. DOI: 10.1016/j.enconman.2018.09.025

Ping, P., Peng, R., Kong, D., Chen, G., & Wen, J. (2018). Thermal management performance of PCM-fin structures for Li-ion modules in high-temperature environments. *Energy Conversion and Management*, *176*, 131–146.

Poggi, F., Macchi-Tejeda, H., Leducq, D., & Bontemps, A. (2008). Refrigerant charge in refrigerating systems and strategies of charge reduction. *International Journal of Refrigeration*, *31*(3), 353–370. DOI: 10.1016/j.ijrefrig.2007.05.014

Poggi, F., Macchi-Tejeda, H., Leducq, D., & Bontemps, A. (2008). Refrigerant charge in systems and strategies for charge reduction. *International Journal of Refrigeration*, *31*(3), 353 ?"370.

Pradelle, F., Quiles, C., Martinez-Hernandez, E., & Feijoo, G. (2019). Recent developments in biodiesel combustion and emission control. *Energy Conversion and Management*, *183*, 1–14. DOI: 10.1016/j.enconman.2019.01.012

Prasad, B.N and Saini, J. S. (1988). *Effect of artificial roughness on heat transfer and friction factor in a solar air heater. 41*(6), 555–560.

Qi, P. C., He, Y. L., Wang, X. L., & Meng, X. Z. (2013). Experimental investigation of the optimal heat rejection pressure for a transcritical CO2 heat pump water heater. *Applied Thermal Engineering*, *56*(1–2), 120–125. DOI: 10.1016/j.applthermaleng.2013.03.045

Qyyum, M. A., Wei, F., Hussain, A., Noon, A. A., & Lee, M. (2018). An innovative vortex-tube turbo-expander refrigeration cycle for performance enhancement of nitrogen-based natural-gas liquefaction process. *Applied Thermal Engineering*, *144*, 117 ?"125.

Rachman, M. A., & Prasetyo, J. (2014). "Electricity from bio-waste-based gas hydrogen as a source of renewable energy biomass." *Proceedings of ICIEV*.

Rachman, M. A., & Prasetyo, J. (2022). Investigating the effects of biodiesel produced from *Crambe abyssinica* on combustion, performance and emissions. *Renewable Energy Research Journal*, *10*(2), 123–134.

Rahmani, A.. (2024). Recent advancements in battery thermal management for electric vehicles. *Batteries*, *10*(2), 112.

Rahmani, A., Sharma, S., & Kumar, R. (2024). Recent advancements in battery thermal management systems for electric vehicles. *Batteries, 10*(8), 265. DOI: 10.3390/batteries10080265

Rajasekaran, R., & Palekar, M. G. (1985). Mixed convection about a rotating sphere. *International Journal of Heat and Mass Transfer, 2*(5), 959–968. https://doi.org/ https://doi.org/10.1016/0017-9310(85)90277-7

Rakopoulos, C. D., Rakopoulos, D. C., & Giakoumis, E. G. (2010). Comparative study of alcohol–diesel fuel blends. *Energy Conversion and Management, 51*(12), 2850–2858. DOI: 10.1016/j.enconman.2010.07.014

Rakopoulos, C. D., Rakopoulos, D. C., Hountalas, D. T., & Giakoumis, E. G. (2008). Combustion and performance characteristics of butanol–diesel blends in CI engines. *Energy Conversion and Management, 49*(5), 1090–1099. DOI: 10.1016/j.enconman.2007.10.007

Rana, B. K., & Senapati, J. R. (2020). *Laminar Mixed Convection Over a Rotating Vertical Hollow Cylinder Exposed in the Air Medium*. Springer., DOI: 10.1007/978-981-15-7831-1

Rana, B. K., Singh, B., & Senapati, J. R. (2021). Thermofluid Characteristics on Natural and Mixed Convection Heat Transfer From a Vertical Rotating Hollow Cylinder Immersed in Air : A Numerical Exercise. *Journal of Heat Transfer, 143*(2), 022601. DOI: 10.1115/1.4048830

Rao, A. P., & Kumar, S. R. S. (2024). A review of integrated battery thermal management systems for lithium-ion batteries of electric vehicles. *E-Prime: Advances in Electrical Engineering. Electronics and Energy, 8*, 100526. DOI: 10.1016/j.prime.2024.100526

Rasheed, T., Hussain, T., Anwar, M. T., Ali, J., Rizwan, K., Bilal, M., Alshammari, F. H., Alwadai, N., & Almuslem, A. S. (2021). Hybrid nanofluids as renewable and sustainable colloidal suspensions for potential Photovoltaic/Thermal and solar energy applications. *Frontiers in Chemistry, 9*. Advance online publication. DOI: 10.3389/fchem.2021.737033 PMID: 34646812

Rashid, F. L., Aljibori, H. S., Mohammed, H. I., Ameen, A., Ahmad, S., Hamida, M. B. B., & Al-Rubaye, A. H. (2024). Recent advances and developments of the application of hybrid nanofluids in parabolic solar collector energy systems and guidelines for future prospects. *Journal of Engineering Research*. Advance online publication. DOI: 10.1016/j.jer.2024.04.023

Rasool, G.. (2024). Recent advancements in battery thermal management for EVs. *Frontiers in Energy Research, 12*, 1402932. DOI: 10.3389/fenrg.2024.1402932

Ravikiran, N., Padya, B., Rao, A., Kali, R., & Jain, P. K. (2023). Classification of phase change materials. In *Multifunctional phase change materials* (pp. 95–112). Woodhead Publishing.

Reader-Harris, M. (2015). *Experimental fluid mechanics, "Orifice plates and venture tubes*. Springer., DOI: 10.1007/978-3-319-16880-7

Rezayan, O., & Behbahaninia, A. (2011). Thermodynamic optimization and exergy analysis of CO_2/NH_3 cascade refrigeration systems. *Energy, 36*, 888–895.

Roul, M. K., & Dash, S. K. (2012). Single-Phase and Two-Phase Flow through Thin and Thick Orifices in Horizontal Pipes. *Journal of Fluids Engineering, ASME, 134*, 091301–1.

Roy, A.. (2024). Reliability analysis of prismatic lithium-ion batteries under varying thermal environments. *Philosophical Magazine, 104*(12), 2135–2154. DOI: 10.1080/14786435.2024.1234567

Ruivo, C. R., Angrisani, G., & Minichiello, F. (2015). Influence of the rotation speed on the effectiveness parameters of a desiccant wheel: An assessment using experimental data and manufacturer software. *Renewable Energy, 76*, 484 ?"493.

Ruivo, C. R., Carrillo-AndrA(c)s, A., Costa, J. J., & DomA-nguez-MuA oz, F. (2013). Exponential correlations to predict the dependence of effectiveness parameters of a desiccant wheel on the air flow rates and on the rotation speed. *Applied Thermal Engineering, 51*(2), 442 ?"450.

Saber, N.. (2025). Battery thermal management strategies for prismatic cells: Challenges and opportunities. *Energies, 18*(5), 2154.

Safdari, M., Ahmadi, R., & Sadeghzadeh, S. (2022). Numerical and experimental investigation on electric vehicles battery thermal management under New European Driving Cycle. *Applied Energy, 315*(March), 119026. DOI: 10.1016/j.apenergy.2022.119026

Saha, B. B., & Rupam, T. H. (2023). Specialty grand challenge: Thermal energy storage and conversion. *Frontiers in Thermal Engineering, 3*. Advance online publication. DOI: 10.3389/fther.2023.1157794

Sahlot, M., & Riffat, S. B. (2016). Desiccant cooling systems: A review. *The International Journal of Low Carbon Technologies, 11*(4), 489 ?"505.

Saini, M. C., & Jakhar, O. P. (2023). CFD simulation and experimental validation of PCM thermal energy storage system for micro trigeneration system application. *International Journal of Refrigeration, 149*, 119–134.

Saini, R. P., & Verma, J. (2008). Heat transfer and friction factor correlations for a duct having dimple-shape artificial roughness for solar air heaters. *Energy, 33*(8), 1277–1287. DOI: 10.1016/j.energy.2008.02.017

Salvi, B. L., & Subramanian, K. A. (2015). Sustainable development of road transportation sector using hydrogen energy system. *Renewable & Sustainable Energy Reviews, 51*, 1281–1296. DOI: 10.1016/j.rser.2015.07.009

SA nchez, D., Andreu-NA cher, A., Calleja-Anta, D., Llopis, R., & Cabello, R. (2022). Energy impact of low-GWP alternatives to R-134a in a beverage cooler: Experimental analysis and optimization. *Energy Conversion and Management, 256*, 115388.

SA nchez, D., Larrondo, R., Vidan-Falomir, F., & Cabello, R. (2024). Experimental evaluation of CO_2/R-32, CO_2/R-1234yf and CO_2/R-1270 in a transcritical plant with IHX. *Applied Thermal Engineering, 236*, 121473.

SA nchez, D., Vidan-Falomir, F., Nebot-AndrA(c)s, L., Llopis, R., & Cabello, R. (2023). Alternative blends of CO_2 for transcritical refrigeration systems: Experimental approach and energy analysis. *Energy*.

Sanghani, C. R., Jayani, D. C., Dobariya, C. R., Sabhadiya, G. K., & Gohil, J. R. (2016). Comparative Analysis of different profile geometrics for pressure drop. *International Journal of Science Technology & Engineering, 2*(10), 494–498.

Sanyal, S. (2016). *Tea Manufacturing Manual* (2nd ed.). Tea Research Association.

Saputra, D. A., Osaka, Y., Tsujiguchi, T., Haruki, M., Kumita, M., & Kodama, A. (2020). Experimental investigation of desiccant wheel dehumidification control method for changes in regeneration heat input. *Energy, 205*, 118 ?"109.

Sarkar, J., Bhattacharya, S., & Lal, A. (2013). Selection of suitable natural refrigerants pairs for cascade refrigeration system. *Institution of Mechanical Engineers, Part A: Journal of Power and Energy, 227*(5), 612–622.

Sarkar, J., Bhattacharyya, S., & Gopal, M. R. (2004). Optimization of a transcritical CO2 heat pump cycle for simultaneous cooling and heating applications. *International Journal of Refrigeration, 27*(8), 830–838. DOI: 10.1016/j.ijrefrig.2004.03.006

Sarkar, J., Bhattacharyya, S., & Ram Gopal, M. (2004). Optimization of a transcritical CO_2 heat pump for simultaneous cooling and heating. *International Journal of Refrigeration*, *27*(8), 830 ?"838.

Sawalha, S. (2008). Safety of CO_2 in large refrigeration systems. Proceedings of the International Refrigeration Conference, Purdue University, 81 ?"93.

Say, D., Ganesan, A. L., Lunt, M. F., Rigby, M., O ?(tm)Doherty, S., Harth, C., & Bauguitte, S. (2019). Emissions of halocarbons from India inferred through atmospheric measurements. *Atmospheric Chemistry and Physics*, *19*(15), 9865 ?"9885.

Sazali, N. (2020). Emerging technologies by hydrogen: A review. *International Journal of Hydrogen Energy*, *45*(23), 12034–12045. DOI: 10.1016/j.ijhydene.2020.02.035

Schaeffler, S.. (2024). In-situ measurement of internal thermal gradients and their effect on thermal runaway onset. *Journal of Power Sources*, *589*, 233523.

Schöberl, J.. (2024). Thermal runaway propagation in large-format automotive Li-ion modules. *Journal of Energy Storage*, *72*, 108619.

Sen, S., Mittal, S., & Biswas, G. (2011). Flow past a square cylinder at low reynolds numbers. *International Journal for Numerical Methods in Fluids*, *67*, 1160–1174.

Sethin, A., Thirumalini, S., & Prakash, S. (2024). Effects of blended diesel–biodiesel fuel on emissions of a CRDI engine. *ACS Omega*, *9*(3), 1005–1015. DOI: 10.1021/acsomega.4c00012

Shahzad, H. M. A., Asim, Z., Khan, S. J., Almomani, F., Mahmoud, K. A., Mustafa, M. R. U., & Rasool, K. (2024). Thermochemical and biochemical conversion of agricultural waste for bioenergy production: An updated review. *Discover Environment*, *2*(1). Advance online publication. DOI: 10.1007/s44274-024-00171-w

Shanfang, H., Taiyi, M., Dong, W., & Zhong, L. (2013). Study on discharge coefficient of perforated orifices as a new kind of flow meter. *Experimental Thermal and Fluid Science*, *46*, 74–83.

Shareef, S. M., & Rao, G. A. P. (2025). Comparative Evaluation of Phase Change Materials and Fins in Battery Thermal Management During High Discharge. *Energy Storage*, *7*(7), e70271. DOI: 10.1002/est2.70271

Sharma, A., & Eswaran, V. (2004). Heat and fluid flow across a square cylinder in the two-dimensional laminar flow regime. *Numerical Heat Transfer Part A*, *45*, 247–269.

Sharma, S. K., & Kalamkar, V. R. (2017). Experimental and numerical investigation of forced convective heat transfer in solar air heater with thin ribs. *Solar Energy, 147*, 277–291. DOI: 10.1016/j.solener.2017.03.042

Sharma, S., Das, R. K., & Kulkarni, K. (2021). Computational and experimental assessment of solar air heater roughened with six different baffles. *Case Studies in Thermal Engineering, 27*(July). Advance online publication. DOI: 10.1016/j.csite.2021.101350

Sharma, V., & Dhiman, K. A. (2012). Heat transfer from a rotating circular cylinder in the steady regime : Effects of prandtl number. *Thermal Science, 16*(1), 79–91. DOI: 10.2298/TSCI1100914057S

Shatnawi, N. A. Q., Aljaberi, M., & Aljaberi, N. (2018). "Hydrogen-based energy storage systems: a review." *Proceedings of ICRERA*.

Shen, H., Suzuki, A., Takata, N., & Kobashi, M. (2024). Enhancing capillary pressure of porous aluminum wicks by controlling bi-porous structure using different-sized NaCl space holders. *Materials (Basel), 17*(19), 4729. PMID: 39410300

Shomali, A., & Abbasi Souraki, B. (2019). Experimental investigation and mathematical modeling of drying of green tea leaves in a multi-tray cabinet dryer. *Heat and Mass Transfer, 55*(12), 3645–3659. DOI: 10.1007/s00231-019-02662-6

Simal, S., Rossello, C., Berna, A., & Mulet, A. (1998). Drying of shrinking cylinder-shaped bodies. *Journal of Food Engineering, 37*(4), 423–435. DOI: 10.1016/S0260-8774(98)00095-8

Singh Yadav, A., & Bhagoria, J. L. (2015). Numerical investigation of flow through an artificially roughened solar air heater. *International Journal of Ambient Energy, 36*(2), 87–100. DOI: 10.1080/01430750.2013.823107

Singh, I., Vardhan, S., Singh, S., & Singh, A. (2019). Experimental and CFD analysis of solar air heater duct roughened with multiple broken transverse ribs: A comparative study. *Solar Energy, 188*(February), 519–532. DOI: 10.1016/j.solener.2019.06.022

Singh, R. P., Das, R. K., & Upadhyay, S. N. (2023). Effectiveness approach for predicting desiccant wheel performance: Correlation utilizing experimental results. *International Journal of Refrigeration, 156*(December).

Skullong, S., Thianpong, C., & Promvonge, P. (2015). Effects of rib size and arrangement on forced convective heat transfer in a solar air heater channel. *Heat and Mass Transfer/Waerme- Und Stoffuebertragung, 51*(10), 1475–1485. https://doi.org/DOI: 10.1007/s00231-015-1515-5

Smith, S. (2012). AHRI Standard 550 ?"590 ?"Overview of latest updates: Performance rating of water-chilling and heat-pump water-heating packages using the vapor-compression cycle. *ASHRAE Transactions*, *118*(1), 317 ?"325.

Srivathsan, B., G, T., Ram, K. V., & R, H. (2023). Multiphase simulation of sustainable nanoenhanced ionic liquid coolants for improved thermal performance in Ti–6Al–4V alloy drilling. *Heliyon*, *9*(12), e23020. DOI: 10.1016/j.heliyon.2023.e23020 PMID: 38076113

Straka, M., Fiebach, A., Eichler, T., & Koglin, C. (2018). Hybrid simulation of a segmental orifice plate. *Flow Measurement and Instrumentation*, *60*, 124–133. DOI: 10.1016/j.flowmeasinst.2018.02.006

Suhara, A., Kumar, P., & Reddy, A. "Biofuel sustainability: Review of progress and performance." *Green Energy and Sustainability*, 2024. URL: https://www.sciencedirect.com

Sun, X.. (2023). Critical review of thermal runaway prediction and mitigation. *Energy Materials Advances*, *2*, 100008. DOI: 10.34133/energymatadv.000008

Sun, X., Shi, L., Tian, H., Wang, X., Zhang, Y., & Shu, G. (2021). A composition-tunable combined cooling and power cycle using CO_2-based binary zeotropic mixture. *Energy Conversion and Management*, *244*, 114419.

Sutar, S., Rout, S. K., Senapati, J. R., Barik, D., Dennison, M. S., & Praveenkumar, S. (2024b). Numerical investigation on the thermohydraulic performance of a solar air heater duct featuring parabolic rib turbulators on the absorber plate. *Case Studies in Thermal Engineering*, *64*, 105399. DOI: 10.1016/j.csite.2024.105399

Sutar, S., Rout, S. K., Senapati, J. R., & Muduli, K. K. (2024b). Numerical investigation of solar air heater for space heating application. *Applied Solar Energy*, *60*(5), 708–726. DOI: 10.3103/s0003701x24602151

Takhar, H. S. A. J. C. & G. N. (2000). Combined heat and mass transfer along a vertical moving cylinder with a free stream. *Heat and Mass Transfer*, *36*(3), 237–246. https://doi.org/https://doi.org/10.1007/s002310050391

Talukdar, S., Afroz, H. M. M., Hossain, M. A., Aziz, M. A., & Hossain, M. M. (2019). Heat transfer enhancement of charging and discharging of phase change materials and size optimization of a latent thermal energy storage system for solar cold storage application. *Journal of Energy Storage*, *24*, 100797.

Tea Board India. (2022). *Global Tea*. https://www.teaboard.gov.in/pdf/Global_Tea_Website_1005_pdf7206.pdf

Teixeira, A. C. R.. (2021). Alternative fuel technologies emissions for heavy-duty trucks: A review. *Environmental Science and Pollution Research International, 28*, 34297–34334. DOI: 10.1007/s11356-021-14013-y PMID: 33686605

Temple, S. J., & Van Boxtel, A. J. B. (1999). Thin layer drying of black tea. *Journal of Agricultural Engineering Research, 74*(2), 167–176. DOI: 10.1006/jaer.1999.0448

Tezduyar, T. E., Mittal, S., Ray, S. E., & Shih, R. (1992). In- compressible flow computations with stabilized bilinear and linear equal-order-interpolation velocity-pressure elements. *Computer Methods in Applied Mechanics and Engineering, 95*, 221–242.

Thi, N., Quynh, N., Kuo, H., Liao, C., & Lin, K. C. (2025). 3D CFD analysis of convective heat transfer from a heated horizontal cylinder rotating in air: From laminar to turbulent flow. *Journal of Heat and Mass Transfer,* https://doi.org/DOI: 10.1115/1.4068377

Tian, Z., Zheng, W., Sun, X., Wang, L., Jiang, Y., & Mi, X. (2024). Research on prediction model of two-phase flow and friction pressure drop in spiral pipe based on flow regimes. *Thermal Science and Engineering Progress, 103018*. Advance online publication. DOI: 10.1016/j.tsep.2024.103018

Tyagi, V., Kaushik, S., & Tyagi, S. (2012). Advancement in solar photovoltaic/thermal (PV/T) hybrid collector technology. *Renewable & Sustainable Energy Reviews, 16*(3), 1383–1398. DOI: 10.1016/j.rser.2011.12.013

Ukoba, K., Olatunji, K. O., Adeoye, E., Jen, T., & Madyira, D. M. (2024). Optimizing renewable energy systems through artificial intelligence: Review and future prospects. *Energy & Environment, 35*(7), 3833–3879. DOI: 10.1177/0958305x241256293

United Nations Environment Programme. (1987). *Montreal Protocol on Substances that Deplete the Ozone Layer*. U.S. Government Printing Office.

United Nations Environment Programme. (2014). *Environmental effects of ozone depletion and its interactions with climate change: 2014 assessment*. United Nations Environment Programme.

Vaccaro, G., Milazzo, A., & Talluri, L. (2024). A non-flammable, fluorine-free, CO_2-based mixture as a low-TEWI refrigerant. *International Journal of Refrigeration, 158*, 157 ?"163.

Velders, G. J. M., Fahey, D. W., Daniel, J. S., Andersen, S. O., & McFarland, M. (2015). Future atmospheric abundances and climate forcings from scenarios of global and regional HFC emissions. *Atmospheric Environment, 123*, 200 ?"209.

Verma, A., Saikia, T., Saikia, P., Rakshit, D., & Ugalde-Loo, C. E. (2023). Thermal performance analysis and experimental verification of lithium-ion batteries for electric vehicle applications through optimized inclined mini-channels. *Applied Energy*, *335*(February), 120743. DOI: 10.1016/j.apenergy.2023.120743

Visvanathan, C., & Kumar, S. (2002). *Small medium scale Industries in Asia (Energy and Environment): Tea Sector*. Regional Energy Resources Information Center (RERIC), Asian Institute of Technology, Pathumthani, Thailand. https://doi.org/ DOI: 10.13140/2.1.3285.0720

Voller, V. R., & Prakash, C. (1987). A fixed grid numerical modelling methodology for convection-diffusion mushy region phase-change problems. *International Journal of Heat and Mass Transfer*, *30*(8), 1709–1719.

Wahab, A.. (2025). Advances and challenges in immersion cooling for electric vehicle batteries. *Renewable & Sustainable Energy Reviews*, *195*, 113281.

Wahab, A.. (2025). Immersion cooling for EV batteries: Recent innovations and challenges. *Renewable & Sustainable Energy Reviews*, *195*, 113281. DOI: 10.1016/j.rser.2025.113281

Wang, G., Gao, Q., Yan, Y., & Wang, Y. (2020). Thermal management optimization of a lithium-ion battery module with graphite sheet fins and liquid cold plates. *Automotive Innovation*, *3*(4), 336–346. DOI: 10.1007/s42154-020-00121-1

Wang, G., Gao, Q., Yan, Y., & Wang, Y. (2020). Thermal management optimization of lithium-ion battery modules with graphite sheet and aluminum fins. *Automotive Innovation*, *3*(4), 336–346.

Wang, P.. (2025). Thermal runaway modeling of large 280 Ah LFP prismatic cells. *Energy*, *280*, 127988. DOI: 10.1016/j.energy.2025.127988

Wang, Y., & Wu, J. J. (2023). Thermochemical conversion of biomass: Potential future prospects. *Renewable & Sustainable Energy Reviews*, *187*, 113754. DOI: 10.1016/j.rser.2023.113754

Wang, Z., Yelishala, S. C., Yu, G., Metghalchi, H., & Levendis, Y. A. (2019). Effects of Carbon Dioxide on Laminar Burning Speed and Flame Instability of Methane/Air and Propane/Air Mixtures: A Literature Review. *Energy & Fuels*, *33*(10), 9403–9418. DOI: 10.1021/acs.energyfuels.9b02346

Warrick, D., Tobalske, B., Powers, D., & Dickinson, M. “The aerodynamics of hummingbird flight.” *AIAA Conference Proceedings*, 2011.

Welzl, M., Heberle, F., & BrA1/4ggemann, D. (2017). Simultaneous experimental investigation of nucleate boiling heat transfer and power output in ORC using R-245fa and R-1233zd(E). *Energy Procedia, 129*, 435 ?"442.

Weng, J., Ouyang, D., Yang, X., Chen, M., Zhang, G., & Wang, J. (2020). Optimization of internal fins in PCM modules for improved battery thermal management. *Applied Thermal Engineering, 167*, 114698.

Weng, J., Ouyang, D., Yang, X., Chen, M., Zhang, G., & Wang, J. (2020). Optimization of the internal fin in a phase-change-material module for battery thermal management. *Applied Thermal Engineering, 167*, 114698. DOI: 10.1016/j.applthermaleng.2019.114698

Worwood, D.. (2019). Experimental benchmarking of fin materials for large pouch cells. *Energies, 12*(23), 4495. DOI: 10.3390/en12234495

Wright, D. R., Garcia-Araez, N., & Owen, J. R. (2018). Review on high temperature secondary Li-ion batteries. *Energy Procedia, 151*, 174–181. DOI: 10.1016/j.egypro.2018.09.044

Wu, J., Chu, Y., Hu, J., & Liu, Z. (2009). Performance of R-152a/R-125/R-32 in domestic air-conditioners. *International Journal of Refrigeration, 32*(5), 1049 ?"1057.

Wu, X. N., Ge, T. S., Dai, Y. J., & Wang, R. Z. (2019). Investigation on novel desiccant wheel using wood pulp fiber paper with high coating ratio as matrix. *Energy, 176*, 493 ?"504.

Xue, Q., & Fox, R. O. (2014). Computational modeling of biomass thermochemical conversion in fluidized beds: Particle density variation and size distribution. *Industrial & Engineering Chemistry Research, 54*(16), 4084–4094. DOI: 10.1021/ie503806p

Xue, Z. L., Xu, Y., Zhang, T., Zhang, J. L., & Chen, Z. H. (2021) Research on numerical simulation of stratified gas-liquid flow through crescent orifice plate, *2021 IEEE International Instrumentation and Measurement Technology Conference (I2MTC)*, 2021,1-6, doi: DOI: 10.1109/I2MTC50364.2021.9459904

Xu, H., Zhang, X., Xiang, G., & Li, H. (2021). Optimization of liquid cooling and heat dissipation system of lithium-ion battery packs of automobile. *Case Studies in Thermal Engineering, 26*(March), 101012. DOI: 10.1016/j.csite.2021.101012

Xu, L.. (2024). Thermal management and heat transfer in lithium-ion batteries: A comprehensive review. *Energies, 17*(16), 3873.

Xu, X., Chen, X., Shen, J., Kong, J., Zhang, H., & Zhou, F. (2023). Thermal management system for prismatic battery module with biomimetic cephalofoil fin and film heater. *Applied Thermal Engineering*, *227*, 120379. DOI: 10.1016/j.applthermaleng.2023.120379

Yadav, A. S., & Bhagoria, J. L. (2013). Heat transfer and fluid flow analysis of solar air heater: A review of CFD approach. *Renewable & Sustainable Energy Reviews*, *23*, 60–79. DOI: 10.1016/j.rser.2013.02.035

Yadav, H. S. (1995). ISH M., Athar M., ISH M., Kaushal D. R., Flow Characteristics of Sector Orifice Plates. *ISH Journal of Hydraulic Engineering*, *1*(2), 84–94.

Yanfeng, G., Zheng, J., & Shi, T. (2006). Study on the metering characteristics of a slotted orifice for wet gas flow. *Flow Measurement and Instrumentation*, *17*, 123–128.

Yang, H., & Pan, M. (2015). Engineering research in fluid power: A review. *Journal of Zhejiang University. Science A*, *16*(6), 427–442. DOI: 10.1631/jzus.a1500042

Yang, X., Cui, Q., You, Z., Liu, J., Zhang, Y., & Wei, J. (2024). Experimental and theoretical study on performance of bi-porous wick for passive phase-change devices. *Physics of Fluids*, *36*(12).

Yang, X., Wang, X., Liu, Z., Luo, X., & Yan, J. (2022). Effect of fin number on the melting phase change in a horizontal finned shell-and-tube thermal energy storage unit. *Solar Energy Materials and Solar Cells*, *236*, 111527.

Yang, Z., & Wu, X. (2013). Retrofits and options for alternatives to HCFC-22. *Energy*, *59*, 1 ?"21.

Yao, S., Wei, Y., Lu, Z., Guo, S., Chen, J., & Yu, Z. (2023). Review of research progress on Concentrated Solar Energy Utilization System. *Renewables*, *1*(4), 397–414. DOI: 10.31635/renewables.023.202200018

Yelishala, S. C. (2019). Thermodynamic and combustion study on blends of hydrocarbons and carbon dioxide as alternative refrigerants.

Yelishala, S. C., Kannaiyan, K., Wang, Z., Metghalchi, H., Levendis, Y. A., & Sadr, R. (2020). Thermodynamic study on blends of hydrocarbons and carbon dioxide as zeotropic refrigerants. *Journal of Energy Resources Technology*, *142*(8), 082304.

Yelishala, S. C., Kannaiyan, K., Wang, Z., Metghalchi, H., Levendis, Y. A., & Sadr, R. (2020). Thermodynamic study on blends of hydrocarbons and carbon dioxide as zeotropic refrigerants. *Journal of Energy Resources Technology, Transactions of the ASME, 142*(8), 1–10. https://doi.org/DOI: 10.1115/1.4045930

Yelishala, S. C., Wang, Z., Metghalchi, H., Levendis, Y. A., Kannaiyan, K., & Sadr, R. (2019). Effect of Carbon Dioxide on the Laminar Burning Speed of Propane-Air Mixtures. *Journal of Energy Resources Technology, Transactions of the ASME, 141*(8). https://doi.org/DOI: 10.1115/1.4042411

Yilmaz, N., & Davis, S. M. (2016). PAH formation tendencies in higher-alcohol biodiesel combustion. *Fuel Processing Technology*, *150*, 123–131. DOI: 10.1016/j.fuproc.2016.02.011

Yu, G., Sun, Y., Zhou, Y., & Xu, H. (2025). Review on heat transfer structures for enhancing thermal performance of latent heat thermal energy storage systems. *Journal of Energy Storage*, *132*, 117908. DOI: 10.1016/j.est.2025.117908

Yusuf, A., Bhatti, M., & Ellahi, R. (2025). Study of ionic water/graphene nanofluids in solar panels under the effects of thermal radiation and slip conditions using experimental data. *International Communications in Heat and Mass Transfer*, *164*, 108845. DOI: 10.1016/j.icheatmasstransfer.2025.108845

Yu, Z.. (2023). BTMS with heat-pipe/PCM hybrids: A review. *Journal of Energy Storage*, *68*, 107532. DOI: 10.1016/j.est.2023.107532

Zamzamian, S. A. H. (2022). A new approach for the heat and moisture transfer in rotary desiccant wheel. *Iranian Journal of Chemistry and Chemical Engineering*, *41*(2), 197 ?"208.

Zeng, Z., Han, C., Li, H., Chen, T., Zeng, J., Fang, Z., Li, B., & Wu, W. (2024). Modeling and evaluation in transient analysis of industrialized counter-flow tea drying system. *Case Studies in Thermal Engineering*, *53*(November 2023), 103781. https://doi.org/DOI: 10.1016/j.csite.2023.103781

Zeng, Z., Han, C., Wang, Q., Yuan, H., Zhang, X., & Li, B. (2024). Analysis of drying characteristic, effective moisture diffusivity and energy, exergy and environment performance indicators during thin layer drying of tea in a convective-hot air dryer. *Frontiers in Sustainable Food Systems*, *8*(April), 1–13. DOI: 10.3389/fsufs.2024.1371696

Zhang, K. (2025). A Review of Biomass Energy: Comparison of utilization methods and future Prospects. *E3S Web of Conferences*, *606*, 05007. DOI: 10.1051/e3sconf/202560605007

Zhang, B., & Guo, X. (2018). Prospective applications of Ranque ?"Hilsch vortex tubes to sustainable energy utilization and energy efficiency improvement with energy and mass separation. *Renewable & Sustainable Energy Reviews*, *89*, 135 ?"150.

Zhang, X.. (2025). Comprehensive analysis of airflow and PCM coupling in Li-ion battery packs. *Energy Reports*, *11*(2), 10921–10935.

Zhang, Z., Ito, M., & Hihara, E. (2024). *Laminar Burning Velocity in R290 / R32, R290 / R1234yf, and R290 / HFO-1123 Binary Mixtures : Comparing Experimental Results to Mixing Rule Calculations AGC Inc. Chemicals Company Research & Development General Division*. The University of Toky.

Zhao, Y.. (2024). Review of PCM-based battery thermal management systems: Convection and conduction effects. *Journal of Energy Storage*, *73*, 108920.

Zhao, Y., Wei, M., & Dan, D. (2024). Modeling, design, and optimization of loop heat pipes. *Energies*, *17*(16), 3971.

About the Contributors

Kamalakanta Muduli, is presently working as Associate Professor and Head of the School of Mechanical Engineering, Papua New Guinea University of Technology, Lae, Morobe Province, Papua New Guinea. He is also a Visiting Researcher at University of Derby, UK. He has obtained PhD from School of Mechanical Sciences, IIT Bhubaneswar, Orissa, India. Dr Muduli has over 19 years of academic experience in Universities in India and Papua New Guinea. Dr Muduli is a recipient of ERASMUS+ KA107 award provided by European Union. He has published 102 papers in peer reviewed international journals most of which are indexed in Clarivate analytics, Scopus and listed in ABDC and more than 37 papers in National and International Conferences. He has been also guest edited 10 Special Issues in journals and 16 books published by Springer, Taylor and Francis, MDPI, CRC Press, Wiley Scrivener and Apple Academic Press. Dr Muduli also has supervised four PhD students. His current research interest includes Manufacturing, Sustainable supply chain management, and Industry 4.0 applications in operations and supply chain management. Dr Muduli is a fellow of Institution of Engineers India. He is also a senior member of Indian Institution of Industrial Engineering, member of American Society of Mechanical Engineers(ASME) and Professional member of Industrial Engineering and Operations Management(IEOM) Society.

Sachindra Kumar Rout has a strong background in refrigeration engineering, CFD and cryogenics. His experience in the three fields is deep and varied. At the moment most of his time and effort are directed towards developing the technology of closed cycle regenerative cryorefrigerators, specifically the low speed GM and GM type Pulse Tube cryocoolers. The cryogenics laboratory at C V Raman Global university, which he founded with generous support and encouragement of the University and is heading today, has achieved remarkable success in recent months. The single stage GM refrigerator designed and built completely at C V Raman has achieved a low temperature of 30 Kelvin, this record being a landmark in India on the road to achieving self-sufficiently in designing cryogenic equipment. Dr. Rout has been supporting his research effort through a series of small and medium size projects secured from national funding agencies, the best among them being a mid sized project on development of GM type cryocooler from the Board of Research on Nuclear Sciences. In addition to the main research target, his team has crafted several small devices - a retort type vacuum furnace, a membrane type nitrogen generator, an Arduino controller based nitrogen purity meter and more. At present in addition to the mission of developing India's first complete closed cycle cryocooler.

Bukke Kiran Naik is an Assistant Professor in Mechanical Engineering Department at NIT Rourkela from 2020. He received both his Ph.D. (2019) and M-Tech (2014) from IIT Guwahati and B-Tech (2012) from JNTU Anantapur in Mechanical Engineering. He worked as a Queen Elizabeth Postdoctoral fellow at Simon Fraser University, Canada during 2019–2020 and as Project Engineer at IIT Guwahati in 2019. He is the recipient of Paired Early Career Fellowship in Applied Research (PECFAR) from Indo-German Science and Technology Centre (IGSTC), Queen Elizabeth Scholars (QES) Fellowship from Universities Canada, CCSTDS Travel fellowship from INSA/CSIR/DAE-BRNS-CCSTDS, and SERB-ITS travel grant from DST, Government of India. Further, he was selected as Young Indian Scientist to the 6th BRICS (Brazil, Russia, India, China, and South Africa) conclave 2021 in the thematic area of Energy Solutions and he also received INAE fellowship for working as a visiting researcher at IIT Kanpur. He has filled three patents till date and secured three funded research projects from ISRO, SERB and NIT Rourkela. He also received funding from several funding agencies (SERB, ATAL and ISHRAE) to organize outreach programs (workshops and short-term courses). He served as a student's activities chair, K-12 chair and CWC member for ISHARE Guwahati-sub chapter from year 2016–2018 and Program chair for ISHARE Bhubaneswar sub-chapter from year 2021–2022. Currently, serving as ISHRAE Bhubaneswar Sub- Chapter President for the year 2022–2023. He has about 50 research articles in reputed international journals and conference proceedings. Moreover, Dr. Naik has been invited by various reputed institutes/societies to deliver talk on his research work carried out in energy-water nexus or sustainable energy and buildings.

Albert Uchenna Ude received the M.Eng. and Ph.D. degrees in mechanical and materials engineering from the National University of Malaysia, in 2006 and 2011, respectively. He then works as a Senior Researcher for some years after graduation at the Department of Mechanical and Materials Engineering, National University of Malaysia. In 2015, he was appointed as a Senior Lecturer and later the Head of the Department Mechanical Engineering, Papua New Guinea University of Technology. He is currently serving as an Associate Professor at the Botswana

International University of Science and Technology (BIUST), Botswana, where he is also the current Research Chairperson of the Advanced Industrial, Materials, Systems (AIMS) Research Group. He has authored two book chapters and many refereed published scholarly/scientific journal articles. His research interests include machine design, PV optimization, materials

***

B Abhiram completed bachelor's in mechanical engineering from Cvr college of engineering in 2025.

Siba Padarabinda Behera is working as technical asssistant in school of mechanical engineering at KIIT.

Umamahesh Boddepalli is a research scholar at IIT Kharagpur

Sajal Chanda is a General Manager I/C (HOS) in Coke Oven of SAIL, Rourkela Steel Plant, Rourkela. I am pursing Ph.D from NIT,Rourkela in Thermal Engineering. I have done research work in thermal engineering, natural convection and solar. In thermal engineering I have carried out research work in charging distribution system of Blast Furnaces and analysis the heat flux with in distribution Gear box and re design the cooling system of distribution gear box. In natural convection I have carried out research work in a square enclosure with insert at the corners of the enclosure. In solar I am pursuing research work on solar air heater duct with rivet rib in the absorber plate of solar air heater duct.

Kashvi Chaturvedi, is studying B.Tech., specialisation in Artificial Intelligence and Data Science at Ajeenkya D Y Patil University. She started solidifying her programming skills by completing various certification courses like Data Structures in C, Tableau functions and Operating Systems along with doing few projects using languages like Python, C, C++ and creating Web Page using HTML, CSS and Java Script. She had also attended an Arduino-Based Automation course at Dolphin Labs to enhance her software knowledge. There, she learned how to work with sensors, microcontrollers, and circuit design. She has also written multiple review papers on topics such as Geographic Information Systems, Comparative Analysis of BJT Configurations, Swiss Bank vs. European Central Bank: Data Analysis. She is also researching on and learning many topics such as Small Modular Reactors, Bibliometric Analysis, UNSDGs etc. Because of her interest in latest technologies, she also participated in Global Hyperloop Competition hosted by IIT Madras representing her college as team Vac-Vectoor where she learnt more about various subtopics from the mentors.

Balaji Choudhury completed his B.E. in Mechanical Engineering from Berhampur University in 2002. He completed his M. Tech in Thermal Engineering from NIT Rourkela in the year 2009 through GATE and continued to complete his PhD from NIT Rourkela in the year 2014 having Cryogenics specialization. His research interests include Low-temperature refrigeration, Heat transfer equipment, CFD, Turbo-machinery, Process design and analysis, etc.

Susanta Das is a seasoned educator with a proven record of teaching and mentoring diverse students, working at multiple institutions, and providing service to the various initiatives of organizations. He received his Ph.D. and M.A. degrees from Western Michigan University (WMU), USA, and M.Sc. degree from Banaras Hindu University (BHU), India, all in Physics. He continued his research as a Marie-Curie post-doctoral fellow at Stockholm University, Stockholm, Sweden on beam diagnostics for the DESIREE (Double ElectroStatic Ion Ring ExpEriment) in the project DITANET (Diagnostic Techniques for particle Accelerators – a Marie-Curie initial training NETwork), at the Indian Institute of Science Education and Research-Kolkata, India on high-pressure physics as a Project Scientist-B, and at the University of Electro-Communications, Tokyo, Japan on ion-surface interactions as a Post-doctoral fellow. Throughout his career, Dr. Das worked and collaborated with many researchers, Ph.D., master, and visiting students from different countries. He further visited many countries to discuss research and an international conference participant (UK, Italy, Germany, Greece, Belgium, Bulgaria, Romania, Brazil etc.). He worked at several institutions before assuming his current role at Ajeenkya D Y Patil University. He received the Marie-Curie post-doctoral fellowship in Sweden, Science Academies' Summer Research Fellowship in India, Gowen Frostic Doctoral Fellowship, Department Graduate Research and Creative Scholar Award by WMU, and Leo R. Parpart Doctoral Fellowship by Dept. of Physics, WMU, among many others, throughout his academic journey. He co-authored several research articles in WoS/Scopus indexed international journals, conference proceedings, and scholarly book chapters and co-edited several books published by renowned international publishing houses (IGI-Global, Springer Nature, Nova Science Publishers, Wiley, Scrivener-Wiley, Notion Press). At present, Dr. Das continues to delve into cutting-edge fields, with a keen interest in nanotechnology, UN SDGs, and data science. His multifaceted contributions, spanning teaching, research, and administrative leadership, showcase a dedicated professional who is instrumental in advancing the vision and mission of the institutions he serves.

Kavya Dashora received her PhD from the prestigious Central Arid Zone Research Institute (CAZRI-ICAR) on developing low-cost biocontrol solutions for pearl millet smut caused by Tolyposporium penicillarie. She has expertise in the areas of nanotechnology, bioprocess engineering, food (plant-based proteins, mock meats), value chain management, and AI in agriculture. She is the recipient of "best innovation award- UNDP-2021" and "innovation in Food award- PETA-2020". Currently, she is serving as an associate professor in the Centre for Rural Development and Technology, Indian Institute of Technology Delhi.

Harshit Gupta is a Mechanical Engineer. He completed his M.Tech in Mechanical Engineering from IIT Kharagpur. He works in air conditioning and ventilation filed. He loves work with new ideas. He loves travel various new places.

Satrajit Halder is currently a Ph.D. scholar in Mechanical Engineering at NIT Durgapur, working on computational fluid dynamics (CFD). He earned his Master's degree in Nuclear Engineering (specialized in Thermal Engineering) from Jadavpur University with an SGPA of 8.77 and a B.Tech in Mechanical Engineering from Government College of Engineering and Textile Technology, Berhampore with a CGPA of 9.01. His core interests include thermal engineering, CFD simulations, and fluid-thermal interaction studies, with a growing focus on integrating data science and artificial intelligence into engineering analysis and simulation.

Sujit Karmakar is currently serving as an Assistant Professor in the Department of Mechanical Engineering at National Institute of Technology (NIT) Durgapur. He holds a Ph.D. in Mechanical Engineering from the Indian Institute of Technology (IIT) Madras, preceded by an M.Tech and B.Tech from NIT Durgapur (formerly Regional Engineering College, Durgapur). With a strong academic background and specialized expertise in thermal sciences, Dr. Karmakar's research interests encompass thermal engineering, heat transfer, power plant engineering, energy systems, and advanced thermal-fluid analysis. He has authored over 30 research articles in reputed international journals, underscoring his ongoing contributions to the global scientific community.

Deepak Kumar is an Assistant Professor in the Department of Mechanical Engineering at the National Institute of Technology (NIT) Durgapur. He holds a B.Tech degree from NIT Jamshedpur, and subsequently earned his M.Tech and Ph.D. from the Indian Institute of Technology (IIT) Guwahati. Dr. Kumar's research interests lie in the field of fluid mechanics, with a particular focus on bluff body flows, laminar and low Reynolds number flows, hydrodynamic stability, and linear stability analysis. His work explores various aspects of flow control—especially passive control approaches involving splitter plates—as well as wakes, jets, flow separation, and vortex shedding phenomena. He actively utilizes computational fluid dynamics (CFD) tools, with a specialization in the finite element method, to conduct his research. Dr. Kumar has published several research papers in reputed international journals, including a notable publication in Physics of Fluids (AIP Publishing). He also serves as a reviewer for multiple peer-reviewed journals, including Physics of Fluids.

K. Ravi Kumar received his Ph.D. from Indian Institute of Technology (IIT) Madras in 2012. At IIT Madras, he worked on the investigation of energy efficient receiver for line-focus concentrating solar power (CSP) systems, such as parabolic trough collector and linear Fresnel reflector for process heating and power generation applications. After his PhD, he worked as a postdoctoral fellow at ABB Corporate Research Center, Bangalore. Since November 2016, he has been part of the Department of Energy Science and Engineering, Indian Institute of Technology Delhi and currently serving as an associate professor. He is interested in working in the areas of solar thermal technology, applications, thermal energy storage, solid state hydrogen storage, and solar radiation forecasting, etc.

Shantanu Kunar is a PhD scholar at the Department of Energy Science and Engineering, IIT Delhi, India. He works in the area of solar thermal technology and its applications. He has completed a master's degree in Renewable Energy Systems from NIT Kurukshetra, Haryana, India. Additionally, he holds an undergraduate degree in Mechanical Engineering from NIT Kurukshetra.

Snehangshu Mondal is a Ph.D. research scholar at the School of Energy Science and Engineering in Indian Institute of Technology Kharagpur, India. He is from Asansol city in West Bengal, India. He works on building energy, air conditioning and ventilation, refrigeration, renewable energy, conventional energy and waste management. He completed his M.Tech in Mechanical Engineering from Indian Institute of Technology (Indian School of Mines) Dhanbad. He has Post graduate diploma in thermal power plant engineering. He has worked in power sector (power generation) as an operation and commissioning Engineer for 4 years. He works in different academic and industrial projects. His hobby is gardening and cooking.

Maddali Ramgopal is a senior faculty member in the Department of Mechanical Engineering at IIT Kharagpur and served as the Dean of the Faculty of Engineering and Architecture. He is a Dean of Infostructure in IIT Kharagpur. His research focuses on refrigeration and air conditioning, heat transfer, energy materials and hydrogen storage, and solar energy. He has contributed significantly to the development of sustainable thermal systems and energy-efficient technologies. Prof. Ramgopal has published extensively in these areas, with over 50 publications and more than 1,200 citations. His work has advanced the understanding of CO_2-based refrigeration cycles, hybrid air conditioning systems, and solar-assisted trigeneration systems, making him a recognized expert in thermal engineering.

G. Amba Prasad Rao is Professor (HAG) in the Department of Mechanical Engineering at the National Institute of Technology (NIT) Warangal, India. He received his Ph.D. in Mechanical Engineering from NIT Warangal (2003) for his work on biodiesel as an alternative fuel for diesel engines, an M.Tech. in Design and Production of IC Engines and Gas Turbines from REC Warangal (1992), and a B.Tech. in Mechanical Engineering from JNTU Anantapur (1988). With more than three decades of teaching and research experience, Dr. Rao's areas of specialization include internal-combustion engines, alternate fuels, emission control, and renewable energy systems. He has supervised 11 Ph.D. scholars, guided over 50 M.Tech. projects, and published more than 80 international-journal papers and 40 international-conference papers. Dr. Rao is the co-author of the book Engine Emission Control Technologies (Apple Academic Press, CRC Press/Taylor & Francis Group, 2020) and has completed several sponsored research projects in sustainable energy and engine performance enhancement. At NIT Warangal, he has served in leadership roles including Dean (Faculty Welfare), Associate Dean (Academics), and Professor-in-charge (Training and Placement), contributing significantly to academic administration and curriculum innovation. He is a Life Member of ISTE, IEI, Combustion Institute (Indian Section), Solar Energy Society of India, and Indian Society of Mechanical Engineers, and a member of SAE India. His biography was featured in Marquis Who's Who in Science and Engineering (2003–2004). Dr. Rao remains actively involved in collaborative research, journal reviewing, and capacity building for sustainable mechanical engineering education.

P S Rao is a senior faculty member in the Department of Agricultural and Food Engineering, Indian Institute of Technology, Kharagpur with more than 25 years of teaching and research experience. His research area includes processing and value addition of millets, post harvest & food process engineering, extraction & encapsulation of bioactive, process equipment design & development and nonthermal methods of food processing with 90+ publications and more than 4500 citations.

P. Srinivas Reddy is a Senior Assistant Professor in the Department of Mechanical Engineering at CVR College of Engineering, Hyderabad. He obtained his Ph.D. in Mechanical Engineering from Osmania University (2025), following an M.Tech in Thermal Engineering from NIT Bhopal (2008) and a B.E. from the University College of Engineering, Osmania University (2004). He has more than seven years of industrial experience with organizations such as Thermal Systems (Hyderabad) Pvt. Ltd., Indian Oil Tanking Ltd. (Mumbai), and KL Crescent Infrastructure Pvt. Ltd., complemented by over a decade of academic service. His research interests span internal-combustion engines, alternative fuels, heat-transfer enhancement, renewable-energy systems, and machine-learning applications in mechanical and thermal analysis. Dr. Reddy has authored multiple papers in SCIE and Scopus-indexed journals including Environmental Science and Pollution Research and the International Journal of Internet Manufacturing and Services. His conference contributions address topics ranging from turbine-blade stress modeling and additive manufacturing to deep-learning-based solar-power prediction. He has also written book chapters on sustainable energy and AI-driven predictive diagnostics and holds three published patents in 2025 on self-lubricating gears and PCM-based thermal-energy-recovery systems. An active member of ISTE, IEI, and IAENG, Dr. Reddy has coordinated more than 50 expert lectures and 20 faculty development programs. His professional endeavors combine experimental engineering and computational intelligence to advance sustainable, data-driven innovations in modern mechanical systems.

Sunil Sankathala is pursuing B. Tech Computer Science and Engineering at Ajeenkya D Y Patil University, with a diverse range of interests that bridge technology, sustainability, and innovation. His academic focus lies in computer science fundamentals like programming, data structures, algorithms, and software development. He actively supplements this with hands-on experience in web development and robotics. Beyond the core curriculum, He has a strong research inclination in emerging and impactful fields. His current research interests include Artificial Intelligence in Healthcare, bioplastics, nano bioplastics, e-waste management, and solid waste management—areas where he believes technology can make a meaningful difference in solving real-world problems. He has been publishing research articles and book chapters in reputed national and international journals and books. He is also part of an ambitious Hyperloop project, which involves interdisciplinary collaboration and cutting-edge engineering aimed at revolutionizing transportation. Working on this project has enhanced his skills in teamwork, innovation, and systems thinking. In the tech domain, he has developed proficiency in web development using technologies such as HTML, CSS, JavaScript, and React. His interests in robotics further reflect building smart, responsive systems that interact with the physical world. Throughout his course, he has helped strengthen problem-solving and collaborative skills. He is proficient in languages such as Python, C, C++, MySQL, Ubuntu, Machine Learning and familiar with frameworks and tools like React, Git. He actively participates in hackathons, coding competitions, and tech events to challenge himself and learn from peers. He is also a quick learner and highly adaptable, with an eagerness to stay updated with the latest trends in the tech world. He aspires to become a skilled software engineer and contribute to building innovative solutions that have a positive impact on society. The long-term goal is to work in a dynamic environment where knowledge, grow professionally, and make meaningful contributions to the field of technology. Driven by a spirit of exploration and sustainability, he aims to contribute to technological solutions that address both human and environmental needs. The long-term goal is to become a versatile engineer and researcher, working at the intersection of computer science, environmental innovation, and human well-being.

Sk. Mohammad Shareef is an Assistant Professor in the Department of Mechanical Engineering at CVR College of Engineering, Hyderabad, India, and is currently pursuing his Ph.D. in Mechanical Engineering at the National Institute of Technology (NIT) Warangal. He received his M.Tech. in Advanced I.C. Engines from Jawaharlal Nehru Technological University, Anantapur (2016) and his B.Tech. in Mechanical Engineering from K.S.R.M. College of Engineering, Kadapa (2013) under Sri Venkateswara University, Tirupati. With more than eight years of academic experience, he specializes in internal-combustion engines, alternative fuels, computational-fluid-dynamics (CFD) simulation, and emission-control technologies. His postgraduate dissertation, "Effect of Exhaust Gas Recirculation on Emissions of a Direct-Injection Diesel Engine Using CFD Simulation," employed STAR-CD modeling integrated with experimental validation to assess combustion and pollutant behavior. Mr. Shareef is proficient in AutoCAD, PRO-E, CATIA, and C-language programming, enabling strong analytical and design skills. He qualified the national GATE-2014 examination and has been recognized for academic excellence since his early education. At CVR College, he mentors student projects, develops laboratory infrastructure, and integrates simulation-based learning in thermal-engineering education. His present research at NIT Warangal focuses on sustainable-energy systems, advanced combustion strategies, and phase-change-material-based thermal management. Dedicated, innovative, and collaborative, Mr. Shareef strives to advance cleaner and more efficient energy-conversion technologies in mechanical engineering.

Aditya Shrivastav is an aspiring technologist currently pursuing a Bachelor of Technology (B. Tech.) in Computer Science and Engineering at Ajeenkya D Y Patil University. He holds solid programming expertise in C, C++, and Python, alongside strong command over data and business tools such as Microsoft Excel and Tally software. Aditya is a member of Team Vac-Vectoor Hyperloop and participated with the team in the Global Hyperloop Competition organized by IIT Madras, contributing to cutting-edge transportation solutions. He actively engages in hackathons and various technical events to expand his skills and contribute to innovative problem-solving. He has authored and co-authored book chapters in the domains of solid waste management, bioplastics, nanotechnology applications, and artificial intelligence, showcasing his multidisciplinary approach and research acumen. Aditya is also working on an innovative project on predictive e-waste management. The project is targeted toward IT departments, schools, and colleges, aiming to support sustainable practices and resource optimization. In parallel with his academic journey, Aditya is serving as a faculty intern at Disha Computer Institute, where he teaches C, C++, basic Python, basic Java, Excel formulas, and Tally software. His dedication and teaching proficiency earned him a Letter of Appreciation and the opportunity to conduct a workshop on Advanced Excel at G. H. Raisoni College of Arts, Commerce and Science. With a strong foundation in technology, a passion for sustainability, and hands-on experience in both teaching and project development, Aditya Shrivastav is committed to driving impact through research, education, and innovation.

Aaditya Singh is currently a fourth-year Aerospace Engineering student at KIIT Bhubaneswar, where I have developed a strong foundation in aerodynamics, aircraft design, and sustainable engineering practices. My academic pursuits include a project on metamorphic wings, focusing on innovative adaptive wing structures that enhance aircraft performance and efficiency. Additionally, I have authored a research paper on sustainable development strategies for electronic waste (e-waste), reflecting my commitment to integrating environmental sustainability within engineering solutions. Through these experiences, I aim to contribute to advancing aerospace technology while promoting responsible and eco-friendly engineering practices.

Abhilas Swain is an accomplished academic and researcher specializing in mechanical and thermal engineering, with a distinct expertise in advanced HVAC systems, heat transfer enhancement, and sustainable building energy solutions. As an Assistant Professor at KIIT, Dr. Swain has contributed extensively to both experimental and modeling research on cutting-edge thermal technologies, regularly integrating innovative approaches such as machine learning and renewable energy into engineering practice. His professional repertoire also encompasses academic writing, conference presentations, and administrative responsibilities, reflecting a commitment to both technical excellence and academic leadership in the field of engineering education and research

Aadhya Vig is a final-year Aerospace Engineering student at Kalinga Institute of Industrial Technology. With a strong academic foundation and a deep curiosity in aerospace systems, her primary interests lie in propulsion, avionics, and aerodynamics. Her current work, titled "Simulation of Desiccant-Based Air-Conditioning System Using Vortex Tube as a Cooling Source," explores sustainable alternatives to conventional cooling methods—reflecting her inclination toward practical, energy-efficient innovations in engineering. Strongly motivated by an interest in cross-disciplinary research and practical application, Aadhya actively seeks out opportunities that blend fundamental aerospace theory with new technologies.

Index

A

B

C

D

E

F

G

H

L

M

N

O

P

R

S

T

V

W

Z

www.ingramcontent.com/pod-product-compliance
Lightning Source LLC
LaVergne TN
LVHW062331300926
849161LV00007B/1649